PROCEEDINGS OF THE INTERNATIONAL CONFERENCE ON LASERS IN THE CONSERVATION OF ARTWORKS XIII (LACONA XIII), 12-16 SEPTEMBER 2022, FLORENCE, ITALY

Lasers in the Conservation of Artworks XIII

Edited by

Salvatore Siano & Daniele Ciofini

Institute of Applied Physics "N. Carrara"- National Research Council - Florence, Italy

CRC Press
Taylor & Francis Group
Boca Raton London New York Leiden

CRC Press is an imprint of the
Taylor & Francis Group, an **informa** business
A BALKEMA BOOK

First published 2023
by CRC Press/Balkema
4 Park Square, Milton Park, Abingdon, Oxon, OX14 4RN

2385 NW Executive Center Drive, Suite 320, Boca Raton FL 33431

CRC Press/Balkema is an imprint of the Taylor & Francis Group, an informa business

Typeset by Integra Software Services Pvt. Ltd., Pondicherry, India

British Library Cataloguing-in-Publication Data
A catalogue record for this book is available from the British Library

Library of Congress Cataloging-in-Publication Data

A catalog record has been requested for this book

ISBN: 978-1-032-47995-8 (hbk)
ISBN: 978-1-032-47996-5 (pbk)
ISBN: 978-1-003-38687-2 (ebk)
DOI: 10.1201/9781003386872

Lasers in the Conservation of Artworks XIII – Siano & Ciofini (eds)
© 2024 The Editor(s), ISBN 978-1-032-47995-8
Open Access: www.taylorfrancis.com, CC BY-NC-ND 4.0 license

Table of contents

Lasers in the Conservation of Artworks XIII – Siano & Ciofini (eds)
© 2024 The Editor(s), ISBN 978-1-032-47995-8
Open Access: www.taylorfrancis.com, CC BY-NC-ND 4.0 license

Preface

Since about twenty-eight years ago, LACONA - *Lasers in the Conservation of Artworks* has gathered together researchers from natural and technological sciences, conservators, and restorers of public institutions and private enterprises for discussing about innovation in conservation-restoration of cultural heritage based on the development of non-invasive optical and spectroscopic techniques for material characterization, dedicated laser systems, and optimized laser irradiation treatments.

Thus, over such a long period, the Conference has represented a stimulating context to advance the international debate on extending the use of light in conservation-restoration of tangible cultural assets. Its mentioned core topics can be conventionally divided in "photonic diagnostics" and "laser treatments", although their inextricable interdependence and indivisibility is quite evident, since no diagnostic assessments can be successfully addressed without focusing on the specific information needed for optimizing the conservation treatments and the latter should never be carried out without exhaustive diagnostic insights before, during, and after the preservation intervention.

Non invasive analytical techniques and laser treatments were introduced several decades ago, but their development and widespread application underwent a significant acceleration during the 1990s. This, mainly thanks to the growing of R&D opportunities driven by the reaction of the research world against the serious damage to cultural heritage caused by natural disasters and anthropogenic factors. LACONA was born during such a very dynamic period and throughout its thirteen editions (see http://www.lacona13.eu/ and https://lacona-conferences.org/) it has collected numerous development and application contributions, concerning successful technological transfer to the industrial production and best practices. At the same time, it has also had positive cross-fertilization effects.

LACONA pursues two fundamental objectives: 1) to favour effective technical exchanges among researchers, conservators, restorers, curators, and stakeholders of the photonic industry; 2) to promote the exploitation of the significant advantages associated with the synergistic potential provided by the combination of portable diagnostics and laser conservation treatments. Rigorous validations, methodological and thematic insights, and hence innovation in the present field were and continue to be strictly dependent on the extent to which such goals are achieved through effective multidisciplinary and multiprofessional symbioses.

LACONA XIII aimed at strengthening such a tradition and its positive follow-up in the profound methodological transformation, which is going on in monitoring and maintenance practices of tangible heritage. In line with the previous editions, the Conference held in Florence last year (12-16 September 2022) included almost balanced numbers of presentations dedicated to the mentioned conventional subjects (photonic diagnostics 42, laser treatments 53), which focused on the experimentation of the most advanced laser systems and sensors, optimization and validations of improved conservation treatments, and discussion of a number of important case studies. Here, a collection of these works is presented, while other contributions can be found in the associated Special Issue of the Journal of Cultural Heritage. These publications show the strong commitment of the present community in exploiting the opportunities offered by novel solid state photonic technologies, as well as the growing knowledge of the degradation and deterioration phenomena. Finally, it is important to emphasize here that the last Conference involved new groups and many young scholars, which represent a very encouraging premise towards the perspective of further growing of this community, long life of LACONA, and wide spreading of the photonic approach in cultural heritage and other fields presenting similar material knowledge and material treatment problems.

Salvatore Siano
Chair of LACONA XIII
Florence, June 2023

Lasers in the Conservation of Artworks XIII – Siano & Ciofini (eds)
© 2024 The Editor(s), ISBN 978-1-032-47995-8
Open Access: www.taylorfrancis.com, CC BY-NC-ND 4.0 license

International steering committee

John F. Asmus	*Center for Advanced Nanotechnology, Department of Physics, University of California, San Diego (UCSD), La Jolla, CA, US*
Marta Castillejo	*Instituto de Química Física Rocasolano, CSIC, Madrid, ES*
Martin Cooper	*Lynton Lasers Ltd., Holmes Chapel, Cheshire, UK*
Vincent Detalle	*Centre de Recherche et de Restauration des Musées de France, FR*
Abdelrazek Elnaggar	*Faculty of Archaeology, Fayoum University, Al-Fayoum, EG*
Wolfgang Kautek	*University of Vienna, Department of Physical Chemistry, Vienna, AT*
Martin Labourè	*MESCLA Patrimoine, Strasbourg, Grand Est, FR*
Marco Leona	*Department of Scientific Research, Metropolitan Museum of Art, NY, US*
Austin Nevin	*Department of Conservation, Courtauld Institute of Art, London, UK*
Johann Nimmrichter	*Federal Office of Historical Monuments (Bundesdenkmalamt), Department for Restoration and Conservation, Vienna, AT*
Vadim Parfenov	*St. Petersburg State Electrotechnical University, RU*
Paraskevi Pouli	*Institute of Electronic Structure & Lasers–FORTH, Heraklion, EL*
Jose Santiago Pozo-Antonio	*Department of Natural Resources and Environment Engineering University of Vigo, ES*
Roxana Rădvan	*National Institute of Research and Development for Optoelectronics INOE, Bucharest- Magurele, RO*
Matija Strlič	*Institute for Sustainable Heritage London, University College London UK*
David Saunders	*International Institute for Conservation, UK*
Salvatore Siano	*Institute of Applied Physics "Nello Carrara" - CNR, Florence, IT*
Alessandro Zanini	*El.En. SpA, Calenzano, IT*

Lasers in the Conservation of Artworks XIII – Siano & Ciofini (eds)
© 2024 The Editor(s), ISBN 978-1-032-47995-8
Open Access: www.taylorfrancis.com, CC BY-NC-ND 4.0 license

International advisory committee

Giorgio Bonsanti	*Former Professor at the University of Florence and Superintendent of the Opificio delle Pietre Dure, Florence, IT*
Klaus Dickmann	*Laser Center, Muenster University of Applied Sciences, DE*
Costas Fotakis	*Emeritus Professor of Physics at the University of Crete and Distinguished Member of the Foundation Organization for Research and Technology (FORTH), Heraklion, Crete, EL*
Renzo Salimbeni	*Former Research Director at CNR and Director of the Institute of Applied Physics "N. Carrara" - CNR, Florence, IT*
Manfred Schreiner	*Emeritus Professorat the Academy of Fine Arts, Institute of Natural Sciences and Technology in the Arts Lecturer, Vienna, AT*
Véronique Vergès-Belmin	*Former Head of Department, Laboratoire de Recherche des Monuments Historiques, Champs sur Marne, FR*
Kenneth Watkins	*Former Professor at the University of Liverpool, UK*

Lasers in the Conservation of Artworks XIII – Siano & Ciofini (eds)
© 2024 The Editor(s), ISBN 978-1-032-47995-8
Open Access: www.taylorfrancis.com, CC BY-NC-ND 4.0 license

Local organizing committee

Chair:
Salvatore Siano
Institute of Applied Physics "N. Carrara"- CNR, Florence, IT

Stefania Agnoletti	*Opificio delle Pietre Dure, Firenze, IT*
Juri Agresti	*Institute of Applied Physics "N. Carrara"- CNR, Florence, IT*
Alessia Andreotti	*DCCI-University of Pisa, Pisa, IT*
Anna Brunetto	*Restauri Brunetto, Vicenza, IT*
Ilaria Cacciari	*Institute of Applied Physics "N. Carrara"- CNR, Florence, IT*
Daniele Ciofini	*Institute of Applied Physics "N. Carrara"- CNR, Florence, IT*
Francesco Colao	*UTAPRAD-ENEA, Frascati, IT*
Roberta Fantoni	*UTAPRAD-ENEA, Frascati, IT*
Raffaella Fontana	*National Institute of Optics - CNR, Florence, IT*
Marco Giamello	*DSFTA- University of Siena, Siena, IT*
Stefano Legnaioli	*Institute of Chemistry of Organometallic Compounds - CNR, Pisa, IT*
Rachele Manganelli del Fà	*Institute of Heritage Science-CNR, Florence, IT*
Andrea Azelio Mencaglia	*Institute of Applied Physics "N. Carrara"- CNR, Florence, IT*
Austin Nevin	*Department of Conservation, Courtauld Institute of Art, London, UK*
Iacopo Osticioli	*Institute of Applied Physics "N. Carrara"- CNR, Florence, IT*
Silvia Rescic	*Institute of Heritage Science - CNR, Florence, IT*
Cristiano Riminesi	*Institute of Heritage Science-CNR, Florence, IT*
Vincenzo Palleschi	*Institute of Chemistry of Organometallic Compounds - CNR, Pisa, IT*
Olga De Pascale	*Institute for Plasma Science and Technology - CNR, Bari, IT*
Paolo Romano	*Institute of Heritage Science - CNR, Catania, IT*
Giorgio Saverio Senesi	*Institute for Plasma Science and Technology - CNR, Bari, IT*
Francesco Taccetti	*National Institute for Nuclear Physics, Florence, IT*
Renato Torre	*Department of Physics and Astronomy, University of Florence, IT*
Alessandro Zanini	*El.En. SpA, Calenzano, IT*

PHOTONICS DIAGNOSTICS

Painted surfaces

Lasers in the Conservation of Artworks XIII – Siano & Ciofini (eds)

Rubens's *Death of Hippolytus* (1610-12): How scanning XRF shed new light on old questions

K. Stonor
Tager Stonor Richardson, London, UK

C. Richardson, S.R. Amato, A. Nevin & A. Burnstock
Courtauld Institute of Art, London, UK

ABSTRACT: A version of Rubens's *Death of Hippolytus* in the Courtauld Gallery collection was examined in 2011, using microscopy, x-radiography, infrared reflectography and dendrochronology. Further examination using scanning x-ray fluorescence spectroscopy in 2022 gave greater insights into the painting technique, condition and conservation history and complemented the earlier study to allow for a better understanding of the status of the picture in Rubens's oeuvre.

1 INTRODUCTION

A traditional technical analysis of the *Death of Hippolytus* by Peter Paul Rubens (Figure 1) was undertaken by Richardson and Stonor as part of a study of early works by the artist in the Courtauld Gallery collection in 2011.

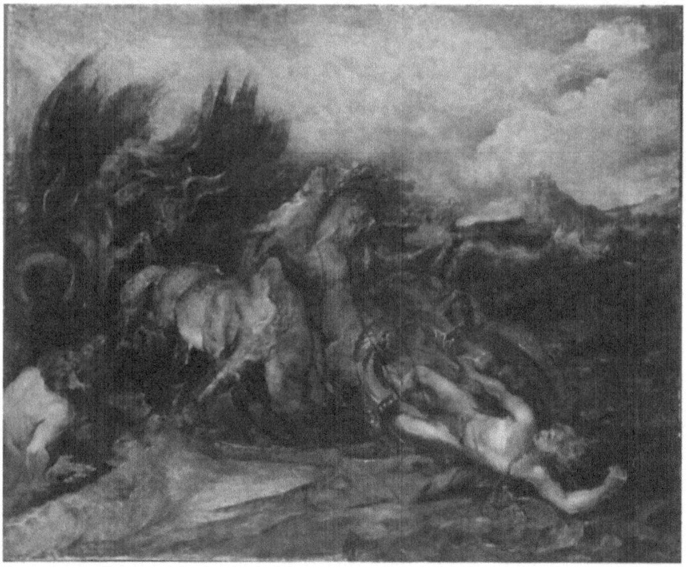

Figure 1. *The Death of Hippolytus*, Peter Paul Rubens, © The Courtauld, London (Samuel courtauld trust).

The heavily discolored varnish and conservation history made the thinly painted work hard to sample and difficult to interpret with surface microscopy and key art historical questions

DOI: 10.1201/9781003386872-1

about the function of the work within Rubens's studio practice, intimately connected to its condition, could not be adequately answered. Through dendrochronology and the identification of key pigments, it was possible to confirm that the painting dated from Rubens's lifetime, but it remained a tantalizing puzzle. With the application of scanning X-Ray Fluorescence spectroscopy (XRF), distribution maps of a wide range of elements could be identified non-invasively. The results gave new insights into the making of the work, as well as a better understanding of its condition, allowing a more informed investigation of the art historical questions around the status of the picture.

2 REPUTATION AND CONSERVATION HISTORY

The picture is one of two painted versions of the subject by Rubens, both of which are of almost the same dimensions. The Courtauld painting, (Figure 1) on a Flemish oak panel, measures 51×65 cm^2, whilst the version now in the Fitzwilliam Museum, on a copper support, is slightly larger at 50.2×70.8 cm^2. Both paintings are linked to a drawing in the Musée Bonnat-Helleu, Bayonne which shows a different treatment of the composition: the action is moved over towards the right and the foreground is occupied by figures fleeing the scene. This group of works depicting the *Death of Hippolytus* has traditionally been dated to between 1610 and 1614 and shows Rubens's preoccupation with the motif of dueling horses that was inspired by his studies of Leonardo da Vinci's *Battle of Anghiari.* Recently, Gifford has reappraised the Courtauld work as part of her study of *The Fall of Phaeton* and sees it as sitting comfortably in a chronology of iconography explored by Rubens after his return to Antwerp from Italy in 1608 (Gifford 2019).

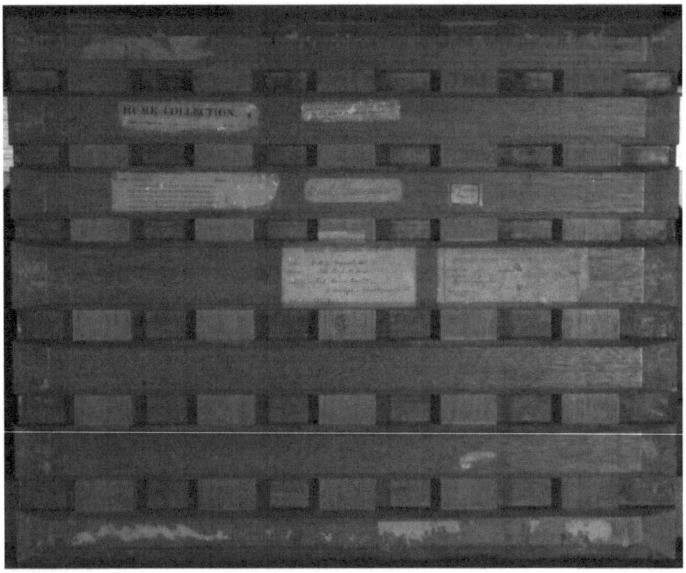

Figure 2. Reverse of *The Death of Hippolytus*, Peter Paul Rubens, © The Courtauld, London (Samuel Courtauld trust).

The Courtauld and Fitzwilliam paintings both depict Hippolytus falling onto a beach from his broken chariot, which is pulled by four horses. The horses have been frightened by a sea monster, that appears within a huge wave. At the bottom left of both pictures, a sea triton is shown blowing a conch shell. In the shallower setting of the Courtauld picture, this is the entire cast of characters and the right-hand side of the picture shows a short stretch of beach before a stormy sea with the suggestion of a distant landscape beyond. The deeper pictorial space of the Fitzwilliam version shows the shoreline continuing to the right-hand side, with two figures fleeing the melee. The muscular figure of Hippolytus in the Fitzwilliam picture is inspired by a drawing after Michelangelo's *Tityus* in the Royal Collection, and draws on an unusual high view of the *Laocoön* sculpture (Jaffé and McGrath, 2005), in contrast to the slight figure of the Courtauld picture.

The Courtauld painting has been the subject of much debate amongst Rubens scholars, enjoying a good reputation in the 19th century, when Rooses considered it the prime version of the subject (Rooses 1886, p.109), agreeing with Smith's 1830 catalogue raisonné which described it as "free, loose and masterly" in comparison to the "dry and careful" Fitzwilliam version (Smith 1830, p.177-178). At that time the work was in the collection of Abraham Hume and was sold by his descendants in 1923 to a private collection in Berlin. It subsequently entered the collection of Count Seilern and is recorded in his catalogue in 1955.

The painting has been thinned and cradled, a typical 19th-century treatment which applies a wooden lattice to the reverse of the panel, thought to support panel joins and prevent warping. The panel join, running horizontally, is associated with paint cracking and loss, and it seems likely that this process was intended to limit movement and further damage to the paint. This intervention can be dated to before or during its ownership by Hume, as revealed by labels applied to the reverse: the earliest label relates to the Hume collection catalogue, whilst there are subsequent labels from exhibitions at the end of the 19th century when it was in the collection of Lord Brownlow (Figure 2).

Seilern's catalogue describes more of the painting's conservation history, comparing it with its engraving by Maria Cosway (1760-1838) made in the late 18th or early 19th-century (Figure 3):

"*As this engraving shows, the present sketch was completely overpainted in the late 18th or early 19th century: the overpainting has been carefully and successfully removed by Mr. S. Isepp*" (Seilern 1955, p. 32).

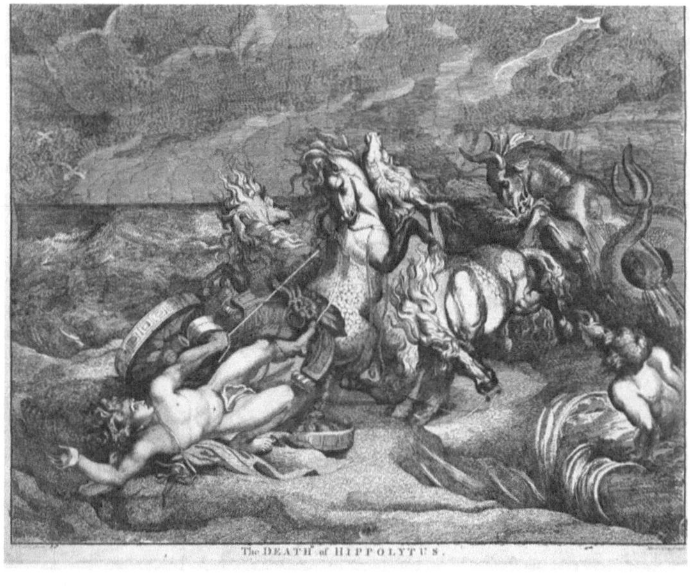

Figure 3. *The Death of Hippolytus*, Etching by Maria Cosway after Peter Paul Rubens © The trustees of the British Museum. Shared under a creative commons attribution-non-commercial-share alike 4.0 international (CC BY-NC-SA 4.0) licence.

Sebastian Isepp was an Austrian émigré restorer who followed Seilern to London from Vienna, fleeing persecution in the late 1930s. Isepp had worked for Seilern from the 1920s, and continued to do so until his death in 1954. Formerly chief restorer at the Vienna Kunsthistorisches Museum, Isepp worked at the Ashmolean Museum and National Gallery, London and was a highly respected restorer. Isepp was one of the first restorers to make use of x-radiography for the study of paintings, having established a laboratory in Vienna with Johannes Wilde for the examination of paintings (Simon, 2015). Unlike Seilern, who had seen the painting firsthand prior to its restoration, we cannot know which of the key differences between the engraving and the painting in its current form are related to the removal of the 18th or early 19th century overpainting campaign, and which are simplifications made by Cosway to render the subject in the graphic medium. For example, the engraving shows a flat horizon, without the dark waves behind the sea monster. The figure of Hippolytus is shown with an open hand flailing behind him, rather than a closed fist and his foot is visible in strong foreshortening, beneath his left leg.

After the cleaning and restoration, Seilern proposed that the painting should be reconsidered as a *modello,* or oil sketch, rather than a finished work. Writing some years later in 1980, Held expressed doubts that the painting belonged in his catalogue raisonné of oil sketches, believing the 'sketchy' appearance was possibly 'due more to its condition than Rubens' intention' (Held, 1980, p. 333-334). By 2011, when the work was re-examined as part of a technical study of Rubens's earliest works in the Courtauld collection, it was rarely displayed thanks to the now heavily discolored varnish and patchy appearance of old restoration campaigns. It was particularly hard to judge whether the extent to which Rubens's typical streaky *imprimatura* was now visible in the sky related to intention or subsequent abrasion of the paint layers. This seemed key to its interpretation as a quickly painted sketch.

3 2011 FINDINGS

In 2011 the painting was examined out of the frame. The painting was imaged using infrared reflectography (IRR) with an OSIRIS InGaAs camera and paint cross sections were taken. Finally, Ian Tyers undertook dendrochronology on the oak panel. Dendrochronology gave a felling date of after 1574 for one of the two boards comprising the panel, and matched ring sequences to a Flemish reference series, rather than the more typical Baltic oak (Tyers, 2011).

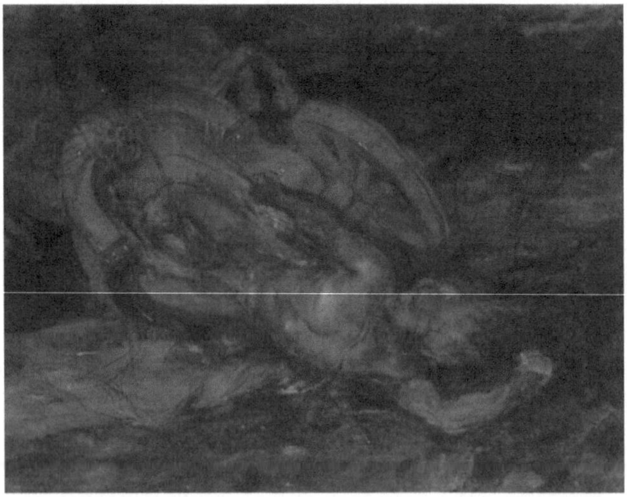

Figure 4. Infrared reflectography detail of *The Death of Hippolytus*, Peter Paul Rubens, © The Courtauld, London (Samuel Courtauld trust).

IRR revealed a comprehensive freehand underdrawing done in a dry, carbon-based medium. The underdrawing includes the legs of the fleeing attendant figures painted in the Fitzwilliam version and marks indicating Hippolytus's left foot, as well as the closed fist of his left hand (Figure 4).

Cross sections confirmed the presence of a chalk (calcium carbonate) ground layer followed by a thin, grey *imprimatura* composed of lead white (lead carbonate) with some chalk and charcoal black. Samples showed a palette of pigments which would be expected for Rubens and typical of the early 17th century including lead white, chalk, lead tin yellow (lead-tin oxide), vermilion (mercuric sulfide), red lake (organic dye typically precipitated onto potash alum - potassium aluminium sulphate), red earth (iron oxide), yellow ochre (iron hydroxide), azurite (basic copper carbonate), charcoal black and bone black (calcium phosphate). However, sampling opportunities were limited and where samples were taken the paint layers were very thin, the friable paint was prone to delamination, and brushy paint application meant layers could be missed in cross section.

The appearance of the sky could not be clarified by sampling – a single sample site was identified and only layers of ground and *imprimatura* could be found. The infrared imaging of the sky revealed a mottled appearance, particularly at the right-hand side, and only the streaks of the *imprimatura* could be seen at the left.

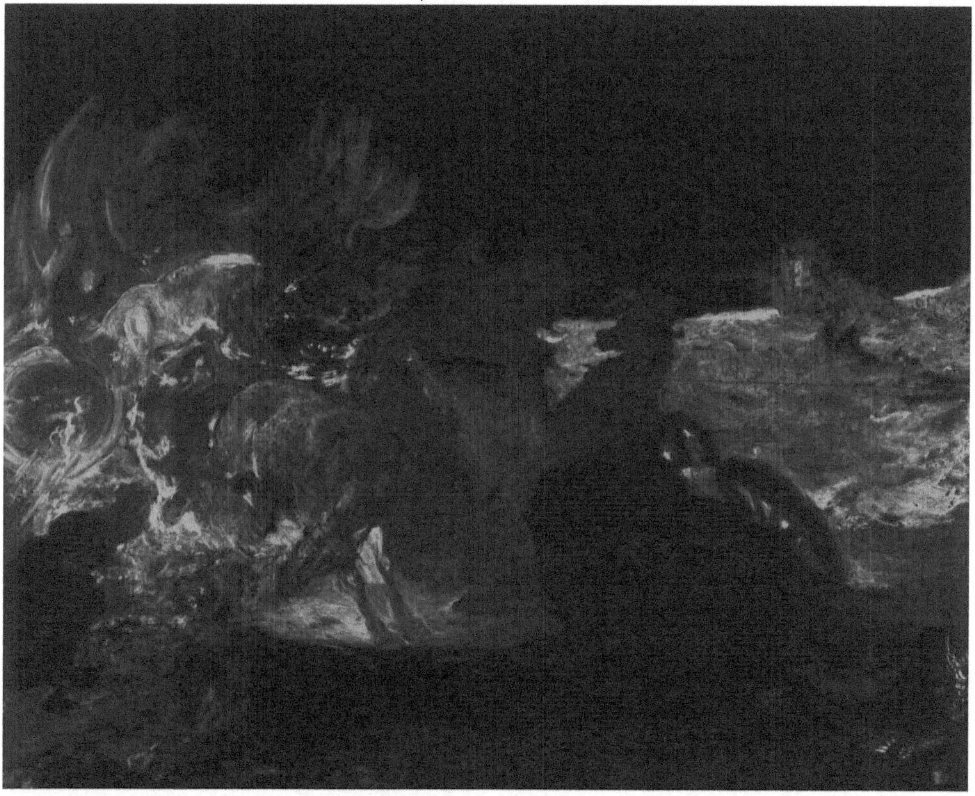

Figure 5. CuK map of *The Death of Hippolytus*, Peter Paul Rubens, © The Courtauld, London (Samuel Courtauld trust).

Evidence for 19th or 20th century interventions could also be found in cross section, with the identification of zinc white (zinc oxide) in combination with cadmium yellow (cadmium sulfide) in a passage of green overpaint. Using surface microscopy, the greener passages of the sea could be seen to be reinforced with a restorer's viridian glazes (hydrated chromium oxide).

4 SCANNING XRF

Scanning XRF was identified as a non-invasive analytical method that could help to answer the outstanding questions about the painting's original appearance by detecting areas of restoration and establishing the condition of the panel. Scanning XRF was carried out using a M6 Jetstream instrument from Bruker Nano GmbH, which is equipped with a 30 W Rh-target micro-focus X-ray tube mounted with two 60 mm^2 SDDs (with energy resolution of < 145 eV for MnKα) on a motorized stage. The tube uses a polycapillary X-ray lens for beam focusing with a maximum voltage of 50 kV and a maximum current of 0.6 mA. The elemental maps were acquired using a voltage of 50 kV and a current of 0.6 mA with 100 μm spot size, 400 μm step size and 30 ms dwell time, and were processed using the Bruker Jetstream software.

The brushy, directional application of the grey *imprimatura*, so typical of Rubens's panels, is readily apparent in both the PbL and CaK maps, as is the horizontal panel join. Scanning XRF confirmed the presence of the pigments identified in cross section with clear maps for HgL showing the use of vermilion in the flesh paints and drapery and the map of CuK (Figure 5) confirming the presence of azurite throughout the sea and landscape. However, as in the cross section, azurite was noticeably absent from the sky.

Overlaying maps for elements known to indicate modern pigments (titanium, zinc and chromium) also usefully indicated the extent of later retouching. It was particularly instructive to compare the shapes of retouching brushstrokes with those of Rubens, as they were quite different. The retouching often appeared patchy and spotty, in comparison to the decisive brushwork of the original paint. As some of these pigments were introduced in the later 19th century, only the titanium white passages can be associated with certainty with Isepp's restoration. Whilst Seilern's account implies the total removal of all earlier overpaints, it is possible that chromium and zinc containing passages could be associated with a 19th century treatment, for example the cradling of the panel support.

Scanning XRF identified the presence of cobalt, something that had been absent from the cross sections. Two explanations are possible here: that cobalt blue (cobalt aluminium oxide) was used as a retouching pigment, or that Rubens had employed the blue glass pigment smalt, a pigment found elsewhere in his oeuvre, in the sky paint which had degraded, losing its blue colour and contributing to the pale appearance of this part of the composition. In order to address this question, cobalt was added to the overlay of titanium, zinc and chromium, which showed some correlation between the presence of cobalt and other retouching pigments (Figure 6a).

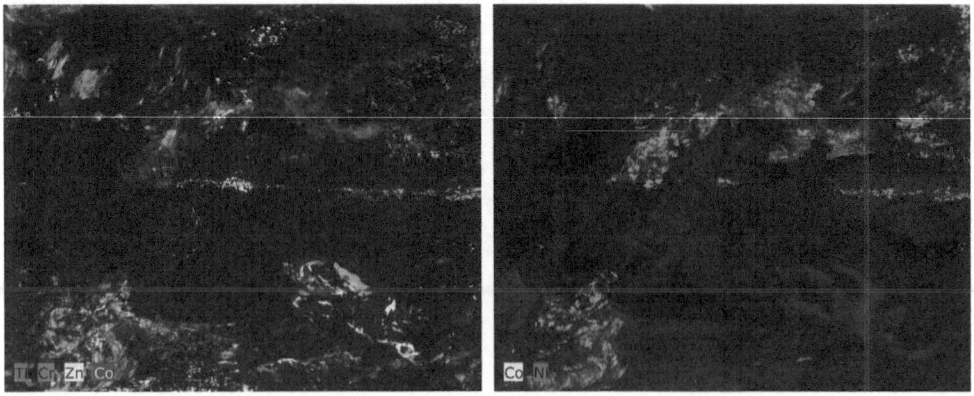

Figure 6. Left to right: (a) Overlay of Ti, Cr, Zn and Co K maps, (b) Overlay of Ni and Co K maps of *The Death of Hippolytus*, Peter Paul Rubens, © The Courtauld, London (Samuel Courtauld trust).

Smalt is usually found in combination with trace elements of nickel, potassium, bismuth and iron. Bismuth was not detected, and iron and potassium were detected all over the painting. However, creating an overlay with the nickel K map (Figure 6b) showed a partial correlation between the passages of cobalt-containing paint, and the nickel associated with the cobalt ore used in smalt production (Spring, Higgit, Saunders, 2005). This raised the possibility that both cobalt blue and smalt may be present.

Subsequent microscope examination of the paint surface confirmed the presence of a modern blue pigment in the retouching (Figure 7a). However, examining the painting under the microscope along the horizon line, where there was the strongest signal for both cobalt and nickel, revealed little evidence of blue pigment particles. In one area, a well-preserved blue could be found, with a particle shape that evoked either azurite or smalt (Figure 7c). With the overlying yellowed varnish, it was impossible to use the hue of the particles to make any further distinction between the two. The only other evidence for smalt was in the presence of now colorless, glassy particles (Figure 7b). However, in the absence of any other observable blue pigment, it seemed that discolored smalt remained the most likely explanation for the detection of cobalt. If the smalt sky paint had completely lost its colour early in the painting's history, this would offer a plausible explanation for the painting having been entirely repainted in this area in the late 18th or early 19th century. However, the overlay of the cobalt and nickel map indicated that smalt remains only as an incomplete layer in the sky, suggesting that the paint layer was also significantly damaged as well as discolored.

Turning to the other differences between the engraving and the painting: fascinatingly, the exposed left foot and extended fingers of the left hand of Hippolytus, recorded by Cosway, also appear to be visible in the PbL map (Figure 8). The figure of Hippolytus also looks quite different and is closer in appearance to the more muscular figure of the Fitzwilliam version. In particular, the rendering of the torso has the rippling flesh inspired by Rubens's study of the works of Tintoretto on his Italian travels, and the Michelangelo *Tityus* drawing that is thought to have been amongst his direct inspirations. Looking closely at the painted surface, it appears that both the open hand and left foot seen in the engraving could have been covered by later retouching. In the case of the hand, the closed fist of the underdrawing, now visible through the thin upper paint, may have been used as a guide to the intended appearance of the hand and the painted open hand could have been interpreted as later restoration. Whilst the damaged left foot may have been covered by red drapery as a means to avoid reconstructing the complex anatomy and foreshortening of the pose in the restoration.

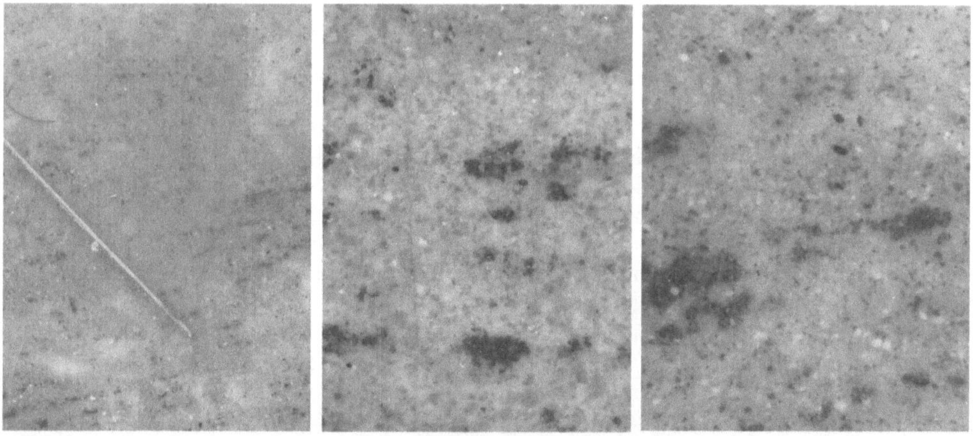

Figure 7. Microscope details of the paint surface, *The Death of Hippolytus*, Peter Paul Rubens, © The Courtauld, London (Samuel Courtauld trust) left to right: (a) modern retouching paint with a brush hair caught in varnish; (b) colourless glassy particles in a matrix of lead white and black pigments in original paint; (c) blue and black pigments in a matrix of lead white in original paint.

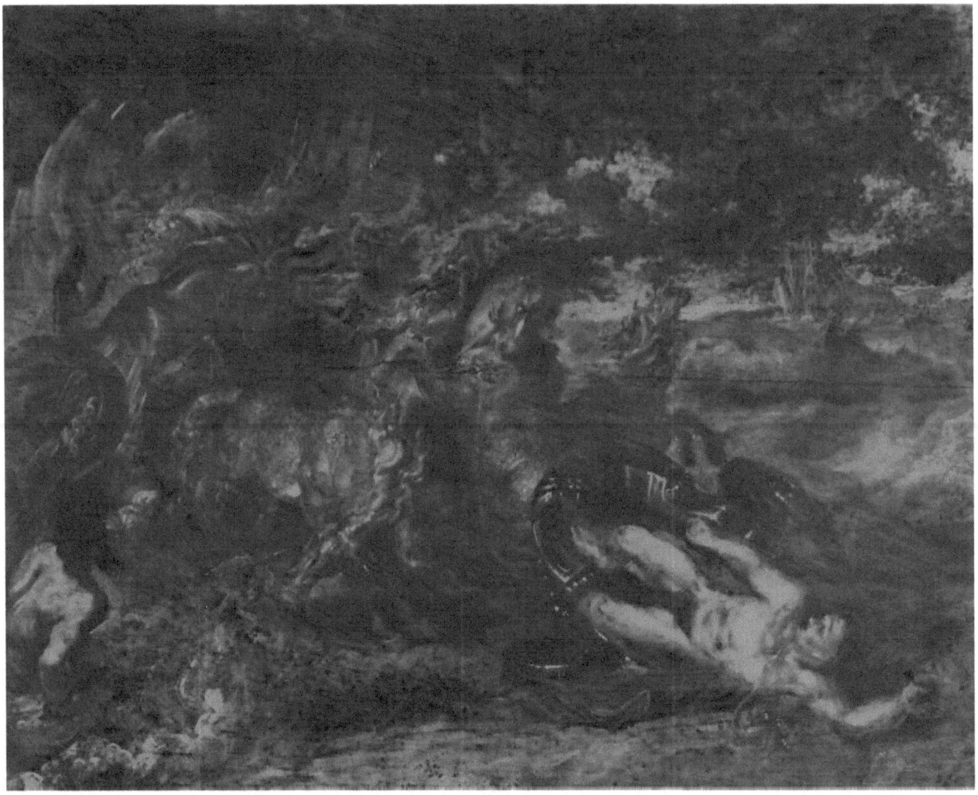

Figure 8. PbL map of *The Death of Hippolytus*, Peter Paul Rubens, © The Courtauld, London (Samuel Courtauld trust).

5 CONCLUSIONS

Scanning XRF has offered considerable insights into the condition of *The Death of Hippolytus*, revealing that, in the foreground, retouching has been used to draw together the thin and abraded paint and also to alter the figure of Hippolytus. The changes to the hand and foot are the most obvious, but comparison with the lead map suggests that the musculature of the torso has also been simplified. The painting may in fact have been much more similar to the Fitzwilliam version than it now appears. In the sky, the presence of a damaged and discolored layer of smalt explains the absence of blue colour and dull, unfinished appearance. As a whole, the results seem to confirm Held's doubts that the painting is a sketch and point to its status as a more finished, early work.

Whilst Scanning XRF is a powerful technique, it has been most fruitful when combined with microscopic surface examination and other methods of technical study, such as infrared examination. This has allowed greater confidence in the implications of the XRF results and given a fuller picture of the artist's technique and the painting's condition history, helping to inform future conservation treatment and understand the painting's art historical status.

ACKNOWLEDGEMENTS

The 2011 study of the painting was part of a Caroline Villers Fellowship for Richardson and Stonor at the Courtauld Institute of Art. Purchase of the scanning XRF instrumentation was made possible by a grant from the Arts and Humanities Research Council (AHRC) Capability for Collections Fund (CapCo).

REFERENCES

Jaffé, D. and McGrath, E. 2005. *Rubens A Master in the Making*. London: National Gallery Company Limited.

Gifford, M. 2019. Rubens's Invention and Evolution: Material Evidence in The Fall of Phaeton. *Journal of Historians of Netherlandish Art* 11:2. DOI: 10.5092/jhna.2019.11.2.1

Held, J. 1980. *The oil sketches of Peter Paul Rubens: a critical catalogue/Vol.1.*, Princeton, N.J: Princeton University Press for the National Gallery of Art.

Rooses, M., 1886. *L'œuvre de P. P. Rubens: Histoire et description de ses tableaux et dessins*, Volume 3, Anvers, J Maes.

Seilern, A. 1955. *Flemish Paintings and Drawings*, London: Shenval Press.

Simon, J. 2015. *British picture restorers, 1600–1950*, https://www.npg.org.uk/research/programmes/directory-of-british-picture-restorers/

Smith, J. 1830. *A Catalogue Raisonné of the Works of the Most Eminent Dutch, Flemish, and French Painters: Peter Paul Rubens*, London.

Spring, M., Higgit, C. & Saunders, D. 2005. Investigation of Pigment Medium Interaction Processes in Oil Paint Containing Degraded Smalt. *National Gallery Technical Bulletin, Volume 26*: 56–70. London: National Gallery Company Limited.

Tyers, I. 2011. *Dendrochronological Consultancy Report 422*, unpublished report

Lasers in the Conservation of Artworks XIII – Siano & Ciofini (eds)
© 2024 The Author(s), ISBN 978-1-032-47995-8

Use of non-invasive methods of analysis applied to the study of Oscar Pereira da Silva paintings

J. Schenatto & M.A. Rizzutto
Laboratory of Archaeometry and Applied Sciences to Cultural Heritage Studies (LACAPC), Nuclear Physics Department, Physics Institute of University of São Paulo, São Paulo, Brazil

ABSTRACT: The application of spectroscopic and imaging techniques, with non-invasive methodologies and portable equipment, to the study of two paintings by the Brazilian artist Oscar Pereira da Silva allowed the characterization of these artworks finalized in 1922. With Energy Dispersive X Ray Fluorescence (ED-XRF) and Raman analysis it was determined that his palette has a large combination of mineral and synthetic pigments, using lithopone and lead white to prepare white layers before depositing the pictorial layer. His palette is similar to both paintings, differing only by the presence of synthetic ultramarine and umber, found only in one of the artworks. With Infrared Reflectography (IRR) it was found creative steps of the artist, such as checkering the canvas, sketching of most of the scene, correction of drafts and *pentimenti*. The characteristics found in this study are aligned with previous research about his work, corroborating to the definition of his artistic style.

1 INTRODUCTION

The use of non-invasive methodologies to the study of paintings has become one of the first choices to the characterization of these artwork objects, due to their priceless cultural value and fragility. (Brooke 2020, Delaney et al. 2020) For a more complete study and comprehension of the cultural heritage objects, different techniques are often used together, including spectroscopic and imaging measurements. The joint analyzes allow to determine and record artwork information, such as its state of conservation, the creative process of the artist, the material that was used in the confection, etc. While the spectroscopic techniques enable researchers to investigate the composition of the pigments present in the paintings, the imaging techniques can reveal regions that have undergone previous interventions, thus preventing restoration pigments from being measured instead of original ones. The techniques that will be presented in this work use portable and easy-to-handle equipment, which allows them to be applied *in situ*, in the museum itself or in the collection where the works are kept.

To this study, two paintings by the artist Oscar Pereira da Silva (1867-1939) belonging to the Paulista Museum collection, linked to the University of São Paulo (USP) in São Paulo city, in Brazil, were selected due to their historical importance. This famous Brazilian artist became known for his neoclassical style and rigor in his creative process, acting as a fundamental character in the construction of historical images of Brazil, illustrating the process of Brazil's independence and other important national events. (Junior 2015)

The analyzes of these paintings were carried out during their restoration process, since the entire Paulista Museum, its building and collection, underwent a renovation and restoration from 2019 to 2022. (USP 2022) The study of the paintings during the restoration process allowed a better definition of the artist's palette, since during this period the paintings had their varnish removed. With this, the Raman spectral analysis detected a better signal, since it had a low fluorescence background, usually caused by the presence of the varnish. Besides

DOI: 10.1201/9781003386872-2

that, the results that were being collected and discussed during the analysis were used to assist in the determination of the proper proceeding to approach the restoration.

2 METHODOLOGY

Two important paintings from the Paulista Museum collection by Oscar Pereira da Silva (OPS) were studied to better understand his style and determine the material he used. The paintings selected were "Sessões da corte de Lisboa" (Lisbon court' session), 315 × 265 cm, henceforward referred to as "COURT", and "Príncipe Dom Pedro e Jorge de Avilez a bordo da fragata União" (Regent prince Dom Pedro and Jorge de Avilez aboard the frigate Union), 315 × 262 cm, henceforward referred to as "REGENT", both oil on canvas, finalized in 1922. These two paintings were chosen based on their historical importance, the great variability of colors, and because both were being restored, which allowed good handling of these large works. As mentioned, during the measurements the paintings were unvarnished, allowing more direct access to the pigments, mainly facilitating Raman spectroscopy analysis. The lack of varnish means that we do not expect interactions with the varnish constituents and/or the restoration materials in the analysis of the pigments.

The non-invasive approaches used in this work combines an imaging examination with Infrared Reflectography (IRR), elemental analysis using Energy Dispersive X-Ray Fluorescence (ED-XRF) spectroscopy, and molecular characterization through Raman spectroscopy. Portable X-ray Fluorescence spectroscopy has proved to be valuable for *in situ* identification of the elemental composition of the pigments and a guide for further investigations with complementary analytical techniques. (Colantonio 2021, Moreau et al. 2023) This method is non-invasive, multi-elemental, fast and low cost (compared with other elemental techniques such as PIXE).

X-ray Fluorescence measurements were performed using a X Ray tube, Mini X, with a silver anode and a 2 mm diameter collimator, operating with 30 kV and 5 μA, and a fast system of a silicon drift detector, XR-100SDD, and the multichannel digital pulse processor PX5, operating during 100 seconds for each point of analysis, all from Amptek. This portable equipment, homemade setup, is particularly efficient for light elements such as S, P, Cl etc, because of the low detection limit. A great advantage of this instrument is its compactness: the setup has almost 2 kg and can be mounted on any light-weight tripod.

The paintings have large dimensions and a big chromatic palette, so in order to have a proper base to the definition of the pigments, 88 points at COURT and 85 points at REGENT were measured, shown in Figure 1. These measuring points were selected according to the color, its distribution across the painting, and the absence of restoration or intervention pigments.

Raman spectroscopic analysis used a EZRamanDual-G, from Enwave Optronics system, equipped with a 785 nm laser operating with approximately 60 mW. The spectrometer is customized to cover 200–2000 cm^{-1} range with spectral resolution of 6 cm^{-1}. The measurement used a contact tip lens to avoid the external light interfering with the signal detection. All the points measured were the same as the ED-XRF, acting as complementary techniques.

Infrared Reflectography (IRR) is an imaging technique that investigates the underlying layer of the painting, due to the different ways in which infrared radiation interacts with the pigments. After a combination of reflection, transmission, and absorption of the radiation by the pigments, the scanned signal, i.e., the resulting black and white image, allows the visualization of details that are hidden from the naked eye, such as the initial sketches in graphite or charcoal, corrections and *pentimenti* performed by the artist.

The analyses are carried out using two 1000 W diffused halogen lamps, positioned at an angle of approximately 45° and a digital camera specialized for the near-infrared detection, OSIRIS, from Opus Instruments, with an InGaAs sensor. In order to obtain an image with clear details, the paintings were scanned in quadrants of approximately 60×60 cm, generating 20 and 24 quadrants for COURT and REGENT, respectively.

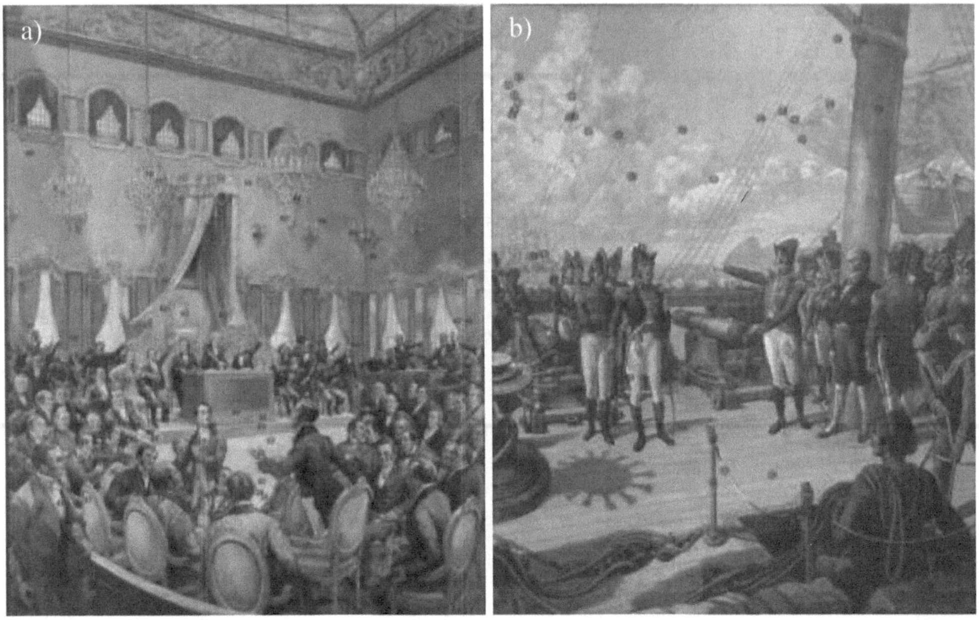

Figure 1. ED-XRF and Raman measurement points on the paintings (a) "Sessões da corte de Lisboa" (Lisbon court' session - COURT), 315 x 265 cm, and (b) "Príncipe Dom Pedro e Jorge de Avilez a bordo da fragata União" (Regent prince Dom Pedro and Jorge de Avilez aboard the frigate Union - REGENT), 315 x 262 cm, by the Brazilian artist Oscar Pereira da Silva, both finalized in 1922 and belonging to the Paulista Museum, linked to the University of São Paulo, in São Paulo city. Photos by J. Rosael/Paulista Museum/USP.

3 RESULTS AND DISCUSSION

The complementary use of ED-XRF and Raman spectroscopies allowed the determination of the most probable palette of pigments used by OPS, showing that the artist used a considerable number of pigments. The presentation of the results and the discussion will be divided in the following colors: white, yellow, red, orange/brown, blue, green, and black pigments.

3.1 *White pigments*

Due to the presence of high amounts of zinc (Zn) and barium (Ba), in the ED-XRF measurements of many points over the paintings, it is possible to suggest the use of lithopone (ZnS, BaSO4), as a white pigment in OPS palette. Figures 2a and 2b show the bar graphs with counts (areas in the XRF spectra) of Zn and Ba, respectively, at each point measured in the painting COURT, to exemplify. Particularly, a direct correlation was verified between these two elements in regions without a pictorial layer, i.e., regions where the preparation layer was exposed. (Squared in red in the Figures 2) This suggests the use of lithopone as a preparation layer, a technique that was very common among classicist painters from the 19th and 20th centuries (Stols 2012) and known to be part of the creative process of OPS. (Campos et al. 2014) Besides that, it is also suggested the use of zinc white (ZnO) and barite (BaSO4), separately, due to the uncorrelated concentration noted at some points, characteristic observed in both paintings.

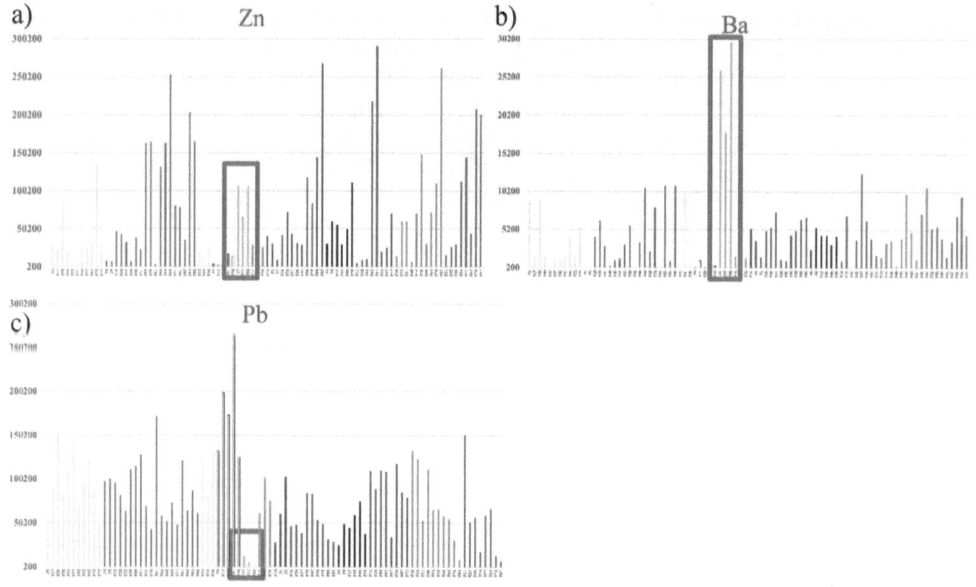

Figure 2. Bar graphs of the XRF counts of the elements (a) zinc, (b) barium, and (c) lead over the measures done in the painting COURT. The points with no pictorial layer (directly in the preparation layer) are indicated by the red rectangle. The correlated countings of Zn and Ba, with the absence of Pb, show that the first layer was made of lithopone (ZnS, BaSO4). The high quantity of Pb in all the points suggests lead white (2PbCO3.Pb(OH)2) as a second white layer.

It was confirmed the presence of the pigment lead white (2PbCO3.Pb(OH)2) in both paintings (COURT and REGENT). The element lead (Pb) was detected by ED-XRF with high quantities in almost every point, except at those points measured over the preparation layer, in regions where it presented a thin pictorial layer, or at points that presented high amounts of other elements, in that case, coming from pigments deposited on top of the lead white. In Figure 2c is shown the bar graph of Pb for the painting COURT to exemplify. In addition, the presence of lead white was verified by Raman, that detected the band of the pigment, at 1050 ± 6 cm^{-1}, in most of the measurement points, reaching higher intensities in light hue regions. The presence of Pb in high quantity in almost every point measured over the paintings suggests that OPS deposited two preparation layers, the first being performed with lithopone, followed by lead white. It is important to highlight that this combination of preparation layers was found in both paintings (COURT and REGENT), revealing a similarity of the construction process by OPS in these artworks.

3.2 *Yellow pigments*

ED-XRF analyses verified the presence of iron (Fe) over yellow regions in both paintings, suggesting the use of the mineral pigment yellow ochre (Fe2O3.nH2O, clay, silica). Another possibility is that the element is coming from iron-based pigments with darker tones, such as orange/brown ochre (Fe2O3.H2O, clay) or burnt sienna (Fe2O3.nH2O, Al2O3), mixed with lighter pigments to produce yellowish hues. This hypothesis is stronger, since the orange and brown regions of the paintings presented higher concentrations of Fe than the yellow regions. To exemplify, the Figure 3a presents the bar graph of Fe counting for the painting COURT, where the counts at brown, black, and red points are significatively higher than at the yellow points.

Also, it was confirmed the use of the synthetic chrome yellow (PbCrO4 or 2PbSO4. PbCrO4) pigment in both paintings by both spectroscopic techniques. While the ED-XRF technique revealed a correlated concentration between chrome (Cr) and lead (Pb) in yellow, green, and orange points, the Raman spectroscopy detected the characteristic set of vibrational bands (with the stronger bands located around 360 cm^{-1} and 843 cm^{-1}) of this yellow pigment at those same points, confirming that the artist had this pigment in his palette and combined it with other pigments to create different colors, such as orange and green. Figure 3b shows the spectra detected at the points yellow P44 and green P60, at the paintings REGENT and COURT, respectively, and the literature spectra of chrome yellow provided by the database of CHSOS. (chsosopensourse.org).

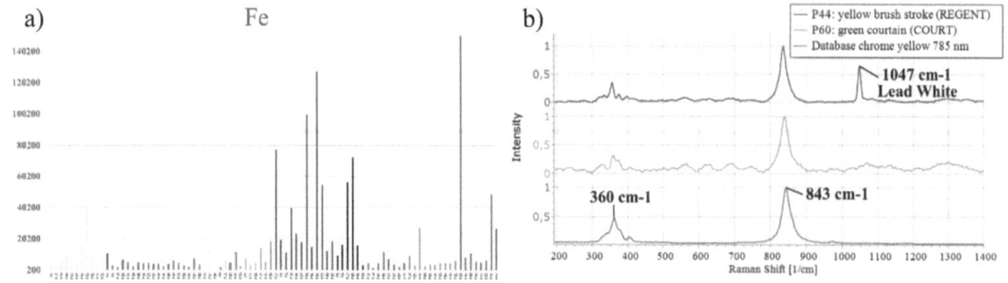

Figure 3. (a) Iron XRF counts over the painting COURT, showing a higher quantity of Fe over brown, black, and red regions, indicating the presence of brown ochre pigments instead of yellow ochre. (b) Raman spectra of the points yellow P44 and green P60 compared with the database spectra of chrome yellow (PbCrO4 or 2PbSO4.PbCrO4) obtained with a 785 nm excitation source, confirming that OPS had this pigment in his palette. It's also identified the band of lead white, at 1047 cm^{-1}, in the point P44.

It is important to highlight that during the comparison of the measured Raman spectra with the database, some bands may seem offset. Since the consulted database is based on a 785 nm excitation source, as are our measures, the main reason for that offsetting might be the fact that the database is composed of pure pigments (with binders), differently than a painting, where the artist can mix numerous pigments to achieve a specific color. This mixture can cause an interference among the response of each pigment present, dislocating the bands, reducing the intensity of the signal or even turning the spectra so noisy that the bands can't be properly read. When evaluating the Raman spectra detected in this study, it's important to take that into account, along with the fact that there can be varnish residual in some regions, since the removing isn't always perfect.

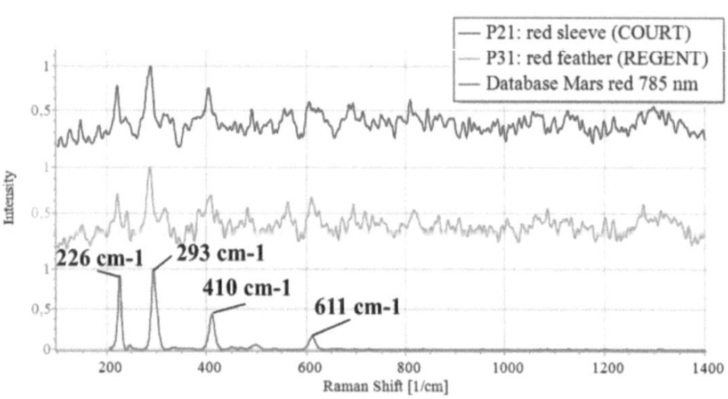

Figure 4. Raman spectra of the red points P21 and P31, measured over the paintings COURT and REGENT, respectively, and the database spectra of the synthetic pigment Mars red (Fe2O3) obtained for a 785 nm excitation source. This confirms the presence of this pigment in OPS palette.

3.3 *Red pigments*

The pigment vermilion (HgS) was confirmed with the spectroscopic results in both paintings, since mercury (Hg) and sulfur (S) were found by ED-XRF and the characteristic Raman spectra of the pigment (with stronger bands in 254 cm^{-1} and 343 cm^{-1}) was detected. The presence of the pigment in red, pink, orange and a few reddish-brown regions in the painting REGENT, and in red, pink, beige, and in a few blue and reddish-brown points in the painting COURT, show that OPS used vermilion to compose other colors.

Another synthetic red pigment confirmed by both spectroscopic techniques was the Mars red (Fe2O3), found in some red regions and, with lower intensity, in some brown regions, both group of color that presented very high concentrations of Fe and that shown the characteristic Raman spectra of the pigment (with stronger bands around 226, 293, 410, and 610 cm^{-1}). Figure 4 presents the spectra of the red points P21 and P31, from COURT and REGENT, respectively, compared to the Mars red spectra provided by the database CHSOS.

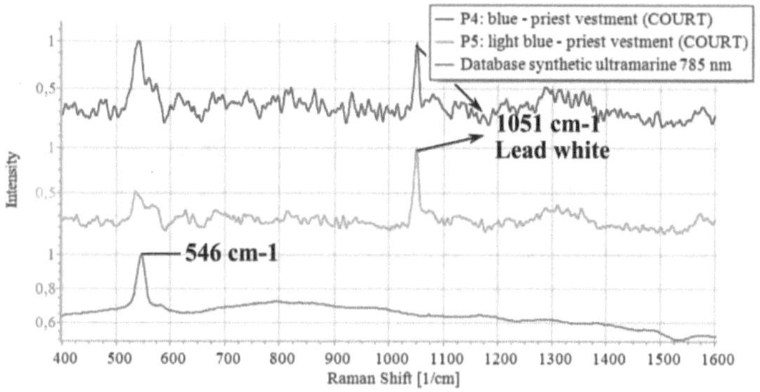

Figure 5. Raman spectra of the blue points P4 and P5 detected exclusively over the painting COURT at a priest's vestment, compared to the database spectra of the synthetic ultramarine pigment (Na8-10Al6Si6O24S2-4), indicating the presence of this pigment in OPS palette. It's also indicated the lead white band, at 1051 cm^{-1}.

3.4 *Orange/brown pigments*

As mentioned before, both paintings present high amounts of Fe, especially in brown, orange, red, black, and yellow regions, indicating that the artist has iron oxides-based pigments in their palettes, like ochre or burnt Sienna. Besides that, it was verified the correlated presence of manganese (Mn) and Fe in some brown and black points of the painting COURT, suggesting the use of the pigment umber (Fe2O3.nH2O, MnO2, Si, Al2O3), while it wasn't detected Mn in the painting REGENT. Unfortunately, these suggested brown pigments didn't respond well to the Raman technique and couldn't be confirmed by both spectroscopic analyses.

3.5 *Blue pigments*

Although the ED-XRF shows counts of Fe in all the blue points measured, which could suggest the use of Prussian blue (Fe4(Fe[CN]6)3), most of the counts have lower values when compared to yellow, orange, brown, black, and red regions. The most probable reason for the detection of Fe in the blue points is the broad use of orange/brown pigments based on Fe over the painting, being present in mixture or even by keeping the brushes a little dirty with the remaining brown pigments. In that sense, the Prussian blue pigment was discarded as a constituent of OPS palette in these paintings.

Both paintings presented significant counts of the element cobalt (Co) in blue and blueish regions, suggesting that the synthetic pigment cobalt blue (CoO.Al2O3) is part of the artist palette. Unfortunately, this pigment couldn't be detected by the Raman spectroscopy.

Finally, the presence of the synthetic ultramarine (Na8-10Al6Si6O24S2-4) was detected only in the painting COURT, in some blue points, by the Raman technique. This pigment, with characteristic band around 546 cm^{-1}, was detected in the priest's vestment at the scene by the measuring points P4 and P5. Figure 5 shows the spectra obtained in these two points and the database Raman spectra of the pigment, indicating that the large band detected in P4 and P5 is due to the presence of the synthetic ultramarine pigment.

3.6 *Green pigments*

For the green pigments, considerable counts of Cr were detected in both paintings with ED-XRF. The presence of this element in green regions, without the detection of yellow pigments based on Cr through the Raman spectroscopy (that usually presents a good response to the technique), indicates that the presence of this element might be due to the green pigment chrome oxide (Cr2O3) or Viridian (Cr2O3.2H2O). Unfortunately, these pigments don't respond well to the Raman spectroscopy and couldn't be detected by this technique.

3.7 *Black pigments*

For the black pigments, the presence of two characteristic vibrational bands in approximately 1325 cm^{-1} and 1580 cm^{-1} were observed by Raman measurements. These bands can suggest the use of carbon black (C), or the pigment bone black (C, $Ca_3(PO_4)_2$), due to its burnt residual indicated by the presence of carbon (C) in the composition. Since it was also verified correlated concentration of phosphorus (P) and calcium (Ca) by ED-XRF mostly in black and dark hue regions, coinciding with the Raman spectra detected, it was possible to confirm the presence of the pigment bone black in both paintings, while the carbon black is only a suggestion.

Table 1. Suggested pallet for the paints COURT and REGENT by OPS, painted in 1922, belonging to Paulista Museum Collection. Color, Principal element detected in ED-XRF, Raman bands detected with a 785 nm excitation source, suggested pigments and their chemical composition.

Color	ED-XRF Principal Element	Raman Bands with 785 nm laser (±6 cm^{-1})	Suggested pigment – Chemical Composition	COURT	REGENT
White	Pb	1050vs	Lead white - 2PbCO3. Pb(OH)2	x	x
	Zn, Ba	988vs	Lithopone - ZnS, BaSO4*	x	x
Yellow	Pb, Cr	339w, 361s, 372m, 403w, 840vs	Chrome yellow - PbCrO4 or 2PbSO4.PbCrO4	x	x
	Fe	not detected	Yellow ochre - Fe2O3. nH2O, clay, silica	x	x
Red	Fe	226vs, 293vs, 410m, 497w, 610m	Mars red - Fe2O3	x	x
	S, Hg	254vs, 284w, 343m	Vermilion - HgS	x	x
Orange/	Fe	not detected	Ochre - Fe2O3.H2O, clay	x	x
brown	Fe	not detected	Burnt sienna - Fe2O3. nH2O, Al2O3	x	x
	Fe, Mn	not detected	Umbra - Fe2O3.nH2O, MnO2, Si, Al2O3	x	
Blue	Co	not detected	Cobalt blue - CoO.Al2O3	x	x
	Si	545vs	Ultramarine - Na8-10Al6Si6O24S2-4	x	
Green	Cr	not detected	chrome oxide or Viridian - Cr2O3/Cr2O3.2H2O	x	x
Black	not detected	1325vs(br), 1580vs(br)	Charcoal - C	x	x
	Ca, P	1325vs(br), 1580vs(br)	Bone black - C, Ca3(PO4)2	x	x

*It's also suggested, for both paintings, the use of barite (BaSO4), with a characteristic Raman band at 988 cm^{-1}, and zinc white (ZnO), that hasn't been detected with Raman.

With the exception of the pigments umber and synthetic ultramarine, detected exclusively in COURT, the palette suggested to both paintings are practically the same, being very similar to a previous discussion of OPS palette. (Campos et al. 2014) A systematization of the pigments suggested for the paintings is listed in Table 1 where the pigments are sorted by color, with its ED-XRF principal elements, and its detected Raman bands indicated.

With the imaging analysis, IRR revealed several stylistic details of the artist that appeared in both paintings. It was observed that several vertical and horizontal lines were drawn with graphite, checkering the canvas, to construct the final image with the correct proportion. Also, almost every single element present in the paintings was drafted with graphite, except for faces in the background or in crowds.

In the Figures 6 it is possible to see very clear lines sketching the contour of the objects, correction of the drafts, changing its dispositions or adding new elements, besides the vertical and horizontal lines of the grid. Figure 6a shows the outline of the cannon made with graphite in the painting REGENT, where it also appears that the ropes were added after painting the cannon. The Figure 6b shows the sketch of windows' ornament and curtains in the painting COURT, where it can also be seen that the artist drew lines connecting the windows, to define

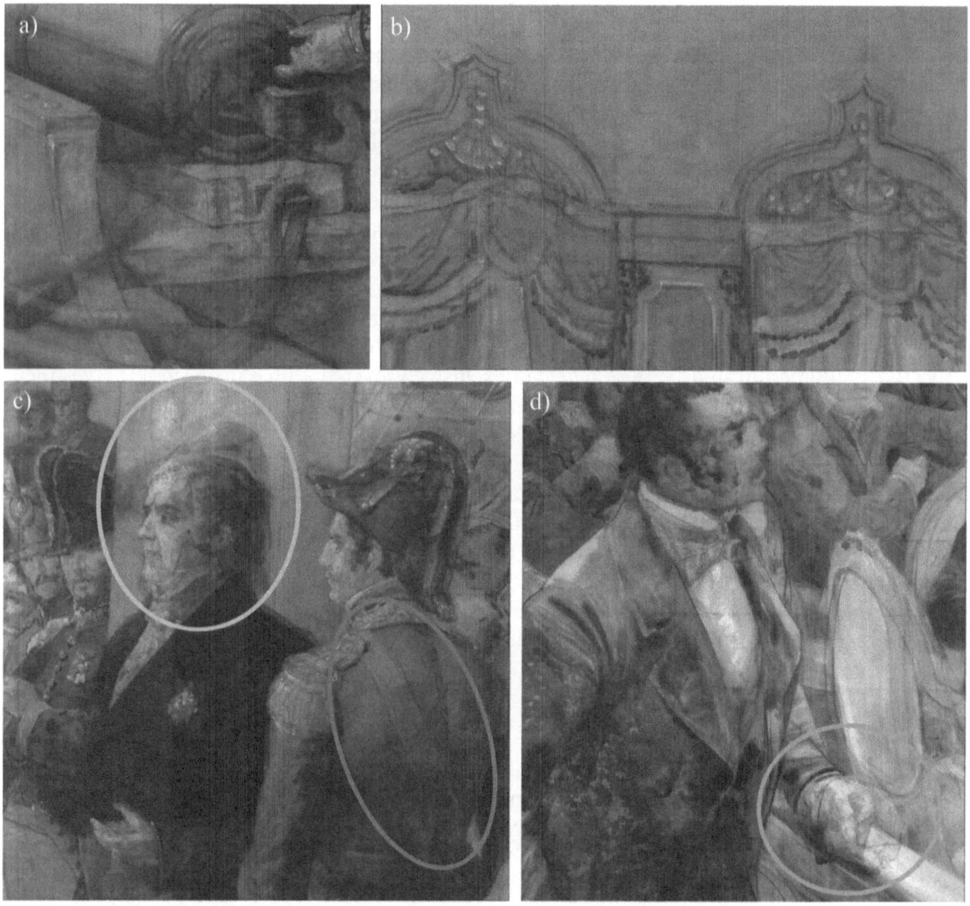

Figure 6. Selection of IRR images showing creative steps of OPS, such as a grid done so the artist could reproduce the scenario with the right proportion, that can be found in all four figures. Also, it can be seen the draft and addition of posterior elements, like in (a) the outline of the cannon and the addition of ropes in the painting REGENT, and (b) the sketching, with corrections, of the windows and curtains in the painting COURT. It was found some *pentimenti* made by OPS, such as (c) removing a hat (blue) and a uniform strip (red) (REGENT), and (d) changing the position of a man's arm and the size of his hand (green) (COURT).

the background perspective of the room. Besides that, it's possible to see that OPS had a few regrets about the sketch and decided to alter the outline of the elements.

The IRR images also allowed the identification of regions that underwent transformations from the initial painting to the final image, that is, the artist´s regrets or *pentimenti.* A few examples of this creative step are presented in the Figures 6c and 6d. In the Figure 6c, selected from the painting REGENT, it can be seen that the artist decided to remove elements of the scene, such as the character's hat (identified by the blue circle) and the man's uniform strip (identified by the red circle). Besides that, it can also be seen that OPS altered the contour of the first man, reducing his shoulders. Figure 6d exemplifies a few repentances identified in COURT such as the modification in the position of the man's arm, that is taken behind his body, and in the size of his left hand, that is drastically reduced (circled in green). These creation steps of the painting were also observed in previous research of this artist (Campos et al. 2014), showing that although he was a very thoughtful artist, creating his works with rigorous style and close to reality details, he wasn't static with his paintings, being able to adapt and recreate if necessary.

4 CONCLUSIONS

The use of non-invasive spectroscopic and imaging techniques of analysis, with equipment that are portable and can be carried to the museums and atelier, offer a great help to the study and characterization of cultural heritage objects. The complementary application of the spectroscopic methods Energy Dispersive of X-ray fluorescence (ED-XRF) and Raman made it possible to verify that, before painting the scenes, OPS prepared both canvas with white pigment layers, composed of a sequence of lithopone and lead white. It was also determined the mutual palette is composed by a combination of four white pigments (lead white, lithopone, zinc white, and barite), iron-based pigments to yellow, orange, and brown colors (ochre and burnt Sienna), chrome yellow, vermilion and Mars red as the red pigments, cobalt blue, chrome based green, lamp and bone black. Besides that, the pigments synthetic ultramarine blue and umber, a combination of iron and manganese oxides, were found only in the painting COURT. Curiously, the synthetic ultramarine was detected exclusively over the blue in the priest's vestment present in the scene. Some of those pigments mentioned were found in previous investigations.

The IRR imaging technique revealed stylistic processes of the artist, where it was found in both paintings the same style of construction of the scenes. OPS systematically used to draw vertical and horizontal lines on top of the preparation layer with graphite, checkering the canvas, to reproduce the scene with the right proportion. Still with graphite, he also sketched almost every element and character of the painting. Many drafts went through small corrections while still in the sketching process, but there were also found repentances in the paintings. These creative steps were also found in previous research of OPS, corroborating to the characterization of his work and making it possible to comprehend his painting style over his professional years.

ACKNOWLEDGEMENTS

This research was supported by the National Council for Scientific and Technological Development (CNPq) and Research Support Foundation of the State of São Paulo (FAPESP). Our thanks to the Paulista Museum organization and the restoration team from Raul Carvalho's atelier, for allowing this study to happen.

REFERENCES

Ainsworth, M. W. 2005. From connoisseurship to Technical Art History: The Evolution of the Interdisciplinary Study of Art. *The Getty Conservation Institute Newsletter* 20(1): 4–10.

Appoloni, C.R., Parreira, P.S. & Blonski, M.S. 2006. Estudo da composição química elementar dos pigmentos de uma pintura atribuída a Gainsborough com um sistema portátil de fluorescência de raios x.

ARC Revista Brasileira de Arqueometria Restauração Conservação, special edition. Olinda: AERPA Editora Resumos do III Simpósio de Técnicas Avançadas em Conservação de Bens Culturais.
Appoloni, C.R., Lopes, F. & Bruno, M.A. 2013. Análise da pintura "Moema" por PXRF, TXRF e espectroscopia Raman. *Moema, restauração/restoration*: 59–78. São Paulo: Comunique Editorial, São Paulo
Artioli, G. 2010. *Scientific methods and cultural heritage - An introduction to the application of materials science to archaeometry and conservation science*. New York: Oxford.
Brooke, C.J. 2020. Raman spectroscopic analysis of an early 20th century English painted organ case by Temple Moore. *Heritage* 3(4): 1148–1161.
Calza, C., Anjos, M.J., Pedreira, A. & Lopes, R.T. 2007. Analysis of the painting "The first mass in Brazil" (Victor Meirelles) using a portable XRF system. *International Nuclear Atlantic Conference* – INAC, Associação Brasileira de Energia Nuclear – ABEN
Campos, P.H.O.V., Kajiya, E.A.M., Rizzutto, M.A., Neiva, A.C., Pinto, H.P.F. & Almeida P.A.D. 2014. X-ray fluorescence and imaging analyses of paintings by the Brazilian artist Oscar Pereira da Silva. *Radiation Physics and Chem*istry 95: 362–367.
Colantonio, C., Clivet, L., Laval, E., Coquinot, Y., Maury, C., Melis, M. & Boust. C. 2021. Integration of multispectral imaging, XRF mapping and Raman analysis for noninvasive study of illustrated manuscripts: the case study of fifteenth century "Humay meets the Princess Humayun" Persian masterpiece from Louvre Museum. *European Physical Journal Plus* 136, Art. n. 958.
Cosentino, A. 2015. Effects of different binders on technical photography and infrared reflectography of 54 historical pigments. *International Journal of Conservation Science* 6(3): 287–298.
Delaney, J.K., Dooley, K.A., Loon, A. & Vandiverde, A. 2020. Mapping the pigment distribution of Vermeer's Girl with a Pearl Earring. *Heritage Science* 8, Art. n. 4.
Gilardoni, A. 1977. *X-rays in art: physics-techniques - applications*. Italy: Gilardoni S.P.A.
Jenkins, R. 1999. *X-Ray Fluorescence Spectrometry*. John Wiley & Sons, Inc.
Junior, C.R.L. 2015. Um artista às margens do Ipiranga: Oscar Pereira da Silva, o Museu Paulista e a reelaboração do passado nacional. Dissertation (Master degree in culture and Brazilian identity). São Paulo: Brazilian studies institute, Unversity of São Paulo.
Lage, M.C.S.M. & Farias Filho, B.B. 2018 Arqueometria aplicada à conservação de sítios de arte rupestre. In: *Cadernos do Lepaarq*, v. XV, n.30., p. 327–43.
Mancia, R. 1944. *L'esame scientifico delle opere d'arte e il loro restauro*. Milano: Ulrico Hoepli.
Moreau, R., Brunel-Duverger, L., Pichon, L., Maignard, B, Gourier, D. & Calligaro, T. 2023. Application of a MA-XRF/RIS/PL scanner to paintwork studies. *European Physical Journal Plus* 138, Art. n.16.
Rizzutto. M.A 2015. Métodos físicos e químicos para estudo de bens culturais, *Arqueometria para Bens Culturais* v. 28, n. 43.
Rosado, A. 2008. História da Arte Técnica e Arqueometria: uma contribuição no processo de autenticação de obras de arte. *19&20* (ISSN 1981-030X) v. 2, n. 1.
Stols, M. 2012. Grounds, 1400-1900. *The Conservation of Easel Paintings: 161-85*. Routledge.
Stuart B.H. 2008. *Analytical Techniques in Material Conservation*. West Sussex: John Wiley & Sons, Ltd
USP. 2022. Museu do Ipiranga conclui obras de restauro de seu edifício-monumento. USP online newspaper.

Lasers in the Conservation of Artworks XIII – Siano & Ciofini (eds)
© 2024 The Author(s), ISBN 978-1-032-47995-8
Open Access: www.taylorfrancis.com, CC BY-NC-ND 4.0 license

Non-destructive investigation of a late panel painting by Lucas Cranach the Elder

I.M. Cortea, L. Ratoiu & R. Rădvan
National Institute of Research and Development for Optoelectronics - INOE 2000, Măgurele, Romania

ABSTRACT: This work presents the results of a complementary diagnostic investigation carried out on a late wooden panel painting attributed to world-famous Renaissance painter Lucas Cranach the Elder. The aim of this study was to characterize the pigments, ground layers and painting technique used in Cranach's painting *Madonna and Child* by means of an integrated non-destructive approach that included Hyperspectral Imaging (HSI), portable X-ray fluorescence (XRF) and Fourier Transform Infrared Spectroscopy (FTIR). Combined XRF and FTIR data allowed identification of a vibrant color palette, with lead white, verdigris, smalt, vermilion, red earth, yellow ochre, and a zinc-rich brown earth among the used pigments. Hyperspectral imaging showed no underdrawings, but various technical features and past restoration interventions were possible to document. The findings provide valuable artistic and technical information regarding the painting materials and style used by this influential 16th-century master, as well as key information that can support the conservation strategy of this remarkable artwork.

1 INTRODUCTION

Lucas Cranach the Elder (1472 - 1553) was one of the most important artist of the 16th century German art world, best known for his portraits of German royalty, biblical scenes and mythological subjects. A prolific artist with over 1000 surviving artworks (Hand 1993; Heydenreich 2007; Lubashevsky 2015), Cranach headed a large workshop that included among the numerous apprentices, his sons Hans, and Lucas the Younger. Cranach's continual search for the most efficient techniques enabled the development of a rapid style that facilitated the workshop production (Heydenreich 1998). Starting from the mid 1520s hundreds of artworks were produced with many works existing in different variations (Frank 2015). After his death, paintings in "Cranach style" and versions of his works were still produced by his studio, giving rise to many problems of attribution. Cranach's impressive legacy - which includes paintings, woodcuts, copper engravings, book illustrations and drawings, inspired numerous studies along the years – from technical studies focused on Cranach's style and workshop practice (Heydenreich 1998, 2007), to stylistic and iconographic analyses (Lubashevsky 2015; Tabutsadze et al. 2021; Zorzin 2020), and, more recently, to interdisciplinary studies focused on technical aspects (Klisińska-Kopacz et al. 2022; Molari and Appoloni 2019). In recent decades, new knowledge on attribution, authenticity, and dating of Cranach's paintings has been gained through systematic study of his painting materials and techniques. However, despite the rich technical documentation available, only a small portion of it has been published (Heydenreich et al. 2014).

In this study, a wooden panel painting (Figure 1a) attributed to the workshop of Lucas Cranach the Elder has been investigated via a combined non-destructive approach with the aim to document the conservation state of the investigated artwork, and to obtain a characterization of the painting materials and techniques.

Taking into account several conservation issues, an integrated non-destructive approach was considered. Hyperspectral imaging (HSI), was carried out in-situ, as a first-step in providing key information in terms of technical features and possible underdrawings, followed by *in situ* portable X-ray fluorescence (XRF) measurements that aimed to obtain a basic chemical

DOI: 10.1201/9781003386872-3

characterization of the employed color palette. For in-depth characterization of both organic and inorganic compounds present in the investigated paint layers, Fourier transform infrared spectroscopy (FTIR) was employed on representative micro-samples.

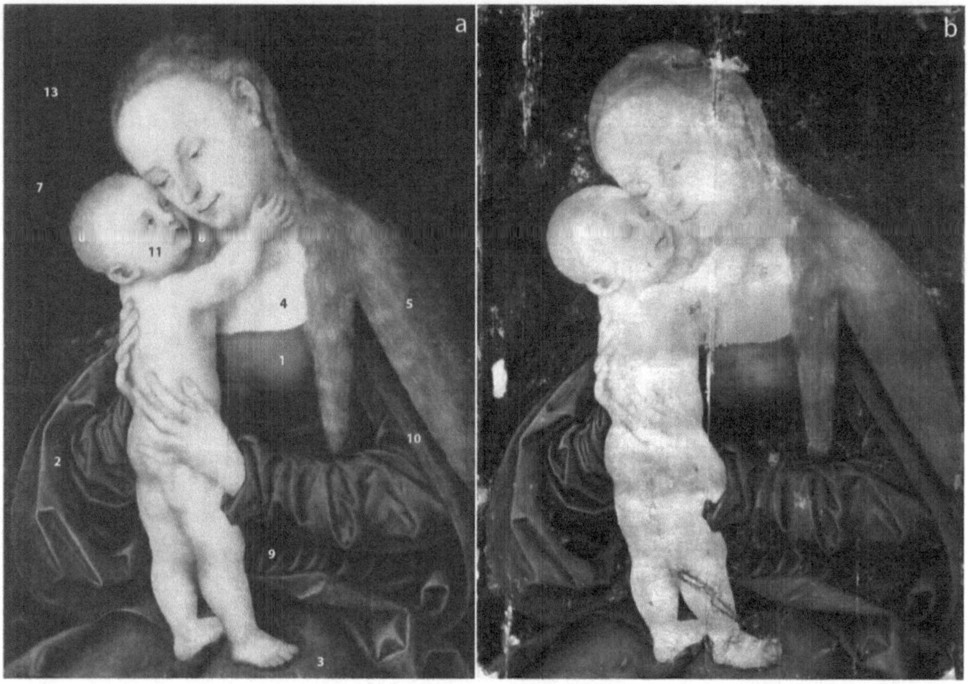

Figure 1. (a) *Madonna and Child* (oil on lime wood panel) attributed to Lucas Cranach the Elder (European Art Gallery, National Museum of Art of Romania, Bucharest). (b) False color infrared image (954, 1600 and 2400 nm) showing various retouches and restoration areas (in blue and white-yellow color).

2 MATERIALS AND METHODS

2.1 *Description of the investigated painting*

The painting investigated in this study, *Madonna and Child*, is attributed to Cranach's late creation period and is currently part of the National Museum of Art of Romania collection. It is signed with Cranach's serpent insignia on the lower left part, with the serpent's wings folded (Figure 2), signature variation adopted after 1537 (Hand 1993). The stylistic features refer to the artist's last years of work when the Madonna or other characters were placed on a dark background having characteristic postures and movements as found in iconography. The same subject is treated the same way in several other versions (Cranach Digital Archive). The painting is in a reasonably good condition, although the paint layer is thin in places and there are some paint losses especially on the edges. The panel painting has undergone several restoration interventions along the years, both at the level of the paint layer and at the level of the wooden support, but only the recent interventions, after the 1950s, are well documented (Corduban, Pătrașcu and Vasile 2016).

2.2 *Hyperspectral imaging*

Hyperspectral images of the entire surface of the painting were registered using a HySpex SWIR-384 hyperspectral camera from NEO equipped with a state of the art mercury cadmium telluride (MCT) detector. HSI data were acquired in push broom technique by scanning the artwork horizontally, the camera being mounted on a translation stage. The spectral range was 950-2500 nm, with 288 spectral channels, with a spectral sampling of 5.4 nm and 16 bit resolution. A 30 cm objective was used. During the investigation, illumination was provided by two custom made lamps which focuses the illumination to a line overlapping with the camera FOV. Data processing was carried in ENVI software package (Harris Geospatial Solutions).

2.3 *Portable X-ray fluorescence*

XRF measurements were performed using a Bruker Tracer III-SD handheld XRF instrument equipped with a Rh-anode X-ray tube and a Si-PIN detector with a typical resolution of 190 eV at 10,000 cps. For this study, the working parameters were set at 10.60 μA current intensity, 40 kV tube voltage, no filtering, air atmosphere, 10 s analysis time. The energy response was recorded between 0-40 Kev, with the S1 PXRF software (Bruker AXS Handheld Inc.). XRF analysis was carried out on 13 different areas (as shown in Figure 1a), in order to ensure that the full color palette is covered. Data analysis was carried in ARTAX software, using standard Bayesian deconvolution, and data post-processing was performed with Microsoft Office Excel.

2.4 *Fourier transform infrared spectroscopy*

FTIR measurements were carried on 6 micro-samples taken from original areas (without any interventions) carefully selected by the restorers. Data were registered in attenuated total reflectance (ATR) mode employing a Perkin Elmer Spectrum Two FTIR spectrometer equipped with a GladiATR accessory (monolithic diamond ATR crystal) from Pike Technologies. Spectra were acquired in the 4000-370 cm^{-1} range, at a resolution of 4 cm^{-1} by averaging 16 scans. Spectra are presented in transmittance with baseline correction applied. Data processing was carried out in Essential FTIR Spectroscopy Software Toolbox.

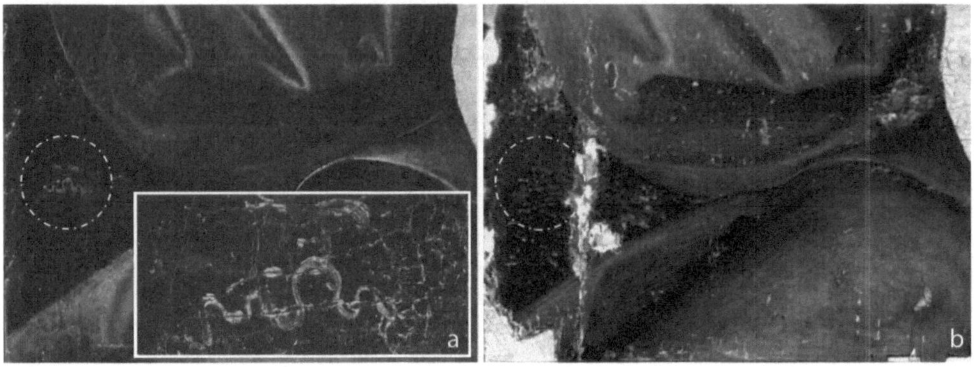

Figure 2. Detail of the signature: (a) visible image, and (b) IR image (2200 nm).

3 RESULTS AND DISCUSSION

3.1 *Imaging investigation*

Hyperspectral imaging investigation allowed identification of various details regarding the technical aspects of execution as well as information related to the actual state of conservation. False color infrared image (IRFC) showed restored areas of the paintings as seen in Figure 1b where interventions from two distinct stages can be clearly distinguished – highlights in in blue and white-yellow color. A series of degradations due to the wooden support could also be highlighted on the child's leg. Small lacunae and abrasions of the original paint layers can be clearly seen especially within the background, shadow areas, Madonna's hair and robe. A fine network of dense cracking spread across the entire surface of the painting can also be seen, denser on the flesh areas and luminous tones in general (Figure 3), a common aspect for Cranach's paintings.

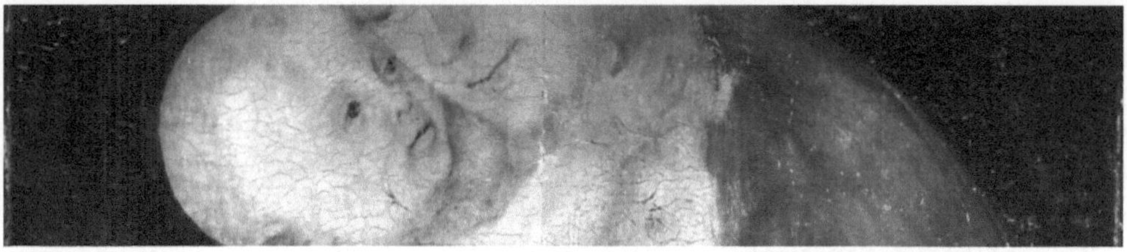

Figure 3. IRFC image (954, 1800, and 2400 nm) showing crack pattern in an area of flesh paint.

No carbon-black underdrawings were found, situation that may be explained either by the absence of any preliminary drawing or to the use of an infrared transparent material. A very fine, liquid medium could have been used over the priming to outline the figures, as suggested in other previous studies carried on Cranach's paintings (Foister 2015). Modeling of the volumes (udermodelling) with greyish washes was frequently documented among Cranach's practice. Other studies (Heydenreich 1998) showed that the linear underdrawings were efficiently integrated in the painting process as final shadows, with underdrawings generally based on a black drawing medium or red chalk in later works. Detailed inspection of the registered infrared images (Figure 4) highlights a masterful technique, and overall a great freedom of execution, with almost no changes in the layout of the composition.

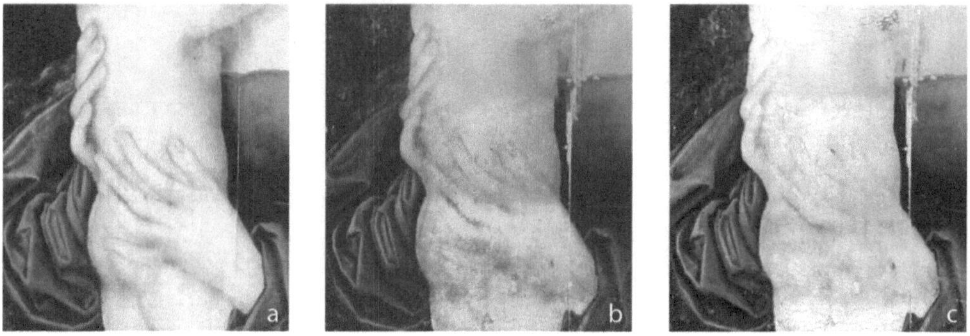

Figure 4. Details on Madonna's hands showing the absence of any preliminary carbon based drawing and a masterful handling of the brush: (a) black and white image, (b) 954 nm, and (c) 1802 nm infrared image.

3.2 *Painting materials*

3.2.1 *Ground layer*

The ground layer was found to be based on calcium carbonate/chalk ($CaCO_3$) and some protein compound (Figure 5a). Chalk, one of the most frequently used mineral fillers for ground layers in North West European oil paintings (Roy 1993; Stols-Witlox 2017) was identified based on the strong absorptions observed at approx. 1400, 872 and 712 cm^{-1} (C–O stretching of carbonate), and the two small peaks centered around 2513 cm^{-1} and 1795 cm^{-1} which are combinations and overtone bands (Gunasekaran, Anbalagan and Pandi, 2006). The weak but broad absorption around 1640 cm^{-1} (amide I band) along the small peaks centered around 2934, 2855 cm^{-1} (C-H stretching vibrations) and 3300 cm^{-1} (N-H group) are clear indicators for protein related content, most probably animal glue. The presence of a calcium based compound at the level of ground layer was also inferred by the intense Ca-K lines registered by XRF analysis (Table 1) on existing lacunas within the paintings. Strontium, also detected for these areas, can be linked to calcium-based-grounds, as Sr frequently accompanies calcium in natural sulfates or carbonates (Rebollo et al. 2013), either under the form of celestine ($SrSO_4$) or under the form of strontianite ($SrCO_3$).

Low intensity Pb and Fe lines were registered on the ground layer as well suggesting the existence of a priming layer (*imprimitura*, a translucent toning of the ground) containing lead white and small amounts of earth pigments. As indicated by previous studies (Heydenreich 1998), Cranach's ground layers were typically white (based on chalk bound in animal glue), with rare exceptions having a reddish color obtained by mixing red lead, white lead and calcium carbonate. The use of a white priming as intermediate layer to adjust the properties of the ground was also documented (Heydenreich 1998) as characteristic for German paintings at the beginning of the 16th century.

3.2.2 *White pigments*

As expected, lead was detected in all XRF investigated areas, indicating the use of lead white (a mixture of cerrusite - $PbCO_3$ and hydrocerussite - $Pb_3(CO_3)_2(OH_2)$ which ratio can change depending on the preparation). The presence of lead white was confirmed by FTIR analysis via the characteristic infrared bands observed c. 1397 and 678 cm^{-1} (strong carbonate stretching vibrations). XRF registered data suggests that lead white was used not only for the white tones but also to give more luminosity to the other pigments. Trace amounts of copper and bismuth, correlated to the lead white pigment, can be ascribed with impurities present within lead ores (Roy 1993). Lead white has been produced artificially since early historical times being by far the most important white pigment in Europe since the Roman period (Eastaugh et al. 2008).

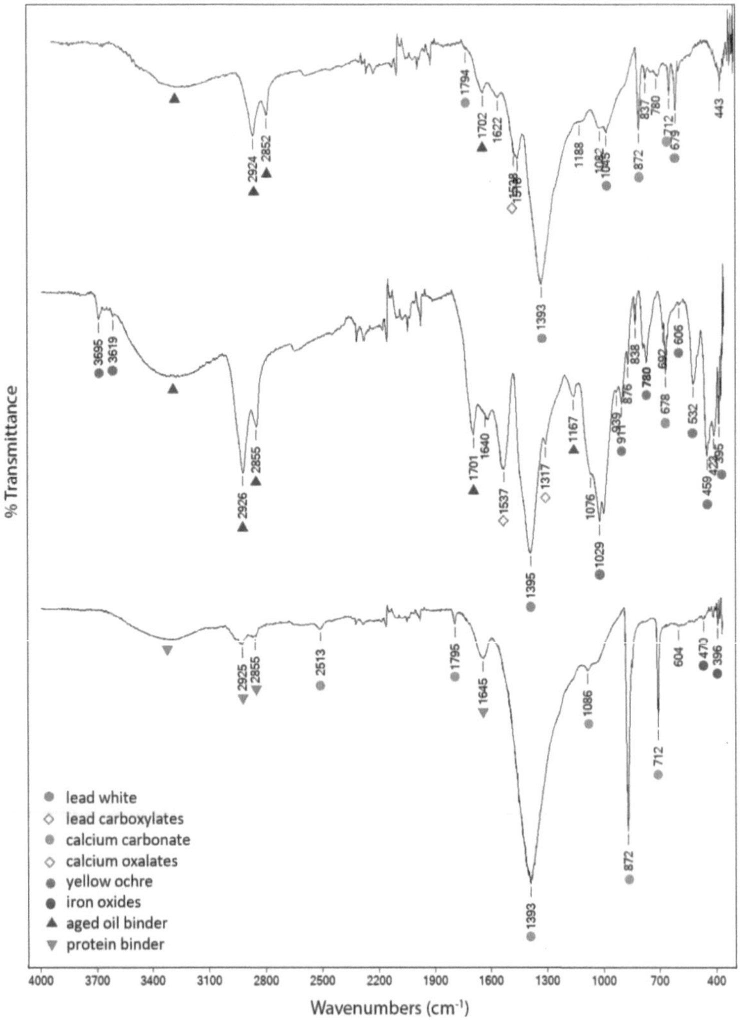

Figure 5. FTIR spectra of: (a) ground layer, (b) ochre paint layer, and (c) blue paint layer.

3.2.3 *Red pigments*

Examination of red areas indicates the use of vermilion (HgS), intense Hg-L lines being registered. No S lines were detected, situation that may be explained by the presence of lead white, Pb-M lines overlapping S-K lines. A widely used term for the synthetic forms of the natural occurring mineral cinnabar or cinnabarite (mercury (II) sulfide), vermilion was the most widely used red pigment until the beginning of the 20th century, when cadmium red was discovered. Small amounts of red iron oxides (red ochre) admixed with vermillion and lead white may also have been used to obtain the pink carnation tones (cheeks, lips), as indicated by the minor inputs of Fe observed in the registered XRF spectra.

3.2.4 *Yellow pigments*

Intense Fe-K peaks were registered on the Madonna's hair, suggesting the use of an iron-rich earth pigment. Higher Ti lines were also registered for these areas, titanium concentration being explained by the terrigenous input (Gil et al. 2007). FTIR spectra (Figure 5b) confirmed the use of a yellow ochre ($FeO(OH)_x$) rich in aluminosilicates. Kaolinite (characteristic peaks at 3695, 3650, 3620 cm^{-1} (hydroxyl ion bands); 1029 cm^{-1} (Si-O-Si band); 1009 cm^{-1} (Si-O-Al band); 938 and 911 cm^{-1} (Al-O-H)), ferric oxides (diagnostic peaks at 535 and 460 cm^{-1}) as well as small amounts of quartz (diagnostic peaks at 1167, 796, 778, 692 cm^{-1}) were identified. The broad peak around 3400 cm^{-1}, ascribed to free hydroxyl ions of kaolinite, suggests the presence of hydrated iron oxides (goethite) (Cortea et al. 2022). To obtain the various hues of the Madonna's hair, yellow ochre was mixed with lead white and small amounts of vermilion, as indicated by XRF data. The red-orange hues within the Madonna's robe may have been obtained by mixing vermilion with a yellow arsenic sulfide pigment such as – realgar (As_4S_4) or orpiment (As_2S_3), as suggested by the registered As lines within this investigated area. Such pigment combination has been previously reported within Cranach's practice (Nöller 2015).

3.2.5 *Green pigments*

The green areas investigated highlighted the most intense Cu-lines, a clear indicator for the use of a copper-based pigment such as malachite or verdigris. FTIR spectrum registered on the green areas could not offer a clear identification in terms of employed pigments, the registered FTIR spectra being mainly defined by the strong absorptions ascribed to lead white and calcite. Some peaks poorly defined within the 1600-1400 cm^{-1} region could be due amorphous copper fatty acid soaps, inferring an alteration of verdigris (Buti et al. 2013). This hypothesis is sustained by previous studies carried out on other Cranach's artworks that showed verdigris to be the first choice pigment for obtaining intense green tones, especially for backgrounds and drapery (Foister 2015; Heydenreich 2004). Verdigris (a collective term for copper acetates of different chemical composition) was commonly used within the 15th to the 17th century to produce intense and pure green tones. Moreover, as shown in previous studies, in Cranach's paintings it was found to be frequently applied over an opaque layer of lead white or mixed with lead-tin yellow to produce desired color effects (Foister 2015).

3.2.6 *Blue pigments*

For the blue areas investigated XRF analysis highlighted the presence of As and Co that suggests the presence of smalt - a potassium glass containing cobalt frequently used as a blue pigment in European oil painting within the 16th century (Roy 1993; Rebollo et al. 2013). It is well documented (Spring 2017) that the only cobalt ores in Europe until the 18th century were found in Saxony, as arsenates and arsenides, with high quality smalt produced within the Netherlands. A rather unstable pigment due the rich potassium level, smalt is prone to deterioration - with formation of soaps and carboxylic acids (Spring, Higgitt, and Saunders 2005), that usually determines a grey-brown appearance of the paint film. FTIR spectra registered on the blue paint sample (Figure 5c) may indicate the presence of potassium soaps, weak but characteristic bands being observed at approx. 1562, 1460, 1403 cm^{-1} (Spring et al. 2005).

Table 1. Results of the XRF analysis obtained on the investigated areas.

Area	Description	XRF detected elements*
1	Intense red hue, Madonna's robe	**Pb**, **Hg**, Ca, Sr, Fe, As (Zn, Ti, Mn, K, Cu, Co, Bi)
2	Dark green, Madonna's robe	**Cu**, **Pb**, **Ca**, Sr, Fe (Zn, Ti, Mn, Hg, Bi, Co)
3	Greenish blue, Madonna's robe	**Pb**, **Cu**, Ca, Co, Fe, As, Sr (Zn, Ti, Mn, Hg, Bi)
4	Light carnation, Madonna's chest	**Pb**, Ca, Hg, Fe (Sr, Zn, Ti, Mn, Cu, Co, Bi)
5	Reddish ochre, Madonna's hair	**Pb**, **Ca**, **Fe**, Sr, Ti, Hg (Zn, Mn, Cu, Co, Bi)
6	Light ochre, Child's hair	**Pb**, Ca, Fe, Sr (Ti, Mn, Cu, Hg, Bi)
7	Dark brown, background	**Ca**, Pb, Sr, Cu, Fe Zn (Ti, Mn, K)
8	Shadow, Child's hand	**Pb**, **Ca**, Fe, Sr, Zn (Cu, Bi, Hg, Ti, Mn, Co)
9	Dark brown, Madonna's robe	**Ca**, **Pb**, Zn, Sr, Fe, Cu (K, Mn, Ti, Hg)
10	Greenish blue, Madonna's robe	**Cu**, **Pb**, Co, As, Fe, Ca, Sr (Bi, Zn, Ti, Mn)
11	Light pink carnation, Child's cheek	**Pb**, Ca, Hg, Fe (Bi, Sr, Cu, Ti, Mn, Co)
12	Pink carnation, Child's lips	**Pb**, Ca, Hg, Fe, Sr (Bi, Cu, Zn, Ti, Mn)
13	Ground layer (lacuna)	**Ca**, Pb, Sr, Fe (Cu, Zn, Mn, K)

* Elements are listed in decreasing order of abundance. With bold - major elements, regular - minor elements, in brackets - trace elements.

3.2.7 *Dark brown and black pigments*

XRF spectra obtained on the dark/black regions are characterized by high calcium content (Ca-K peaks). For the same areas Pb-L, Sr-K, Fe-K, Cu-K and Zn-K peaks are also present. At least partially, the high Ca-K peaks are due to the calcium carbonate present within the ground layer. Another fraction could be due to the use of a bone black pigment, or to the use of some earth pigments. Trace amounts of Mn and K were also detected for the black areas, which along with Fe and Zn, suggests the use of a raw umber. As shown in previous studies (Amadori et al. 2016), Zn-rich brown earths were found in dark brown areas of 16th century Italian paintings, the zinc content being ascribed to zincite. Another possibility that would explain the Zn content would be the presence of zinc sulfate ($ZnSO_4$) also known as zinc vitriol, an additive documented in various historical sources (Spring 2017) used to modify the properties of the paint. Although the presence of zinc as a sulfate was not possible to establish, the use of zinc vitriol may be suggested as previous studies have found zinc in other Cranach's paintings dated after 1520 (Foister 2015; Spring 2017) in similar locations of the painting such in this case – dark background, black underpaint.

3.2.8 *Organic materials*

FTIR analysis showed characteristic bands for an oil binder (lipids) – peaks at 2924, 2852 cm^{-1} (C–H stretching), 1740 cm^{-1} (C=O) and 1167 cm^{-1} (C–O stretching) (Cortea et al. 2020). The carbonyl band is in most cases poorly resolved (shifting and broadening of the band is observed), due to the formation of carboxylic acids – peak at 1708 cm^{-1} (carboxylic group). The large absorption band between 3000 and 3200 cm^{-1} region can also be ascribed to carboxylic acids resulted from the ageing of oil (van der Weerd, van Loon and Boon 2005). Metal carboxylates (metal soaps) commonly present in oil paintings were also detected. The peaks centered around 1538, 1513, 1460 and 1416 cm^{-1} can be ascribed to lead carboxylates (Otero et al. 2014). Calcium oxalates formed in the surface layers (peak at 1320 cm^{-1}) can be related to the atmospheric exposure of old varnish. At the level of the ground layers proteinaceous compounds (characteristic amide absorptions) assigned most probably to animal glue were identified, as already mentioned.

4 CONCLUSIONS

In this study, a wooden panel painting assigned to the late creation period of renowned old master Lucas Cranach the Elder has been investigated by a non-destructive in-situ approach that included hyperspectral imaging, portable XRF and ATR-FTIR spectroscopy. The employed methodology provided detailed scientific documentation of the painting *Madonna and Child*, the key information obtained in terms of painting materials and technique leading to an enhanced comprehension of this remarkable artwork, currently in the collection of the National Museum of Art of Romania.

Combined XRF and FTIR analysis provided information on regard the ground layers, pigments, organic materials and several degradation compounds in specific locations of the paintings surface. Cranach's typical white ground layer based on calcium carbonate (chalk) and protein glue was confirmed. A priming layer consisting in lead white and small amount of iron oxides was also detected. The color palette was found to be mainly based on traditional pigments such as: lead white, verdigris, smalt, vermilion, red earth, yellow ochre, and a zinc-rich brown earth. Realgar and orpiment (red and yellow pigments of arsenic), and zinc vitriol, an additive used to modify the properties of the paint, may also have been used. The use of smalt, a blue pigment introduced during the 16th century, proves Cranach's constant interest in new materials and his willingness to experiment with new techniques. The overall findings are in agreement with existing literature on Cranach's color palette.

Hyperspectral imaging led to the identification of various details regarding the painting process as well as on previous restorations, and overall current conservation state of the painting. No carbon-black underdrawings were detected, but a fine liquid medium may have been used to outline certain details. Changes in composition were found to be rare, revealing a masterful technique and supporting his reputation as a "quick painter".

Reported results show the potential and limits of selected analytical techniques for the characterization of complex paint layers, with minimum amount of sampling. The study demonstrates the key role of a multidisciplinary study and the advantages of using portable devices, in-situ, especially for situation when artworks cannot be moved outside the museum. Not least, we believe that the findings are of great interest for art historians, conservators and restorers, by offering new insights and greater understanding on the artist' working methods as well as key information for the restoration procedure and long term conservation strategy of this valuable 16th-century artwork.

ACKNOWLEDGEMENTS

The authors gratefully acknowledge The National Museum of Art of Romania who gave permission for the in-situ analysis of the paintings. We express our gratitude to all restorers involved within the Oil Painting Conservation Studio, among all Ioan Sfrijan, Sorina Gheorghiţă, Horaţiu Costin and Rareş Pătraşcu, for their helpful discussion of the results.

FUNDING

The financial support of this work has been provided by the Romanian Ministry of Research, Innovation and Digitization under grant no. 18PFE/30.12.2021 and grant no. PN-III-P4-PCE -2021-1605.

REFERENCES

Amadori, M.L., Poldi, G., Barcelli, S., Baraldi, P., Berzioli, M., Casoli, A., Marras, S., Pojana, G. & Villa, G.C.F. 2016. Lorenzo Lotto's Painting Materials: An Integrated Diagnostic Approach. *Spectrochimica Acta Part A: Molecular and Biomolecular Spectroscopy* 164: 110–122.

Buti, D., Rosi, F., Brunetti, B.G. & Miliani, C. 2013. In-Situ Identification of Copper-Based Green Pigments on Paintings and Manuscripts by Reflection FTIR. *Analytical and Bioanalytical Chemistry* 405: 2699–2711.

Corduban, A., Pătraşcu, R. & Vasile, M. 2016. Reversibile and Irreversibile Interventions. Case Study on 'Madonna and Child' by Lucas Cranach. In *Caietele restaurarii (Restoration notebooks)*: 142–155. Bucharest: Art Conservation Support.

Cortea, I.M, Ghervase, L., Rădvan, R. & Seriţan, G. 2022. Assessment of Easily Accessible Spectroscopic Techniques Coupled with Multivariate Analysis for the Qualitative Characterization and Differentiation of Earth Pigments of Various Provenance. *Minerals* 12(6): 755.

Cortea, I.M., Ghervase, L., Ratoiu, L., Dinu, M. & Rădvan, R. 2020. Uncovering hidden jewels: an investigation of the pictorial layers of an 18th-century Taskin harpsichord. *Heritage Science* 8: 55.

Cranach Digital Archive. http://lucascranach.org. Accessed April 2023.
Eastaugh, N., Walsh, V., Chaplin, T. & Siddall, R. 2008. *Pigment Compendium*. London: Routledge.
Foister, Susan. 2015. *National Gallery Catalogues - The German Paintings before 1800*. https://www.nationalgallery.org.uk/research/research-resources/national-gallery-catalogues/the-german-paintings-before-1800. Accessed Apr 2023.
Frank, K. 2015. Reconsidering Cranach: Digital Approaches to the Cranach Œuvre. *Visual Resources* 31 (3-4): 211–215.
Gil, M., Carvalho, M.L, Seruya, A., Candeias, A.E., Mirão, J. & Queralt, I. 2007. Yellow and Red Ochre Pigments from Southern Portugal: Elemental Composition and Characterization by WDXRF and XRD. *Nuclear Instruments and Methods in Physics Research, Section A: Accelerators, Spectrometers, Detectors and Associated Equipment* 580(1): 728–731.
Gunasekaran, S., Anbalagan G. & Pandi, S. 2006. Raman and infrared spectra of carbonates of calcite structure. *Journal of Raman Spectroscopy* 37(9): 892–899.
Hand, J.O. 1993. *German Painting of the Fifteenth through Seventeenth Centuries*. Washington: National Gallery of Art.
Heydenreich, G. 1998. Artistic Exchange and Experimental Variation: Studies in the Workshop Practice of Lucas Cranach the Elder. *Studies in Conservation* 43(supl): 106–111.
Heydenreich, G. 2004. A Note on Schifergrün. *Studies in Conservation* 48(4): 227–236.
Heydenreich, G. 2007. *Lucas Cranach the Elder: Painting Materials, Techniques and Workshop Practice*. Amsterdam: Amsterdam University Press.
Heydenreich, G., Smith-Contini, H., Stahlmann, J., Görres, D., Herrschaft, J., & Sandner, I. 2014. The Cranach Digital Archive: objectives and opportunities for interdisciplinary and inter-institutional research resources. In Janet Bridgland (ed.), *Preprints ICOM-CC 17th triennial conference "Building strong culture through conservation", Melbourne, 15-19 September 2014*. Paris: International Council of Museums (ICOM).
Klisińska-Kopacz, A., Obarzanowski, M., Frączek, P., Moskal-del Hoyo, M., Gargano, M., Goslar, T., Chmielewski, F., Dudała, J. & del Hoyo-Meléndez, J.M. 2022. An Analytical Investigation of a Wooden Panel Painting Attributed to the Workshop of Lucas Cranach the Elder. *Journal of Cultural Heritage* 55: 185–194.
Lubashevsky, R. 2015. 'Cranach Inc.' A Case Study Determining the Nature and Extent of Lucas Cranach the Elder's Involvement in His Industrious Workshop Using Image Processing. In Luca Pezzati & Piotr Targowski (eds.), *Optics for Arts, Architecture, and Archaeology V; Proc. SPIE 9527, Munich, 24–25 June 2015*. Washington: SPIE.
Molari, R. & Appoloni, C.R. 2019. A PXRF and TXRF Study of The Portrait of a Young Gentleman (1539), by Lucas Cranach the Elder. *Radiation Physics and Chemistry* 165: 108413.
Nöller, R. 2015. Cinnabar Reviewed: Characterization of the Red Pigment and Its Reactions. *Studies in Conservation* 60(2): 79–87.
Otero, V., Sanches, D., Montagner, C., Vilarigues, M., Carlyle, L., Lopes, J.A. & Melo, M.J. 2014. Characterisation of Metal Carboxylates by Raman and Infrared Spectroscopy in Works of Art. *Journal of Raman Spectroscopy*: 45(11-12): 1197–1206.
Rebollo, E., Nodari, L., Russo, U., Bertoncello, R., Scardellato, C., Romano, F., Ratti, F. & Poletto, L. 2013. Non-Invasive Multitechnique Methodology Applied to the Study of Two 14th Century Canvases by Lorenzo Veneziano. *Journal of Cultural Heritage* 14(3): e153–e160.
Roy, A. (ed.) 1993. *Artists' Pigments. A Handbook of Their History and Characteristics*, Vol. 2. Washington: National Gallery of Art.
Spring, M., Higgitt, C. & Saunders, D. 2005. Investigation of Pigment-Medium Interaction Processes in Oil Paint Containing Degraded Smalt. *National Gallery Technical Bulletin* 26: 56–70.
Spring, M.. 2017. New Insights into the Materials of Fifteenth- and Sixteenth-Century Netherlandish Paintings in the National Gallery, London. *Heritage Science* 5: 40.
Stols-Witlox M. 2017. *A perfect ground: preparatory layers for oil paintings, 1550–1900*. London: Archetype Publications.
Tabutsadze, N., Kiknadze, E., Managadze, N., Sordia, S. & Kalandadze., N. 2021. "The Procuress": Unlocking the Painting "The ill-Matched Couple" by Lucas Cranach the Elder. In Eka Kiknadze, Nino Tabutsadze, Mariam Gveleisani & Nino Kalandadze (eds.), *The Integrated Study of the Western European Fine Arts Collections*. Tbilisi: Georgian National Museum.
van der Weerd, J., van Loon, A. & Boon, J.J. 2005. FTIR Studies of the Effects of Pigments on the Aging of Oil. *Studies in Conservation* 50(1): 3–22.
Zorzin, A. 2020. A Portrait of Andreas Bodenstein von Karlstadt by Lucas Cranach the Elder. *Reformation and Renaissance Review* 22(3): 238–252.

Lasers in the Conservation of Artworks XIII – Siano & Ciofini (eds)

The contribution of scanning X-ray fluorescence to the investigation of easel paintings: Examples of applications from the Courtauld Gallery

S.R. Amato & A. Burnstock
Department of Conservation, The Courtauld Institute of Art, Somerset House, Strand, London, UK

ABSTRACT: In recent years, a range of spatially resolved imaging techniques to examine paintings has become integrated into the arsenal of analytical methods used in many museum laboratories worldwide. An example is scanning X-ray fluorescence (XRF), a non-invasive method that provides distribution maps of a wide range of elements in materials used for paintings. Scanning XRF has recently been used together with conventional methods for technical study to investigate paintings from the Courtauld Gallery, including *Christ and the Woman taken in Adultery*, dated to 1565, one of the three surviving *grisaille* paintings by Pieter Bruegel the Elder, a large-scale *Portrait of Don Francisco de Saveedra*, painted by Francisco de Goya in 1798, and *Le Déjeuner sur l'herbe* by Édouard Manet, a smaller an undated version of the large work of the same title painted by the artist in 1863 and on display at the Musée d'Orsay in Paris.

1 INTRODUCTION

A description of the painting techniques used for Pieter Bruegel the Elder's *Christ and the Woman Taken in Adultery* panel and Francisco de Goya's *Don Francisco de Saveedra* portrait, both from the Courtauld Gallery, is summarised in this paper and it is based on conventional methods of technical study including light microscopy, X-radiography, infrared reflectography and elemental analysis of paint samples (Burnstock et al. 2017, Burnstock, Serres 2019). This is followed by a discussion of the interpretation of X-ray Fluorescence (XRF) scans and how in each case the information gained from the application of scanning XRF provided further insights into the technique and condition of the paintings. An example of results of scanning XRF that revealed a key aspect of Édouard Manet's painting process is provided by a study of the Courtauld's *Le Déjeuner sur l'herbe* (Amato et al. 2019, Amato et al. 2020).

2 METHODOLOGY

2.1 *XRF analysis*

Scanning XRF was carried out using an M6 Jetstream spectrometer from Bruker Nano GmbH (Berlin, Germany). The equipment consists of a 30 W Rh-target micro-focus X-ray tube mounted with a 30 mm^2 silicon drift detector (SDD, with energy resolution of < 145 eV for MnKα) on a motorised stage that moves across the painting approximately 2 cm from the surface, with a maximum travel range of 80 x 60 cm^2. The tube is equipped with a polycapillary X-ray lens for beam focusing with a maximum voltage of 50 kV and a maximum current of 0.6 mA.

DOI: 10.1201/9781003386872-4

The elemental distribution maps of the paintings were collected with 1 mm spatial resolution and a dwell time of 50 ms per point, with the exception of those acquired from Pieter Bruegel the Elder's Christ and the Woman taken in Adultery, where 50 ms per point and 0.5 mm step size were used. The elemental distribution maps were processed using the data analysis software packages PyMca and Datamuncher (Alfeld, Janssens 2015).

3 RESULTS

3.1 *Pieter Bruegel the Elder, Christ and the Woman taken in Adultery, 1565, 34.4 x 24.1 cm*

Christ and the Woman taken in Adultery (Figure 1) by Pieter Bruegel the Elder is painted on a single oak board prepared with a layer of chalk and animal glue, followed by a priming layer consisting of lead white and possibly red lead in oil: white aggregates of lead soaps are visible in areas of thin paint (Burnstock, Serres 2019).

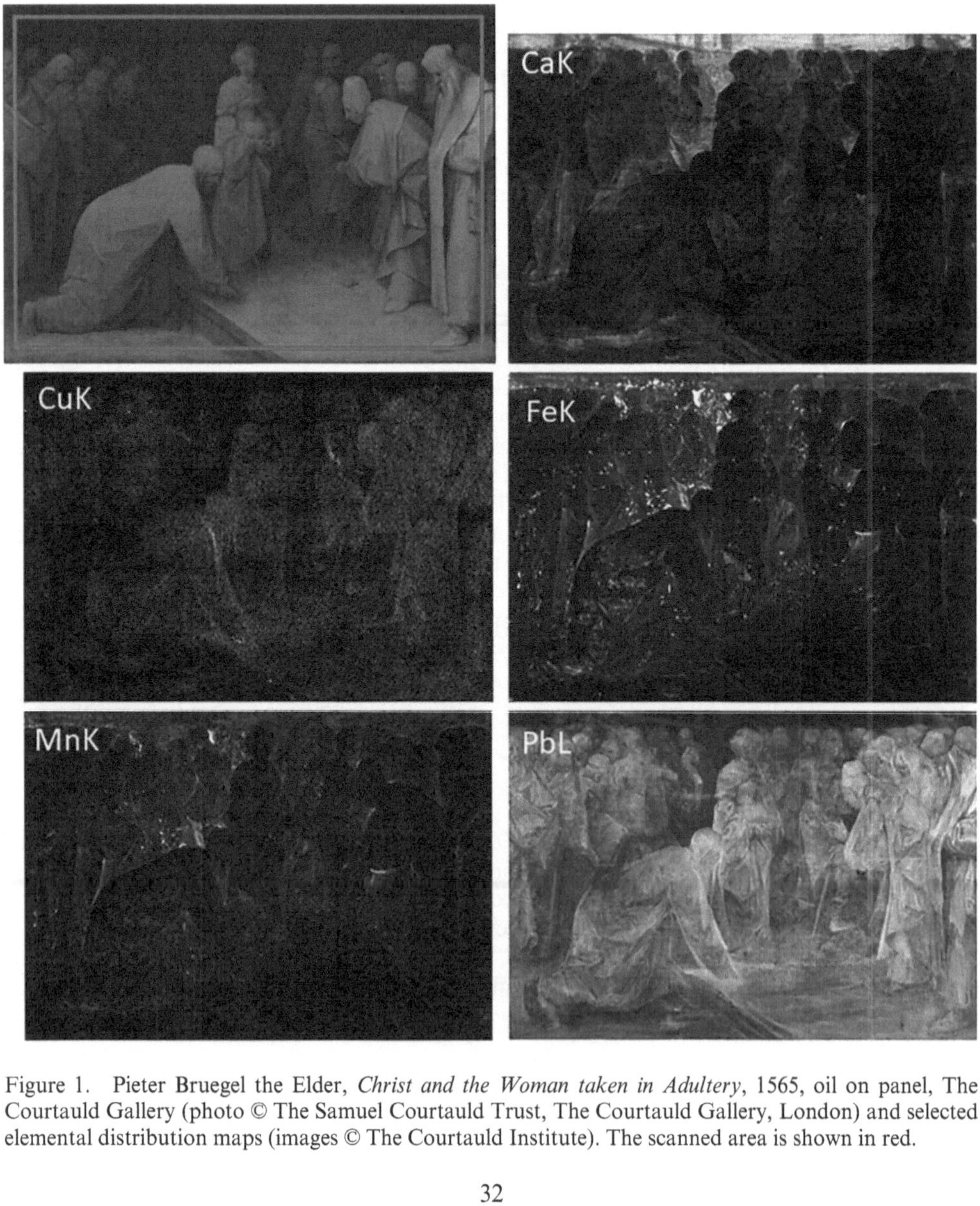

Figure 1. Pieter Bruegel the Elder, *Christ and the Woman taken in Adultery*, 1565, oil on panel, The Courtauld Gallery (photo © The Samuel Courtauld Trust, The Courtauld Gallery, London) and selected elemental distribution maps (images © The Courtauld Institute). The scanned area is shown in red.

The composition was first outlined freehand in red that was overlaid with a more detailed drawing in black, including hatching to indicate areas of shadow. The four central figures were painted directly onto the white priming, using dilute washes of monochrome paint composed of black and white pigments, with a low proportion of red ochre to create shadows and highlights. A low proportion of azurite was added to the paint mixture for the robe and flesh tones of the main figures to impart a cooler tone.

Scanning XRF provided more information about the order of application of the paint (Figure 1). The CaK map shows areas of the chalk ground that are sparsely covered by superficial paint that contains lead; for example, the area below Christ's legs, the background and the dark passages between the figures. The PbL map registers streaky marks running horizontally and vertically across the composition that are likely to indicate the lead present in the priming layer. These marks appear dark in the CaK map, confirming that the Pb signal is coming from a layer above the chalk ground, shielding the Ca signal. The PbL map also indicates changes in the composition in the painting stage using paint containing lead, for example in the area above Christ's shoulders and to the left of the bearded Pharisee, the modelling for the folds of the draperies now overlaid with darker paint, and the addition of extra figures' heads above the Pharisee on the right.

Further information about the condition of the paint layers is evident in the FeK and MnK maps, that show areas of retouching and the use of iron-containing pigments in the original paint layers. These signals are consistently low in the areas corresponding to shadowed passages of the composition, suggesting that the dark passages consist mainly of carbon black, the elemental marker of which is carbon, not detectable with scanning XRF.

The low CuK signal in the map could indicate the presence of azurite in the paint, however the distribution does not clearly relate to the compositional elements in which the copper-containing pigment was observed using light microscopy. Alternatively, copper might be related to a copper-containing material added to the oil medium used for painting to act as a drying agent, or the storage of the oil in a copper vessel.

The first red underdrawing of the composition could not be detected using scanning XRF. This may have been due to the settings used for the mapping, which included a dwell time of 50 ms per point and a step size of 0.5 mm. Future research could involve using longer dwell time and higher spatial resolution to increase the sensitivity to XRF signals stemming from underlaying paint layers to investigate whether the red underdrawing can be visualised.

3.2 *Francisco de Goya, Portrait of Don Francisco de Saveedra, 1798, 200.2 x 119.6 cm*

An X-radiograph of *Portrait of Don Francisco de Saveedra* by Francisco de Goya shows that the canvas support is composed of four pieces of fabric sewn together at the left of the painting (Burnstock et al. 2017). Lead white containing paint was used to conceal the seam, and the sweeping marks visible in the X-radiograph are related to the application of the ground layer that contains lead white, chalk, calcium sulphate, charcoal, aluminosilicates and iron-containing ochre coloured pigments. Highlights in the coat and neck scarf painted with a lead white-containing paint are also visible.

An infrared reflectogram of the painting shows a strongly absorbing underlayer used to block out the sitter's face and the area around it. Elemental analysis of a paint sample taken from this area of the composition indicated that a mixture of Prussian blue, bone black and lead white was used. Technical examination also indicated that the brown paint of the background was applied up to the perimeter of the face covering the underlayer.

The elemental distribution maps gathered by scanning XRF provided additional information about the pigments used for the portrait (Figure 2). Lead white and vermilion were used to create the warm flesh tones of the sitter's face and hands, as indicated by the distribution of lead and mercury illustrated by the PbL and HgL maps. The distribution of manganese and iron visualised in the MnK and FeK maps is indicative of the use of iron-containing pigments, including umber. These were used for the brown background, and in addition to the pigments identified using other methods, a mixture of ochre coloured iron-containing pigments and bone black was indicated in the paint used to block out the face. Bone black was used for the black paint of the trousers, the eyes, and the eyebrows, as illustrated by the distribution of Ca in the

CaK map, indicative of the calcium hydroxyapatite content of this pigment. The CaK map indicates the use of bone black in the paint mixtures used for the portrait, but also areas where the chalk and calcium sulphate containing ground was left visible between brushstrokes.

Phosphorus, related to the phosphate content of bone black, could not be mapped by scanning XRF. Elements with low energy radiation such as phosphorus (Z=15, PK_{α}=2.010 keV) are difficult to detect due to absorption of their X-ray emission in the air path between the painting and the instrument detector.

Scanning XRF confirmed the use of Naples yellow and ochre to paint the tablecloth, as indicated by the distribution of antimony and iron in the SbL and FeK maps respectively. Comparison of these two elemental distribution maps with the PbL map indicative of lead white also made it possible to visualise changes in composition made in paint that were not visible using other imaging methods. For example, the tablecloth and tassels were originally longer and their length was reduced in the final version of the composition.

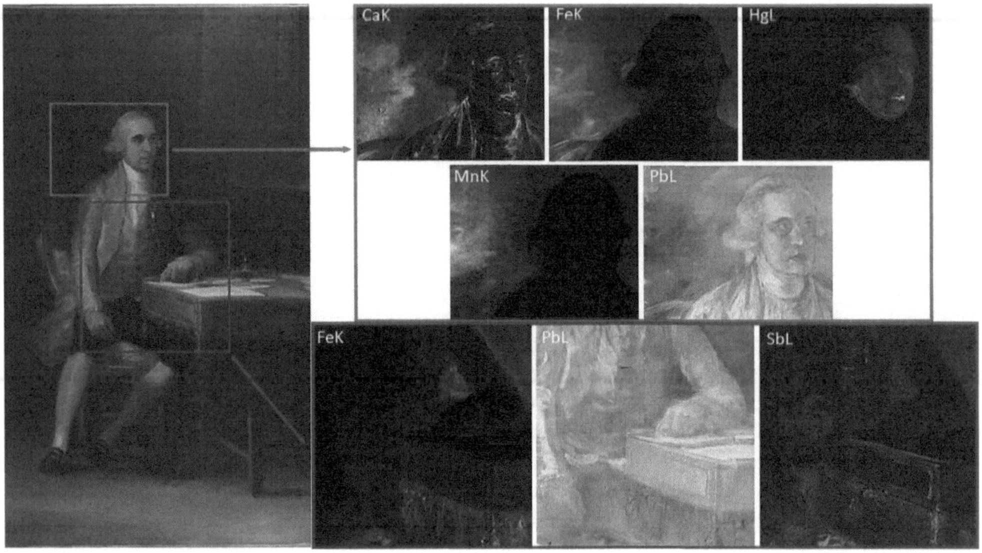

Figure 2. Francisco de Goya, *Don Francisco de Saveedra*, 1798, oil on canvas, The Courtauld Gallery (photo © The Samuel Courtauld Trust, The Courtauld Gallery, London) and selected elemental distribution maps of selected areas (images © The Courtauld Institute).

3.3 *Édouard Manet, Le Déjeuner sur l'herbe, ca. 1863-68, 89.5 x 116.5 cm*

The Courtauld's *Le Déjeuner sur l'herbe* by Édouard Manet has recently been treated in the Department of Conservation of The Courtauld, which coincided with examination of the painting using a combination of imaging and portable spectroscopic methods such as reflectance imaging spectroscopy, scanning XRF, fibre optics reflectance spectroscopy, portable reflection mid-FTIR spectroscopy and Raman spectroscopy (Amato et al. 2019, Amato et al. 2020).

The results indicated the use of both traditional and modern pigments introduced in the nineteenth century. These are consistent with those used by Manet throughout his career, as suggested by comparison with technical studies of many of his works published to date.

Scanning XRF of the painting provided new evidence for discussion of the relationship between the Courtauld painting and the largest version of the composition now at the Musee D'Orsay (Amato et al. 2019, Amato et al. 2020) (Figure 3). The Ca distribution map revealed a system of grid lines used to transfer the design to the canvas, which had never been visualised using other techniques. The presence of calcium in the grid lines suggested the use of either bone black (calcium phosphate) or chalk (calcium carbonate). Complementary reflectance imaging spectroscopy in the IR range suggested that the lines were applied using chalk, which is transparent in the IR range, rather than bone black, which would have been detected in the IR images.

Visual examination of the painting showed a set of marks at smaller intervals as the grid lines readable in the calcium map (Figure 3). These marks are limited to the edges of the canvas and were drawn using a diluted red paint. The presence of these marks around the edges of the canvas suggests that a second squaring system, possibly involving the use of string lines, was also used to transfer the composition (Amato et al. 2020).

Reflectance imaging spectroscopy in the IR range revealed changes made by the artist to paint the group of figures, including adjustments to the shape of the naked woman's back and her drapery, the pose of the woman in the background and the position of the cane and cap worn by the figure on the right.

The changes to the composition visualised using reflectance imaging spectroscopy in the IR range and the identification of a method for transferring the composition that possibly involved the use of two different sets of gridlines provided a hypothesis about the status of the work in relation to the version at the Musée d'Orsay. The presence of two sets of gridlines at different intervals, which could be related to different stages of the development of both compositions,

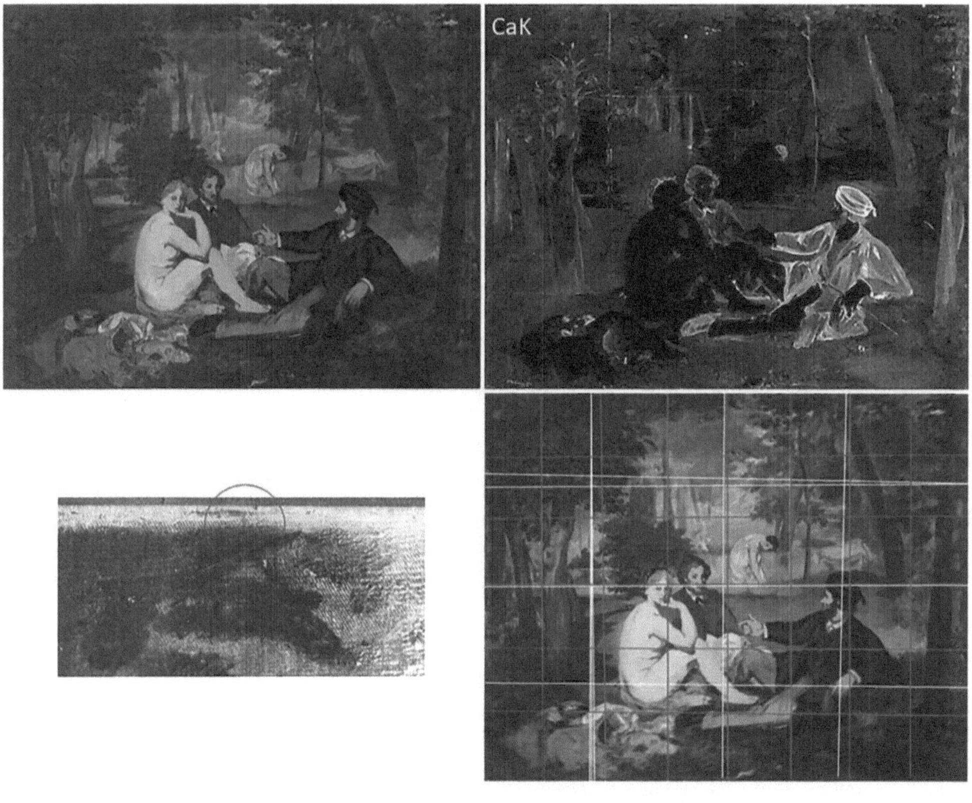

Figure 3. (Top left) Édouard Manet, *Le Dèjeuner sur l'herbe*, 1863-1868 (circa), oil on canvas, The Courtauld Gallery (photo © The Samuel Courtauld Trust, The Courtauld Gallery, London); (top right) calcium distribution map (image © 2018 John Wiley & Sons, Ltd); (bottom left) detail of one of the red marks on the top edge of the canvas (image © The Courtauld Institute); (bottom right) reconstruction of the grid lines based on the distribution map of calcium (white lines) and on the red marks on the edges (red lines, © The Courtauld Institute, image Silvia Rita Amato). Modified from Amato 2020.

combined with evidence of the painting technique used for the Courtauld picture and contextualised within Manet's wider *oeuvre*, suggested that this canvas might be a 'working sketch' for the larger version, thus representing an important contribution to the long-lasting scholarly debate on the relationship between the two paintings (Amato et al. 2020).

4 CONCLUSIONS

The outcome of this research showed the advantages of combining data from scanning XRF with established methods of technical analysis of paintings, for the characterisation of materials and painting techniques demonstrated by examples of selected works from the 16th to the 19th centuries in the Courtauld Gallery.

Scanning XRF provided significant additional information about both the artists' techniques and the condition of the paintings investigated, complementing traditional methods of technical study including light microscopy and X-radiography. Elemental mapping has clear advantages over site-specific analyses that may not be representative of the material composition of the whole painting. It also is an advantage where the opportunity for taking microsamples is limited to the edges of a painting or areas of damage. However, the data obtained from scanning XRF is influenced by the composition of the paint layers and their stratigraphy, and therefore complementary evidence from microsamples is critical for interpretation. The results are also determined by the instrumental and experimental parameters used to acquire the XRF scans.

While the scanning XRF data discussed in this paper were acquired with an XRF scanner equipped with a single 30 mm^2 silicon drift detector, the Courtauld Institute has now acquired an updated version equipped with a dual 60 mm^2 silicon drift detector, which provides improved sensitivity for the same dwell time. Re-examination of paintings using the new and improved scanner and the use of longer dwell time and higher spatial resolution may in future reveal additional technical features and materials.

The results of this research will inform future study into the material composition and the development of compositions of other paintings by Bruegel, Goya and Manet, and by other artists who used similar materials and techniques.

ACKNOWLEDGMENTS

The authors would like to thank Prof. Joris Dik (University of Delft), for the loan of the M6 scanner in 2015. The authors would also like to thank Dr Karen Serres (The Courtauld Gallery) and Graeme Barraclough (The Courtauld Gallery).

REFERENCES

Alfeld, M. & Janssens, K. 2015. Strategies for Processing Mega-Pixel X-Ray Fluorescence Hyperspectral Data: A Case Study on a Version of Caravaggio's Painting Supper at Emmaus. *J. Anal. At. Spectrom.* 30 (3): 777–89.

Amato, S.R., Burnstock, A., Cross, M., Janssens, K., Rosi, F., Cartechini, L., Fontana, R., Dal Fovo, A., Paolantoni, M., Grazia, C., Romani, A., Michelin, A., Andraud, C., Tournié, A. & Dik, J. 2019. Interpreting techincal evidence from spectral imaging of paintings by Édouard Manet in the Courtauld Gallery. *X-Ray Spectrometry* 48: 282–292.

Amato, S.R., Cross, M., Burnstock, A., Janssens, K., Dik, J., Cartechini, L., Michelin, A. & Tournié, A. 2020. Examining Édouard Manet's *Le Déjeuner sur l'herbe* from the Courtauld Gallery using spectral imaging techinques. In M. Kempski, J. Kirby, V. Leanse & K. Mandy (eds), *Tales of the Unexpected': In Paintings Conservation*: 98–110. London: Archetype Publications.

Burnstock, A., Cross, M. & Serres, K. 2017. Insoluble Surface Spots, Metal Soaps and Challenges in the Conservation of Goya's Portrait of Francisco de Saavedra. In J. Bridgland (ed), *ICOM-CC 18th Triennal Conference Preprints, Copenhagen, 4-8 September 2017.* Paris: Pulido & Nunes; ICOM Committee for Conservation.

Burnstock, A. & Serres, K. 2019. Pieter Bruegel the Elder's Grisaille Paintings. In E. Oberthaler, A. Hoppe-Haranoncourt, S. Penot, R. Spronk & M. (eds), *The Hand of the Master. Materials and Techniques of Pieter Bruegel the Elder*: 82–95. Vienna: Kunsthistorisches Museum Publications in collaboration with Hannibal Publishing.

Lasers in the Conservation of Artworks XIII – Siano & Ciofini (eds)

INFRA-ART spectral library: A new open access infrastructure for heritage science

I.M. Cortea & L. Angheluță
National Institute of Research and Development for Optoelectronics - INOE 2000, Măgurele, Romania

A. Chiroșca
University of Bucharest/Networks SRL, Bucharest, Romania

G. Serițan
University Politehnica of Bucharest, Bucharest, Romania

ABSTRACT: Web-based open-access spectral databases relevant to conservation are nowadays a real necessity for heritage scientists and other heritage-related professionals that work with spectroscopic techniques. Fast and easy access to reliable high-quality databases is amplified by the fact that portable analytical techniques are becoming increasingly more used for in situ investigation of objects of art and archaeology. To address this need, within the frame of the postdoctoral project INFRA-ART, an open-access spectral library exclusively dedicated to art and cultural heritage materials has been developed. The INFRA-ART Spectral Library is an ongoing compilation of spectra that now contains over 1300 ATR-FTIR, Raman, and XRF spectra linked to over 680 known reference materials, and is currently part of the services offered by the Romanian hub within E-RIHS (European Research Infrastructure for Heritage Science) DIGILAB. This paper presents the key features of the INFRA-ART database, along some aspects regarding database accessibility, and upcoming developments.

1 INTRODUCTION

Identification and characterization of the various constituent materials in works of art and archaeological artifacts is of crucial importance as it can provide valuable historical, artistic and technical information. A clear diagnosis of the materials present can grant insights into the cultural, social, and technological contexts in which an artwork was created, can yield evidence for dating, attribution and authentication (Craddock 2009), and can support the restoration process. The identification and characterization of materials can also inform curatorial decisions regarding the display, storage, and transportation of works of art (Stoner and Rushfield 2012), essential aspects for their preservation.

To date, a wide range of advanced analytical methods have been successfully employed to investigate cultural heritage objects - from well-established spectroscopic, chromatographic and imaging methods or nuclear beam techniques (Adriaens and Dowsett 2021; Artioli 2010) to emerging approaches such as synchrotron techniques (Bertrand et al. 2016), immunochemical micro imaging methods (Sciutto et al. 2016) or, more recently, hybrid techniques (Bai et al. 2019; Osticioli et al. 2009). The current general trend is to use non- or minimal-invasive methods that can provide targeted information of cultural heritage materials ideally without sampling or any contact with the surface of the object, directly onsite. The necessity of the cultural heritage community for a non-invasive in situ approach has determined a rapid development of miniaturized/portable instrumentation over the recent years, especially of

DOI: 10.1201/9781003386872-5

spectroscopic techniques (Pozzi et al. 2021; Li et al. 2022). X-ray fluorescence (XRF), fiber optic reflectance spectroscopy (FORS), Raman and Fourier-transform infrared (FTIR) spectroscopy, stand among the most important mobile techniques developed over the last years, and at the same time among the most widely used analytical methods in the heritage science field (Drake and MacDonald 2022; Pozzi et al. 2021). Laser-based techniques such as laser-induced breakdown spectroscopy (LIBS) or laser-induced fluorescence (LIF) have also seen an increasing interest following the development of portable compact equipment operable onsite (Bai et al. 2019; Caso, Caneve, and Spizzichino 2021; Ortiz et al. 2015). Although such portable spectrometer systems have lower performance capabilities as compared to benchtop equipment, they represent the best option for the study of objects which may not be moved to a laboratory.

Due to the complexity of heritage samples a "one-size-fits-all" approach is usually unfeasible. A combination of techniques is most frequently needed for a complete characterization as the capabilities of each techniques depends on the material under investigation and on the type of information to be retrieved. In heritage sciences, it has become normal practice to combine elemental data from a handheld XRF analyzer with chemical and structural insights from vibrational spectroscopies (such as FTIR and Raman) (Delgado Robles et al. 2015; Smith, Thompson, and Lennard 2017; Želinská et al. 2018). Commonly found in museum's laboratories and research infrastructures, these techniques stand among the conventional analytical methods available to most heritage scientists today. These easily accessible spectroscopic techniques are probably at the moment the most frequently used analytical tools in conservation and heritage sciences as they offer a series of advantages: relatively low-cost, non- or minimal-invasiveness, high sensitivity and complementary information (on both organic and inorganic compounds). Not least, such a combined approach maximizes the amount of information about the investigated artifacts, at the same time minimizing the damage.

When working with spectroscopic data of unknown materials quite often there is a need for reference spectra. Thus, the availability of databases with high-quality reference spectra is a key aspect to the use of these analytical techniques. Spectral libraries are an important tool for heritage science researchers and conservators, as they allow not only easier identification and characterization of the studied materials, but also access and support to various research directions including provenance studies (by comparing and analyzing the spectral data from different sources), or to new research approaches such as automatic pigment identification/classification in painted works of art using chemometric tools (Cortea et al. 2022; Festa et al. 2022) or training machine learning models (Fiorucci et al. 2020; Vermeulen et al. 2022). Spectral reference data can also be used to monitor changes in the materials over time, helping to inform conservation and preservation strategies.

For interpretation of FTIR or Raman spectra, typically, fingerprinting is used, meaning that the spectrum of an unknown sample is compared against a database with reference spectra. The fingerprint region of a vibrational spectrum typically consists of a complex unique pattern of peaks that can provide information on the molecular structure and functional groups present in the sample. An XRF spectrum will reflect the atomic variation within a sample and although proof positive characterization of a given material cannot be accomplished, strong inferences can however be made based on the spatial co-occurrence of certain elements. Moreover, elemental analysis, especially trace elemental analysis, can identify chemical patterns/chemical fingerprints, or indicate geochemical trends, that can sustain provenance studies.

Currently, there are several commercial spectral databases available for IR and Raman data –KnowItAll® Spectral Database Collection (former Bio-Rad Sadtler) by Wiley being among the largest. Despite the generous number of available commercial spectral libraries, dedicated databases for the cultural heritage field are scarce and the range of artists' materials included is relatively small (Cortea et al. 2023; Vahur, Virro, and Leito 2005). Another drawback is the large institutional costs that these digital products require, most of these libraries being currently available only under an annual subscription. The lack of relevant resources for the cultural heritage field has led over the years to the development of several specialized spectral databases, most of these the result of the efforts of several research groups active in the area of heritage science. An overview of the main published collections of reference spectra of art-related materials can be found in Table 1. As can be easily observed, most of the existing resources are specialized on Raman data, followed by IR and to a lesser extent by XRF and other spectral data. In terms of reference materials, the vast majority of these spectral collections are relatively small (under 100 reference spectra) and cover largely mineral pigments. In terms of accessibility, most of the spectra can be found only within the original publications, some online as images, and, to a lesser extent, as interactive spectra. A limited number of resources allows access to the raw spectral data files for download. At this moment the RRUFF database developed by The University of Arizona (Lafuente et al. 2016), and the Pictorial Materials Database developed by the Foundation Centre for Conservation and Restoration of cultural heritage La Venaria Reale (CCR) in collaboration with the National Institute of Metrological Research (INRIM) and Laboratorio Analisi Scientifiche of Regione Autonoma Valle d'Aosta (LAS) (Cavaleri et al. 2017), are the only integrated online databases that include multiple analytical data for a given sample. However, in terms of material classes relevant for art and conservation, the Infrared and Raman Users Group (IRUG) Spectral Database (Lomax et al. 2013; Price et al. 2002, 2009) represents the most extensive database freely available online, with over 3600 FTIR and Raman spectra of various artists' and cultural heritage materials.

Each of these resources presents advantages and disadvantages, with some of them being significantly more useful, depending on the user's needs. A closer look at the existing spectral collections points out a clear lack in terms of integrated libraries dedicated to the cultural heritage field, the vast majority of these resources covering only one type of spectral data (ensuring thus only a certain degree of characterization of a given material). Another existing gap is represented by the degree of accessibility and re-use of the data, as in most cases the spectra are displayed only as images, offering thus limited use. To address this need, within the frame of the postdoctoral project INFRA-ART, an open-access integrated spectral library exclusively dedicated to art and cultural heritage materials has been developed. The INFRA-ART Spectral Library was designed as a digital support tool for research specialists and other heritage-related professionals that work with non- or minimally invasive spectroscopic techniques, and can be freely accessed online - https://infraart.inoe.ro/. Developed as an easy-to-use web-based resource that offers open-access to high-quality spectral data for the scientific analysis of pigments and other art-related materials, the INFRA-ART library contains at this moment over 1300 ATR-FTIR, Raman, and XRF spectra linked to over 680 known reference materials. The aim of this paper is to present an overview of the database's key features, accessibility, and use, in view of dissemination of this new open-access infrastructure to the cultural heritage scientific community.

Table 1. Overview of the main published collections of reference spectra of art-related materials.

Collection	Description	Access to data	References
UCL - Raman spectroscopic library	Raman spectra (at 514 and 632 nm) of over 60 pigments, both natural and synthetic	Spectra available for download: http://www.chem.ucl.ac.uk/resources/raman/	Bell, Clark, and Gibbs 1997
Raman spectra of synthetic organic azo pigments	Raman spectra (at 780 nm) of 21 azo pigments	Spectra displayed in the published paper	Vandenabeele et al. 2000
UCL- FT-Raman spectroscopic library	FT-Raman spectra (at 1064 nm) of mineral pigments, pigment media and varnishes.	Spectra displayed in the published paper	Burgio and Clark 2001

(*Continued*)

Table 1. (*Continued*)

Collection	Description	Access to data	References
Collection of minerals concerning research in art history or archaeology	Raman spectra (at 514 and 632 nm) of 45 different mineral species	Spectra displayed in the published paper	Bouchard and Smith 2003
ColoRaman database	Raman (at 532, 632 and 780 nm) and fluorescence spectra of 99 oil, tempera and fresco paint pigments	No longer available online	Burrafato et al. 2004
e-VISART database	243 FTIR spectra, 130 FT-Raman spectra (at 1064 nm) and 246 Raman spectra (at 785 nm) of 130 different materials	No longer available online	Castro et al. 2005
Collection of synthetic organic pigments	FTIR and Raman (at 514, 633 and 785 nm) spectra collection of some 170 synthetic organic pigments of the 20th and 21st centuries	Spectra displayed in the published paper. Data available on request in digital format from the corresponding author.	Scherrer et al. 2009
Romanian Database of Raman Spectroscopy	Raman spectra (at 532 nm) collection of 72 individual mineral species + 77 pigment samples	Interactive spectra available online: http://www.rdrs.ro/	Buzgar and Apopei 2009
Collection of synthetic organic pigments	Raman spectra (at 785 nm) of over 270 different synthetic organic pigments	Interactive spectra available online along file download option: https://soprano.kikirpa.be/	Fremout and Saverwyns 2012
Kimmel Center for Archaeological Science Infrared Standards Library	Collection of over 380 FTIR spectra of various standards and archaeological materials	Spectra available for download: https://centers.weizmann.ac.il/kimmel-arch/infrared-spectra-library	Weiner 2010; Weizmann Institute of Science
IRUG database	Extensive collection of FTIR and Raman spectra of artists' and cultural heritage materials	Interactive spectra available online: http://irug.org/search-spectral-database	Lomax et al. 2013; Price et al. 2002, 2009
Collection of black pigments	Raman spectra (at 532 and 785 nm) of various carbon-based black pigments	Spectra displayed in the published paper	Coccato et al. 2015
Collection of green minerals	Raman spectra (at 532 and 785 nm) of various green minerals of interest for cultural heritage research	Spectra displayed in the published paper. Data available on request in digital format from the corresponding author.	Coccato et al. 2016
University of Tartu - Database of ATR-FTIR	Collection of over 150 ATR-FTIR spectra of materials		Vahur et al. 2016

(*Continued*)

Table 1. (*Continued*)

Collection	Description	Access to data	References
spectra of various materials	related to cultural heritage and conservation science	Spectra displayed online: https://spectra.chem.ut.ee/	
CHSOS - Pigments Checker	A collection of various spectroscopic data on the most used pigments from prehistory to contemporary art	Spectra available for download: https://chsopensource.org/pigments-checker/	Caggiani, Cosentino and Mangone 2016; Cosentino 2014; Larsen, Coluzzi and Cosentino 2016
RRUFF database	Extensive collection of Raman, FTIR and X-ray diffraction data of minerals	Spectra available for download: https://rruff.info/	Lafuente et al. 2016
Centre de Recherche et de Restauration des Musées de France Database	Collection of FORS (400 nm -1000 nm) reflection spectra of over 180 historical pigments and colorants	Spectra displayed online: https://copa.hypotheses.org/552	Boust and Wohlgelmuth 2017
Pictorial Material Database	Extensive collection of ATR-FTIR, FORS, XRF and multispectral imaging of 1200 mock-ups with 173 different pigments and/or dyes, four types of binders, two types of varnishes and four different materials for underdrawings	Spectra displayed online: https://webimgc.inrim.it/Hyperspectral_imaging/Database.aspx	Cavaleri et al. 2017
Collection of medieval pigments	Raman spectra (at 488, 532, 633, 785 and 830 nm) of over 30 pigments, dyes and inks commonly used in medieval manuscripts	Spectra displayed in the published paper	Marucci et al. 2018
Collection of precious gemstones	Raman spectra (combination of two excitation wavelengths in near infrared region) of over 42 different precious gemstones and minerals used as cut gems and synthetic crystals	Spectra displayed in the published paper	Culka and Jehlička 2019

2 THE INFRA-ART SPECTRAL LIBRARY

The INFRA-ART Spectral Library integrates at this moment primary ATR-FTIR and XRF spectra, and a preliminary dataset of Raman spectra, of a wide range of art-related materials (Figure 1). Some key features are discussed and presented in the following subsections, along with some aspects regarding database accessibility, and upcoming developments. Topics related to both the design and development of the database, as well as on the collection and characterization of data samples, among other related aspects, have been already discussed in a previous publication (Cortea et al. 2023).

2.1 *Key features*

Currently, the database offers open-access to a wide range of spectral data acquired via easily accessible spectroscopic techniques commonly found in research institution's and museums' laboratories. As mentioned earlier, vibrational spectroscopies such as FTIR and Raman are two of the most frequently used molecular techniques applied to the study of cultural heritage materials due to their versatility that allows the analysis of a wide range of organic and inorganic materials, both crystalline and amorphous. Handheld XRF on the other hand is presently the go-to technique for elemental analysis in the heritage sciences given its non-invasiveness and easy approach. The main strength of the INFRA-ART Spectral Library is the fact that it offers complementary spectral data for each of the investigated samples. More exactly, at this moment most of the investigated reference materials included in the database are accompanied by both FTIR and XRF data, and, to a lesser extent, by Raman data as well. This combined approach allows a multilevel characterization of the investigated samples providing both elemental and molecular information. Moreover, the implementation of a bottom-up architecture allows further extensions of the database with other types of analysis, LIF and hyperspectral data registered on well know historical pigments as well as on paint mockups, being among the new uploads planned to be carried out by the end of this year.

Compared to other existing databases, INFRA-ART covers a wide selection of materials, including some that are not included in other libraries, such as: a large dataset of earth pigments of various provenance, special effect pigments (metal effect pigments, pearlescent pigments, daylight fluorescent pigments, etc.), other rare or hard-to-find materials - from traditional iwa-enogu Japanese watercolor paints to acrylic colors or graffiti spray paints. In addition, new reference samples (from various artist materials manufacturers and suppliers) will be integrated within the database on an ongoing basis as new materials are being analyzed.

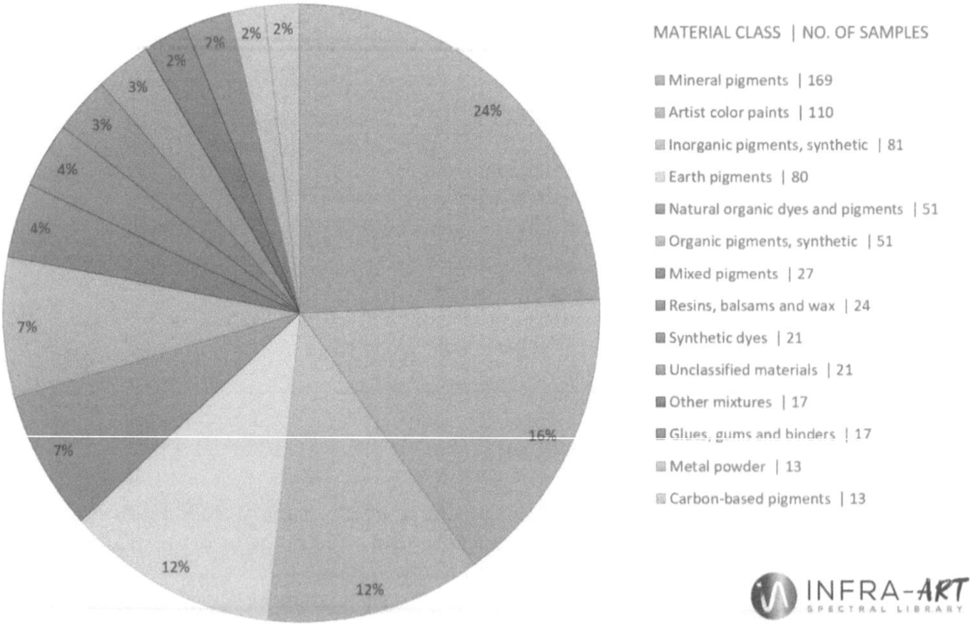

Figure 1. The distribution of the reference materials across the 14 classes of materials included in the database (access date Apr 10, 2023).

In addition to providing access to high quality data, the INFRA-ART database also comprises a comprehensive collection of metadata (such as descriptive, administrative, and structural information) for each reference material that has been analyzed. The metadata contains descriptions associated not only with each of the investigated samples (sample source, origin, sample description, alternative names, chemical information, history of use, etc.) but also with the employed techniques and experimental parameters.

Another strength point is given by the fact that all data are being curated before entering the database to ensure that it is accurate and relevant for research. Furthermore, in order to achieve high-quality spectra, with minimal noise, all the spectra included in the database are measured using optimal experimental conditions. For greater functionality and access, a search engine was incorporated within the database which allows users to search by keywords, material class or data type, and, in terms of basic spectra analysis, an interactive spectra-viewer tool was also integrated within the webpage in order to allow the user to interact with the loaded spectrum - pick points, label peaks, zoom in and out, pan, etc.

Not least, the database is aligned with the European research landscape that supports the uptake of open science practices. To support universal access and the reuse of scientific data, the database follows the European Commission's recommendation on access to scientific information as well as the FAIR (Findable, Accessible, Interoperable, Reusable) Guiding Principles on research data that result from publicly funded research (Burgelman et al. 2019; European Commission 2018; Wilkinson et al. 2016).

2.2 *Open science for increased accessibility*

As already mentioned, the implementation of the INFRA-ART database follows the 2021 UNESCO Recommendation on Open Science, as well as the FAIR Guiding Principles aimed at promoting transparent and accessible knowledge. Adopting the FAIR principles can help to maximize the impact and value of scientific research by making it easier to discover, access, and reuse research data. The INFRA-ART Spectral Library supports cross-disciplinary research and reliable re-use of scientific data for research, innovation, and educational purposes. In order to maximize accessibility, external users can request access to the spectral data files via email and subsequent completion of a File Access Request Form. The database is targeted to the cultural heritage research community as well as to other specialists in the field of art history, conservation, and materials science.

Since September 2022 the INFRA-ART Spectral Library is registered as a resource within the Open Science Cloud (EOSC) Portal for higher visibility and increased FAIRness (Mons et al. 2017). The EOSC Portal is part of the EOSC implementation roadmap and has been developed as an open multi-disciplinary environment for hosting and processing research data to support EU science. Likewise, recently, the INFRA-ART Spectral Library became part of the services offered by the Romanian hub within E-RIHS (European Research Infrastructure for Heritage Science) DIGILAB (Hermon 2021; Striova and Pezzati 2017). The DIGILAB infrastructure facilitates online access to tools and data hubs for heritage research – including measurement results, analytical data and documentation, and was designed in accordance with the policies and strategies concerning scientific data, including the FAIR data principles, the Open Research Data policy, and the EOSC strategy.

2.3 *Upcoming and future developments*

In terms of future development directions, on one hand, we plan to expand the database entries (both in terms of the number of samples and of the spectral data types), and, secondly, to develop a set of online analytical tools that will allow an advanced analysis of the spectral data. More exactly, the database will be updated on an ongoing basis with new reference materials, a special attention being placed to enrich the existing dataset with more types of mixtures (binder-pigment, pigment-pigment, and pigment-dye mixtures). We also plan to integrate a higher number of Raman spectra within the database (at his moment this technique is poorly represented), and complement the current characterization of some of the reference materials with other emerging techniques, such as Laser Induced Fluorescence (LIF) and hyperspectral imaging.

LIF spectroscopy has been used in heritage studies for several decades as it offers a series of advantages for the study of cultural materials: high sensitivity and selectivity (most organic or bio-organic materials as well as a wide range of inorganic materials contain chromophores that upon excitation exhibit characteristic fluorescence emission); high spatial resolution; versatility and minimal interference with the sample (can be used in laboratory or in-situ without the need of sampling or surface pretreatment allowing non-contact and non-invasive analysis). Despite the fact that in recent years this technique experienced an increasing trend, probably due to recent cutting-edge developments in hybrid techniques, at the moment there are no online freely available resources that could aid in spectral interpretation (Ghervase and Cortea 2023). A lack of publically available resources (reference pigment databases) can also be seen in terms of hyperspectral data. By capturing high-resolution images across hundreds of narrow spectral bands, HSI can provide valuable information about the chemical and physical properties of materials, in recent years an increased interest being placed on data treatment procedures (Alfeld and de Viguerie, 2017; Grabowski et al. 2018). As data post-processing methods evolve, there is a clear need for suitable training datasets that can be freely used by researchers, more so as recent years have seen a growing interest in the development of automatic classification/predictive models from large datasets using machine and deep learning techniques (Fiorucci et al. 2020; Kleynhans et al. 2020; Jones et al. 2022).

Taking into account the current research trends, another upgrade that we have taken into consideration is the development of some online analytical tools tailor-made to fit the necessities of the heritage science specialists that work with spectroscopic techniques. Implementation of a peak search option (for faster screening of the possible matches within the library) and the development of a spectral library searching algorithm for the identification of an unknown spectrum are among the key analytical tools that we plan to include within the INFRA-ART website in the upcoming years. Another direction that we would like to explore is the implementation of chemometrics and machine learning tools (neural network analysis) for data mining, including the use of artificial intelligence (AI) for the treatment of spectral data.

3 CONCLUSIONS

Identification and characterization of materials in works of art and archaeological artifacts are essential for historical, artistic, and technical purposes, such as dating, attribution, authentication, restoration, and preservation. Miniaturized and portable spectrometers such as Raman, FTIR, FORS, XRF, or LIBS have been successfully used to investigate cultural heritage objects and in the last decade these techniques have become must-haves thanks to their relatively low-cost, non- or minimal-invasiveness, in-situ operability, and immediate response in terms of results. A combination of techniques is usually required for a complete characterization of artefacts and heritage materials, and the availability of high-quality reference spectra is a key aspect for the use of these spectroscopic techniques. Spectral libraries allow for easier identification and characterization of materials and, at the same time, by providing access to large curated datasets, support various research directions, including recent trends such as machine learning and data mining methods. Despite the large number of available commercial spectral databases, dedicated integrated databases for the cultural heritage field are scarce, and the accessibility and re-use of the data is limited. To assist the greatly increasing number of applications of non- and micro-invasive (mobile) spectroscopic techniques an integrated spectral library, exclusively dedicated to cultural heritage and art-related materials has been developed.

The INFRA-ART Spectral Library is a user-friendly online resource that provides open-access to a diverse collection of spectral data obtained via complementary spectroscopic techniques, commonly found in research institutions and museum labs. The data collection comprises at this moment primary ATR-FTIR and XRF spectra, along with an initial dataset of Raman spectra. Compared to other existing databases, INFRA-ART covers a wide selection of materials, including some that are not included in other libraries, such as earth pigments of various provenance, special effect pigments, and other rare or hard-to-find materials. A rich set of descriptive metadata accompanies each entry for enhanced findability, interoperability, and reusability of the data. The implementation of the INFRA-ART database adheres to the FAIR Guiding Principles, aimed at promoting universal access and reuse of scientific data. Currently, the database is registered as a resource in the European Open Science Cloud (EOSC) Portal, and is featured among the services provided by the Romanian hub in E-RIHS (European Research Infrastructure for Heritage Science) DIGILAB. Expansion of the database with new reference samples and new types of spectral data can be listed among the upcoming upgrades.

ACKNOWLEDGEMENTS

This work was supported by a grant from the Romanian Ministry of Education and Research, CNCS - UEFISCDI, project no. PN-III-P1-1.1-PD-2019-1099, within PNCDI III, and through the Core Program within the National Research Development and Innovation Plan 2022-2027, conducted with the support of MCID, project no. PN 23 05.

REFERENCES

Adriaens, M. & Dowsett M. (eds.) 2021. *Spectroscopy, Diffraction and Tomography in Art and Heritage Science*. Amsterdam: Elsevier.

Alfeld, M. & de Viguerie, L. 2017. Recent developments in spectroscopic imaging techniques for historical paintings - A review. *Spectrochimica Acta Part B: Atomic Spectroscopy* 136: 81–105.

Artioli, G. 2010. *Scientific Methods and Cultural Heritage: An Introduction to the Application of Materials Science to Archaeometry and Conservation Science*. New York: Oxford University Press.

Bai, X., Oujja, M., Sanz, M., Lopez, M., Koch Dandolo, C.L., Castillejo, M. & Detalle, V. 2019. Integrating LIBS LIF Raman into a Single Multi-Spectroscopic Mobile Device for in Situ Cultural Heritage Analysis. In H. Liang, R. Groves & P. Targowski (eds.), *Optics for Arts, Architecture, and Archaeology VII; Proc. SPIE 11058, Munich, 24-26 June 2019*. Washington: SPIE.

Bell, I.M., Clark, R.J.H. & Gibbs, P.J. 1997. Raman Spectroscopic Library of Natural and Synthetic Pigments (Pre-~ 1850 AD). *Spectrochimica Acta Part A: Molecular and Biomolecular Spectroscopy* 53 (12): 2159–2179.

Bertrand, L., Bernard, S. Marone, F., Thoury, M., Reiche, I., Gourrier, A., Sciau, P. & Bergmann, U. 2016. Emerging Approaches in Synchrotron Studies of Materials from Cultural and Natural History Collections. *Topics in Current Chemistry* 374: 7.

Bouchard, M., & Smith, D.C. 2003. Catalogue of 45 Reference Raman Spectra of Minerals Concerning Research in Art History or Archaeology, Especially on Corroded Metals and Coloured Glass. *Spectrochimica Acta Part A: Molecular and Biomolecular Spectroscopy* 59(10): 2247–2266.

Boust, C. & Wohlgelmuth A. 2017. Scientific photography for cultural heritage: pigment image database under UV and IR radiations, https://copa.hypotheses.org/552. Accessed April 2023.

Burgelman, J.C., Pascu, C., Szkuta, K., Von Schomberg, R., Karalopoulos, A., Repanas, K. & Schouppe, M. 2019. Open Science, Open Data, and Open Scholarship: European Policies to Make Science Fit for the Twenty-First Century. *Frontiers in Big Data* 2: 43.

Burgio, L. & Clark, R.J.H. 2001. Library of FT-Raman Spectra of Pigments, Minerals, Pigment Media and Varnishes, and Supplement to Existing Library of Raman Spectra of Pigments with Visible Excitation. *Spectrochimica Acta Part A: Molecular and Biomolecular Spectroscopy* 57(7): 1491–1521.

Burrafato, G., Calabrese, M., Cosentino, A., Gueli, A.M., Troja, S.O. & Zuccarello, A. 2004. ColoRaman Project: Raman and Fluorescence Spectroscopy of Oil, Tempera and Fresco Paint Pigments. *Journal of Raman Spectroscopy* 35(10): 879–886.

Buzgar N., Apopei A.I. & Buzatu, A. 2009. Romanian Database of Raman Spectroscopy, http://www.rdrs.ro. Accessed April 2023.

Caggiani, M.C., Cosentino, A. & Mangone, A. 2016. Pigments Checker Version 3.0, a Handy Set for Conservation Scientists: A Free Online Raman Spectra Database. *Microchemical Journal* 129: 123–132.

Caso, M.F., Caneve, L. & Spizzichino, V. 2021. Improvement of ENEA Laser-Induced Fluorescence Prototypes: An Intercalibration between a Hyperspectral and a Multispectral Scanning System. *Acta IMEKO* 10(1): 70–76.

Castro, K., Pérez-Alonso, M., Rodríguez-Laso, M.D., Fernández, L.A. & Madariaga, J.M. 2005. On-Line FT-Raman and Dispersive Raman Spectra Database of Artists' Materials (e-VISART Database). *Analytical and Bioanalytical Chemistry* 382: 248–258.
Cavaleri, T., Buscaglia, P. Migliorini, S., Nervo, M., Piccablotto, G., Piccirillo, A., Pisani, M., Puglisi, D., Vaudan, D. & Zucco, M. 2017. Pictorial Materials Database: 1200 Combinations of Pigments, Dyes, Binders and Varnishes Designed as a Tool for Heritage Science and Conservation. *Applied Physics A: Materials Science and Processing* 123: 419.
Coccato, A., Bersani, D., Coudray, A., Sanyova, J., Moens, L. & Vandenabeele, P. 2016. Raman Spectroscopy of Green Minerals and Reaction Products with an Application in Cultural Heritage Research. *Journal of Raman Spectroscopy* 47(12): 1429–1443.
Coccato, A., Jehlicka, J., Moens, L. & Vandenabeele, P. 2015. Raman Spectroscopy for the Investigation of Carbon-Based Black Pigments. *Journal of Raman Spectroscopy* 46(10): 1003–1015.
Cortea, I.M., Chiroşca, A., Angheluţă, L. & Seriţan, G. 2023. INFRA-ART: An Open Access Spectral Library of Art-Related Materials as a Digital Support Tool for Cultural Heritage Science. *Journal on Computing and Cultural Heritage*, http://dx.doi.org/10.1145/3593427.
Cortea, I.M., Ghervase, L., Rădvan, R. & Seriţan, G. 2022. Assessment of Easily Accessible Spectroscopic Techniques Coupled with Multivariate Analysis for the Qualitative Characterization and Differentiation of Earth Pigments of Various Provenance. *Minerals* 12(6): 755.
Cosentino, A. 2014. FORS Spectral Database of Historical Pigments in Different Binders. *e-conservation Journal* 2: 53–65.
Craddock, P. 2020. *Scientific Investigation of Copies, Fakes and Forgeries*. London: Routledge.
Culka, A., & Jehlička, J. 2019. A Database of Raman Spectra of Precious Gemstones and Minerals Used as Cut Gems Obtained Using Portable Sequentially Shifted Excitation Raman Spectrometer. *Journal of Raman Spectroscopy* 50(2): 262–280.
Delgado Robles, A., Ruvalcaba Sil, J.L., Claes, P., Manrique Ortega, M.D., Casanova González, E., Maynez Rojas, M.A., Cuevas García, M. & García Castillo, S. 2015. Non-Destructive in Situ Spectroscopic Analysis of Greenstone Objects from Royal Burial Offerings of the Mayan Site of Palenque, Mexico. *Heritage Science* 3: 20.
Drake, L. & MacDonald, B. 2022. *Advances in Portable X-Ray Fluorescence Spectrometer: Instrumentation, Application, and Interpretation*. London: Royal Society of Chemistry.
European Commission. 2018. Commission Recommendation (EU) 2018/790 of 25 April 2018 on Access to and Preservation of Scientific Information, https://eur-lex.europa.eu/eli/reco/2018/790/oj. Accessed April 2023.
Festa, G., Scatigno, C., Armetta, F., Saladino, M.L., Ciaramitaro, V., Nardo, V.M. & Ponterio, R.C. 2022. Chemometric Tools to Point Out Benchmarks and Chromophores in Pigments through Spectroscopic Data Analyses. *Molecules* 27(1): 163.
Fiorucci, M., Khoroshiltseva, M., Pontil, M., Traviglia, A., Del Bue, A. & James, S. 2020. Machine Learning for Cultural Heritage: A Survey. *Pattern Recognition Letters* 133: 102–108.
Fremout, W. & Saverwyns, S. 2012. Identification of Synthetic Organic Pigments: The Role of a Comprehensive Digital Raman Spectral Library. *Journal of Raman Spectroscopy* 43(11): 1536–1544.
Ghervase, L. & Cortea, I.M. 2023. Lighting Up the Heritage Sciences: The Past and Future of Laser-Induced Fluorescence Spectroscopy in the Field of Cultural Goods. *Chemosensors* 11: 100.
Grabowski, B., Masarczyk, W., Głomb, P. & Mendys, A. 2018. Automatic pigment identification from hyperspectral data. *Journal of Cultural Heritage* 31: 1–12.
Hermon, S. 2021. Building DIGILAB – Towards a Data-driven Research in Cultural Heritage. In *VIRTUAL ARCHAEOLOGY - Revealing the Past, Enriching the Present and Shaping the Future, Proceedings of the Forth International Scientific Conference, Krasnoyarsk, 20–22 September 2021*. Krasnoyarsk: SFU.
Jones, C., Daly, N.S., Higgitt, C. & Rodrigues, M.R.D. 2023. Neural network-based classification of X-ray fluorescence spectra of artists' pigments: an approach leveraging a synthetic dataset created using the fundamental parameters method. Heritage Science 10: 88.
Kleynhans, T., Schmidt Patterson, C.M., Dooley, K.A., Messinger, D.W. & Delaney, J.K. 2020. An alternative approach to mapping pigments in paintings with hyperspectral reflectance image cubes using artificial intelligence. Heritage Science 8: 84.
Lafuente, B., Downs, R.T., Yang, H. & Stone, N. 2016. The Power of Databases: The RRUFF Project. In Thomas Armbruster & Rosa Micaela Danisi (eds), *Highlights in Mineralogical Crystallography*: 1–30. Berlin, München, Boston: De Gruyter (O).
Larsen, R., Coluzzi, N. & Cosentino, A. 2016. Free XRF Spectroscopy Database of Pigments Checker. *International Journal of Conservation Science* 7(3): 659–668.

Li, A., Yao, C., Xia, J., Wang, H., Cheng, Q., Penty, R., Fainman, Y. & Pan, S. 2022. Advances in cost-effective integrated spectrometers. *Light: Science & Applications* 11: 174.

Lomax, S.Q., Price, B.A., Lins, A., Davis, C., Pretzel, B., Picollo, M., Richards, G. & Rice, S. 2013. The IRUG Raman Spectral Web Database: Objectives, Progress and Plans. *e-Preservation Science* 10: 38–41.

Marucci, G., Beeby, A., Parker, A.W. & Nicholson, C.E. 2018. Raman Spectroscopic Library of Medieval Pigments Collected with Five Different Wavelengths for Investigation of Illuminated Manuscripts. *Analytical Methods* 10: 1219–1236.

Mons, B., Neylon, C., Velterop, J., Dumontier, M., da Silva Santos, L.O.B. & Wilkinson, M.D. 2017. Cloudy, Increasingly FAIR; Revisiting the FAIR Data Guiding Principles for the European Open Science Cloud. Information Services and Use 37(1): 49–56.

Ortiz, R., Ortiz, P., Colao, F., Fantoni, R., Gómez-Morón, M.A. & Vázquez, M.A. 2015. Laser Spectroscopy and Imaging Applications for the Study of Cultural Heritage Murals. *Construction and Building Materials* 98(15): 35–43.

Osticioli, I., Mendes, N.F.C., Nevin, A., Zoppi, A., Lofrumento, C., Becucci, M. & Castellucci, E.M. 2009. A New Compact Instrument for Raman, Laser-Induced Breakdown, and Laser-Induced Fluorescence Spectroscopy of Works of Art and Their Constituent Materials. *Review of Scientific Instruments* 80(7): 076109.

Pozzi, F., Rizzo, A., Basso, E., Angelin, E.M., Sá, S.F., Cucci, C. & Picollo, M. 2021. Portable Spectroscopy for Cultural Heritage. In R. Crocombe, P. Leary & B. Kammrath (eds.), *Portable Spectroscopy and Spectrometry*: 499–522. Hoboken: Wiley.

Price, B., Pretzel, B. & Lomax, S. (eds) 2009. *Infrared and Raman Users Group Spectral Database*. ed. Vol. 1 & 2. Philadelphia: IRUG.

Price, B., Pretzel, B., Carlson, J., Ehrman, K. & Lins, P.A. 2002. Web-Based Exchange of Infrared and Raman Spectra: A New IRUG Initiative. In Rene van Grieken (ed.), *Art2002, Proceedings of the 7th International Conference on Non-destructive Testing and Microanalysis for the Diagnostics and Conservation of the Cultural and Environmental Heritage, Antwerp, 2-6 June 2002*. Antwerp: University of Antwerp.

Scherrer, N.C., Zumbuehl, S., Delavy, F., Fritsch, A. & Kuehnen, R. 2009. Synthetic Organic Pigments of the 20th and 21st Century Relevant to Artist's Paints: Raman Spectra Reference Collection. *Spectrochimica Acta Part A: Molecular and Biomolecular Spectroscopy* 73(3): 505–524.

Sciutto, G., Zangheri, M., Prati, S., Guardigli, M., Mirasoli, M., Mazzeo, R. & Roda, A. 2016. Immunochemical Micro Imaging Analyses for the Detection of Proteins in Artworks. *Topics in Current Chemistry* 374: 32.

Smith, M., Thompson, K. & Lennard, F. 2017. A Literature Review of Analytical Techniques for Materials Characterisation of Painted Textiles — Part 2: Spectroscopic and Chromatographic Analytical Instrumentation. *Journal of the Institute of Conservation* 40(3): 252–266.

Stoner, J.H. & Rushfield, R. 2012. *Conservation of Easel Paintings*. Abingdon: Routledge.

Striova, J. & Pezzati, L. 2017. The European Research Infrastructure for Heritage Science (E-RIHS). *International Archives of the Photogrammetry, International Archives of the Photogrammetry, Remote Sensing and Spatial Information Sciences* XLII-2/W5: 661–664.

Vahur, S., Teearu, A., Peets, P., Joosu, L. & Leito, I. 2016. ATR-FT-IR Spectral Collection of Conservation Materials in the Extended Region of 4000-80 cm^{-1}. *Analytical and Bioanalytical Chemistry* 408: 3373–3379.

Vahur, S., Virro, K. & Leito, I. 2005. Web-Based Infrared Spectral Databases Relevant to Conservation. *Journal of the Canadian Association for Conservation* 30: 10–17.

Vandenabeele, P., Moens, L., Edwards, H.G.M. & Dams, R. 2000. Raman Spectroscopic Database of Azo Pigments and Application to Modern Art Studies. *Journal of Raman Spectroscopy* 31(6): 509–517.

Vermeulen, M., McGeachy, A., Xu, B., Chopp, H., Katsaggelos, A., Meyers, R., Alfeld, M. & Walton, M. 2022. XRFast and New Software Package for Processing of MA-XRF Datasets using Machine Leaning. *Journal of Analytical Atomic Spectrometry* 37: 2130–2143.

Weiner, S. 2010. *Microarchaeology: Beyond the Visible Archaeological Record*. New York: Cambridge University Press.

Weizmann Institute of Science. Kimmel Center for Archaeological Science Infrared Standards Library, https://centers.weizmann.ac.il/kimmel-arch/infrared-spectra-library. Accessed April 2023.

Wilkinson, M., Dumontier, M., Aalbersberg, I. et al. 2016. The FAIR Guiding Principles for Scientific Data Management and Stewardship. *Scientific Data* 3: 160018.

Želinská, J., Kopecká, I. Svobodová, E., Milovská, S. & Hurai, V. 2018. Stratigraphic EM-EDS, XRF, Raman and FT-IR Analysis of Multilayer Paintings from the Main Altar of the St. James Church in Levoča (Slovakia). *Journal of Cultural Heritage* 33: 90–99.

Metal artefacts

Lasers in the Conservation of Artworks XIII – Siano & Ciofini (eds)

LIBS vs XRF on underwater heritage: The silver coins of "Nuestra Señora de las Mercedes"

I. Donate, S. Díaz, E. García, M. Bueso & María Martín
Instituto del Patrimonio Cultural de España, Ministerio de Cultura y Deporte, Madrid, Spain

M. Oujja, M. Martínez-Weinbaum & M. Castillejo
Instituto de Química Física Rocasolano, CSIC, Madrid, Spain

ABSTRACT: This work presents a comparative investigation using two non-invasive elemental analytical techniques, X-ray fluorescence (XRF) and laser-induced breakdown spectroscopy (LIBS), for the chemical characterization of a real case study of underwater-corroded metals: the silver coins of the Spanish frigate "Nuestra Señora de las Mercedes". A ship loaded with funds and products from the Viceroyalty of Peru, well known to had been sunk by the British Army in 1804 off the coast of Algarve and plundered more than two hundred years later by the commercial company Odyssey Marine Exploration. The combination of both techniques allows overcoming the complexity of the thick patinas covering the coins, compound by corrosion products and the deposits from the seabed. While XRF helps to answer inquiries associated with the conservation and restoration of the metallic pieces, LIBS can solve questions about their core composition and metallurgical techniques employed for their production.

1 INTRODUCTION

Underwater contexts are aggressive environments for cultural heritage materials, especially for metallic artefacts. Its threatening nature lies in the chemical and electrochemical reactions detonated by seawater, the mechanical effects of waves and sediments and the possible microbiological colonization (Bethencourt et al. 2018, Gregory 2009). Consequently, under marine aerobic conditions metals suffer from severe corrosion that produces thick and complex patinas, in addition to the deposition of concretions of the seabed. Later, when the metal objects are extracted from the sea, although they are kept in a medium as similar as possible to that from which they were taken, the slight change in environmental conditions can cause further corrosion and even cracking (Memet 2008).

Despite the above, the nature of the degradation layers does not only depend on the environmental conditions and the geological composition of the seabed. Other factors also intervene in the processes of degradation including the original composition of the metallic alloys, along with the minor elements and traces, the metallurgical processes for obtaining metals and manufacturing, and the previous deterioration processes during the time of use (Angelini et al. 2013). This complexity makes especially challenging the conservation and study of these metal objects.

Particularly, in the case of coins, it is very important to preserve the integrity of the original surfaces to ensure an accurate interpretation (Kotoula & Kyranoudi 2013). For this reason, sampling is usually not allowed and only non-destructive analysis techniques can be applied to them. In most of the times, and considering the small size of coins, scanning electron microscopy combined with energy dispersive spectrometry (SEM-EDS) has been the preferred technique to perform elemental analysis (Ciarlo 2015, Di Fazio et al. 2019, Lei et al. 2003, Mata et al. 2010, Mousser et al. 2011). Its versatility allows for characterising at the same time the superficial topography and detecting the distribution of the chemical elements (Frahm 2014). Nevertheless, SEM-EDS only gives superficial data and to obtain information about the core, that preserves the original alloy composition (free of corrosion products), it is necessary to polish intrusively

DOI: 10.1201/9781003386872-6

the surface of the coins or even take their cross-section (Inberg et al. 2018, Rezk et al. 2022). As mentioned before, this is forbidden in most cases. In addition, it is common that the great numismatic sets cannot be moved from museums or cultural institutions to laboratories, where scanning electron microscopes are, due to the high cost of insurances.

Hence, this work aims to improve a portable and non-invasive methodology for analysing underwater numismatic collections that permits to differentiate between the composition of the metallic core and of the degradation layers. For this purpose, a comparative investigation using two non-invasive elemental analytical techniques, X-ray fluorescence (XRF) and laser induced breakdown spectroscopy (LIBS), was carried out on a real case study, the silver coins of the Spanish frigate "Nuestra Señora de las Mercedes".

XRF is a spectroscopic technique widely applied on cultural heritage objects, whose use dates back to the 50s (Hall 1960) due to its non-invasiveness, speed and high portability (Yatsuk et al. 2022). It offers almost instantaneous results that are easy to interpret on diverse materials, such as minerals, metals, ceramics, glasses, pigments, etc. (Ferreti 2000). In the field of numismatic studies there are numerous works in which XRF is a valuable tool for characterizing the production of mints (Gorghinian et al. 2013), establishing hypotheses about the origin of the metals used in the coinage (Baldassarri et al. 2014) or evaluating the compositional changes on the surface caused by corrosion processes (Felix et al. 2020).

Meanwhile, LIBS is a more recent technique, since its development has been subject to the evolution of laser technology and it was not invented until 1960. However, in a few decades it has been gaining presence in Heritage Science. Its applications in the study of glass (Muller & Stege 2003, Oujja et al. 2021), ceramics (Genc Oztoprak et al. 2016, Melessanaki et al. 2002), pictorial or polychrome layers (Anglos et al. 1997), stone materials (Castillejo et al. 2020) and of course metals (Corsi et al. 2005, Fortes et al. 2005, Siano & Agresti 2015) are already considerable. Restricted to numismatic, although few, there are researches that demonstrate the effectiveness of LIBS in the study of the evolution of alloys composition (BArtoli et al. 2011, Pardini et al. 2012), the detection of additives to alloys to improve their properties (Awasthi et al. 2016) or in the elemental depth profiling of surface coatings (Gaudiuso et al. 2019).

In addition, the combination of both techniques has proven its success in the study of metals (Alberghina et al. 2011, Arafat et al. 2013, Ferreti et al. 2007, Lazic et al. 2018, Lorenzetti et al. 2018), maximizing the information and reducing at the minimum the impact on the objects under study (Botto et al. 2019). On account of that, its use was chosen for this specific numismatic set.

2 MATERIALS AND METHODS

2.1 *The silver coins of the frigate "Nuestra Señora de las Mercedes"*

The Spanish frigate "Nuestra Señora de las Mercedes" was a warship of the Royal Spanish Navy sunk by the British Army on October 5th, 1804, off the coast of Algarve (Portugal). It came from the Viceroyalty of Peru loaded with funds and products for the Spanish crown, when the British squadron attacked by surprise, causing the death of at least 275 people, most of those who were on the ship. Since the aggression took place in a period of peace, it was the trigger for the subsequent declaration of war between Spain and England (García 2020, Marcos 2014).

The wreck "Las Mercedes" remained lost under the sea until 2007, when the private company Odyssey Marine Exploration located it and extracted more than half a million of coins and a few objects, most of which were transferred to Tampa (Florida). The Spanish State decided to reclaim the sovereign property. After 5 years of litigation in a US court, the cargo was returned to Spain, establishing an international precedent in the defense of Underwater Archaeological Heritage (Huang 2012).

The recovered treasure consists of 212 gold coins, 309,396 silver coins, 265,157 silver coins concretized forming blocks and some remains of objects belonging to the ship and to the crew, such as cannonballs, pulleys, snuffboxes, cufflinks, belt and shoe buckles, teaspoons, and textile debris, among others (Nieto 2014).

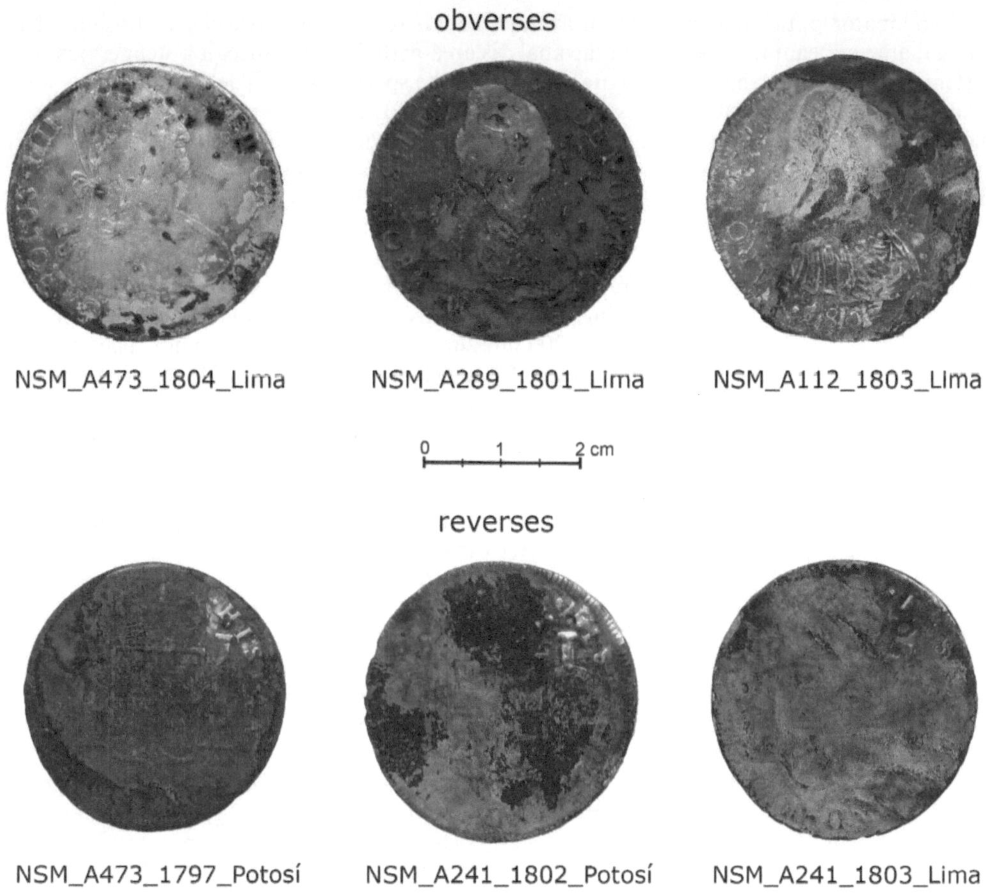

Figure 1. Silver coins from the frigate "Nuestra Señora de las Mercedes" selected for the research. The labels indicate the years of minting, between 1797 and 1804, and the mints of origin, Lima (Perú) or Potosí (actually Bolivia), of each exemplar.

The set of pieces selected in this work was formed by six sterling silver (Ag-92 wt % and Cu-8 wt %) coins, called "*reales de a ocho*" (pieces of eight) from the Bourbon period (Figure 1). They were produced between 1797 and 1804, during the reign of Carlos IV in two Spanish-American mints, Lima and Potosi. They had been manufactured by machine, with flywheel presses, and carved in the edge for greater safety. They display the effigy of the King on the obverse and the crowned shield flanked by the *Pillars of Hercules* with the motto *PLVS VLTRA* on the reverse.

The state of conservation of the coins is common for underwater archaeological metals. In general, they present thick patinas made up of corrosion products and calcium concretions. Thus, they show heterogeneous dull surfaces with greyish, blackish, greenish, blueish and even reddish tones. In addition, the coins were subjected to unknown restoration treatments by the company Odyssey Marine Exploration (Valera y Carrasco 2014), which provoked the acidification of the solutions in which they were immersed. Currently, the coins are subjected to a desalination process.

2.2 *Methods*

Due to the impossibility of moving the coins from the custody institution; all the analyses were developed in-situ at the Spanish Cultural Heritage Institute.

The XRF analyses were carried out using the own system of the IPCE, a Bruker ARTAX^{TM} 400 XRF spectrometer. It is endowed with a mini focus Rh tube (1.2-0.1 mm^2)

with collimator optics that provide an average lateral resolution between 0.2 -1.5 mm. The 1 mm diameter beam was focused on the analysis area with the assistance of a stepper position system, a laser and a camera, which are attached to the spectrometer. The measurements were made in the air to prevent further sample damage and foster the portability of the proposed method. The X-ray generator was operated at 50 keV and 700μA, while the exposure time was 30s. This instrument has a silicon drift detector (SDD) with an active area of 10 mm^2. Acquisition and evaluation of XRF spectra were carried out using the software Spectra (Bruker).

The LIBS setup of the IQFR-CSIC was moved and set up at the IPCE. The excitation source in the LIBS system is a Q-Switched Nd:YAG laser operating in its 4th harmonic at 266 nm delivering 6 ns pulses at 1Hz repetition rate. The spectral analysis and the detection of plasma emission were performed using a 0.2 m Czerny-Turner spectrograph (Andor, Shamrock Kymera-193i-A), equipped with a 1200 lines/mm grating blazed at 500 nm. The spectrograph was coupled to an Andor Technology ICCD detector (iStar CCD 334, 1024x1024 active pixels, 13 μm x 13 μm pixel size) with temporal resolution. The surfaces of coins were irradiated with laser pulses in the same areas explored by XRF, also in air conditions, with a fluence of 6 J/cm^2. The spectra, obtained with a resolution of 0.2 nm, resulted from the accumulation of 5 individual spectra that were collected with a delay time of 500 ns and a temporal observation window of 3 μs. A 300 nm cutoff filter was used to avoid surface scattered light and 2nd order of emissions at shorter wavelengths.

3 RESULTS AND DISCUSSION

The evaluation and comparison of XRF and LIBS results show noteworthy qualitative and quantitative differences between the chemical elemental compositions detected by each technique in the same areas of the silver coins.

Firstly, on the one hand, in the XRF spectra of corroded areas, labeled as black and green crusts, the intensity of copper (Cu) lines is higher than that of silver (Ag) lines, while in the LIBS spectra, Ag signal is stronger (Figures 2-3). This is due to the higher penetration depth of the LIBS probe able to reach the metallic core (Anglos 2019) whereas the depth of the XRF analysis is limited to the surface layer (Beck et al. 2004, Karydas 2007), richer in copper corrosion products.

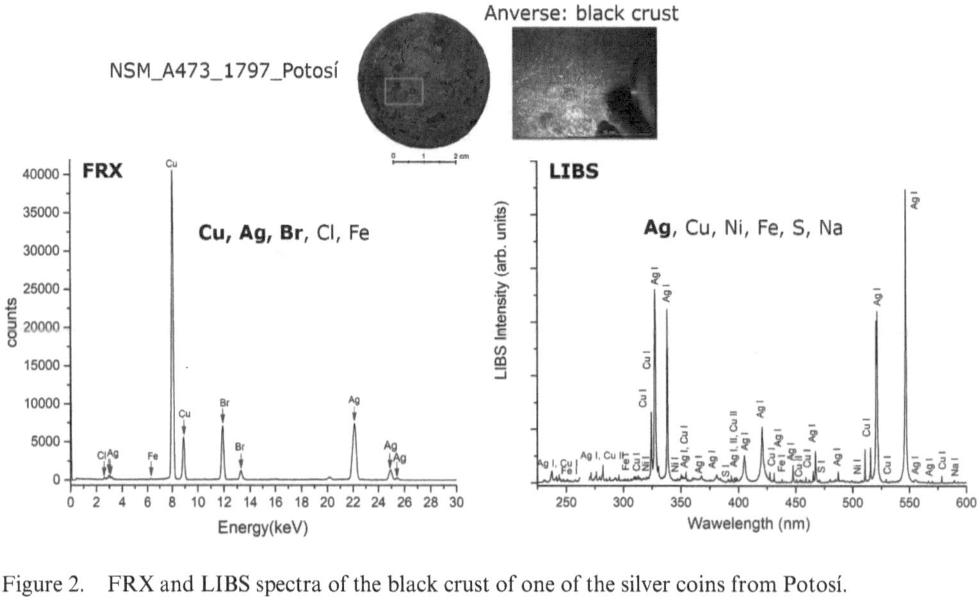

Figure 2. FRX and LIBS spectra of the black crust of one of the silver coins from Potosí.

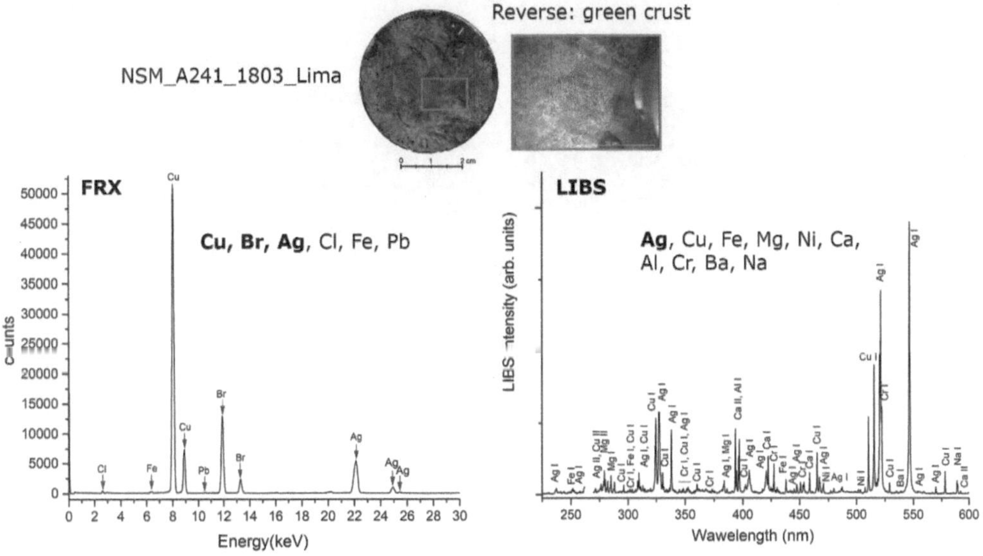

Figure 3. FRX and LIBS spectra of the green crust of one of the silver coins from Lima.

On the other hand, the surface enrichment of copper is caused by the galvanic corrosion boosted by the different redox potentials of metals combined in the alloy. In seawater, Ag has a higher reduction potential than Cu. Thus, Cu corrodes faster than Ag (Selwyn 2004) and a thicker crust of copper corrosion products is formed on the surface. For this reason, to not misguide the original alloy composition of corroded metallic pieces the analysis should reach the core (Carl & Young 2016).

Secondly, chlorine (Cl) and bromine (Br), elements due to the main corrosion products of the alloy under seawater conditions (silver chloride and bromide and copper chloride) are only detected by XRF (Figure 4). The free excited atoms of very reactive elements rapidly recombine with other elements in the air atmosphere or in the surface sample and therefore do not emit the atomic signal necessary for their identification by LIBS (Cremers & Radziemski 1983).

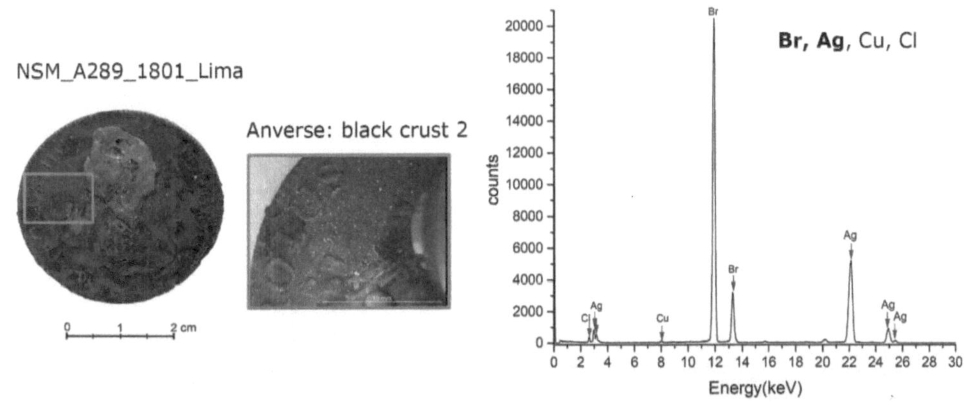

Figure 4. FRX spectrum of the black crust of one of the silver coins from Lima.

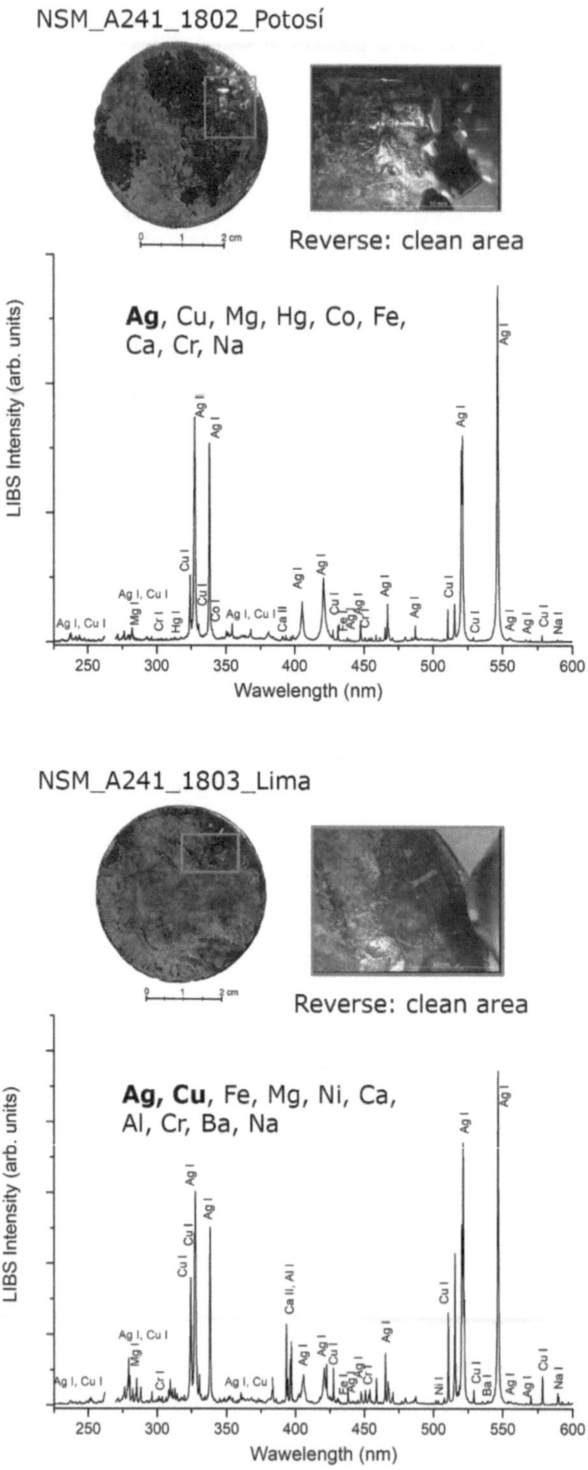

Figure 5. LIBS spectra of clean areas of two of the silver coins.

Nevertheless, LIBS analyses are able to reveal the presence of another element of corrosion compounds, sulphur (S), and of some elements composing the deposits from the seabed, such as sodium (Na), magnesium (Mg) and aluminum (Al) (see LIBS spectra of Figures 2-3). Elements that are not easily observed by XRF due to their lesser sensitivity to lighter elements (Giakoumaki et al. 2007, Shackley 2011).

Additionally, the higher sensitivity of LIBS measurements allows better identification of minor and trace elements, such as nickel (Ni), cobalt (Co), mercury (Hg), chrome (Cr), barium (Ba), strontium (Sr) and zirconium (Zr) (Figure 5) that could have their origin in impurities of alloying elements. Ni and Co usually appear in copper ores (Scott 2002). Moreover, some elements can reveal valuable information about metallurgical technology. For example, in the new territories of America Hg was commonly employed in the amalgamation process to obtain refined silver (Bakewell 2020, Borges et al. 2018). While other elements, such as Cr, Ba, Sr and Zr inform about the seabed composition.

Thereby, the results indicate that both techniques provide a complete characterization of the deterioration layers and of the metallic alloy (Table 1).

Table 1. Elemental composition detected in the analyzed areas.

	Laser-induced breakdown spectroscopy		X-ray fluorescence	
	Major elements	Minor elements	Major elements	Minor elements
NSM_A473_1797_Potosí				
Obverse: black crust	Ag	Cu, Ni, Fe, S, Na	Cu, Ag, Br	Cl, Fe
Reverse: clean area	Ag	Cu, Ni, Fe, S, Na	Ag, Br, Cu	Cl, Pb, Ni
Reverse edge	Ag	Cu, Ni, Fe, Na	Br, Ag	Cu, Cl, Ca, Pb
Reverse: black crust	Ag	Cu, Na	Br, Ag	Cu, Cl, Ca
NSM_A241_1802_Potosí				
Obverse: black crust	Ag, Ca	Cu, Mg, Hg, Co, Ni, Fe, Cr, Ba, Na	Cu, Br, Ag	Cl, Fe, Pb
Obverse: grey crust	Ag	Cu, Mg, Hg, Co, Fe, Ca, Cr, Na	Br, Ag	Cu, Cl, Fe, P
Reverse: clean area	Ag	Cu, Mg, Hg, Co, Fe, Ca, Cr, Na	Cu, Br, Ag	Cl, Pb
Reverse: green crust	Cu, Ca	Ag, Mg, W, Co, Mo, Ni, Fe, Al, Cr, Ba, Na	Cu, Br, Ag	Cl, Ca, Fe, Pb
Reverse: black crust	Cu, Ca	Ag, Mg, As, Cr, Ba, Na	W, Co, Ni, Fe, Al, Cl, Fe, Pb	Cu, Br, Ag
NSM_A289_1801_Lima				
Obverse: black crust 1	Ag	Cu	Br, Ag	Cu, Cl, Ca
Obverse: black crust 2	Ag	Cu, Ca	Br, Ag	Cu, Cl
Reverse: clean area	Ag	Cu, Ni, Fe, Na	Cu, Ag, Br	Cl, Pb
Reverse: black crust	Ag	Cu, Ca, Fe, Na	Br, Ag	Cu, Cl
NSM_A241_1803_Lim				
Obverse: black crust	Cu	Fe, Mg, Ni, Ca, S, Cr, Ba, Na	Cu, Br, Ag	Cl, Fe, Pb

(*Continued*)

Table 1. (*Continued*)

	Laser-induced breakdown spectroscopy		X-ray fluorescence	
	Major elements	Minor elements	Major elements	Minor elements
Obverse: clean area	Ag, Cu	Fe, Mg, Ni, Ca, Cr, Ba, Na	Cu, Ag, Br	Cl, Ca, Fe, Pb, Ni
Reverse: clean area	Ag, Cu	Fe, Mg, Ni, Ca, Al, Cr, Ba, Na	Br, Ag, Cu	Cl, Ca, Fe, Pb
Reverse: green crust	Ag	Cu, Fe, Mg, Ni, Ca, Al, Cr, Ba, Na	Cu, Br, Ag	Cl, Fe
NSM_A112_1803_Lima				
Obverse: orange crust	Fe, Cu, Ag	Zr, Mn, Ca, Na	Fe, Ag	Cu, Br, Ca, Pb, Sn, Ti
Obverse: clean area	Fe, Cu, Ag	Zr, Mn, Ca, Na	Ag, Fe, Cu	Br, Cl, Ca, Mg, Pb
Reverse: white crust	Ag	Cu, Mg, Fe, Zr, Ca, Cr, Sr, Na	Ag, Cu, Fe	Br, Cl, Ca, Mg, Pb
Reverse: orange crust	Ca	Ag, Cu, Mg, Fe, Cr, Sr, Na	Fe, Ag	Cu, Ca, Br, Pb
Reverse: clean area	Ag	Cu, Fe, Ca, Cr, Na	Ag, Cu, Fe	Br, Cl, Ca, Pb
NSM_A473_1804_Lima				
Obverse: uncleaned area	Ag	Cu, Fe, Na	Ag, Cu	Br, Cl, Ca
Reverse: clean area	Ag	Cu, Fe, Na	Ag, Br, Cu	Cl, Pb, Ni

4 CONCLUSIONS

The study carried out on a selection of samples from the frigate "Nuestra Señora de las Mercedes" leads to identifying the different capabilities of XRF and LIBS techniques for the elemental characterization of the silver coins. While XRF is more effective in detecting the main corrosion products of the surface of the coins, such as silver chloride and bromide, copper chlorides and carbonates, LIBS analyses allow better identification of the original composition of the metallic core, including minor elements such as Hg, Ni and Co, regardless of the patina components. These results add information about the state of conservation of these coins affected by underwater corrosion and unknown treatments, and point out to the use of mercury amalgamation technique for the refining of silver, confirming the historical documentation. Therefore, the complementarity of both techniques in providing comprehensive insights into artefacts is demonstrated: XRF can address queries related to the conservation and restoration of the metallic pieces and LIBS can resolve questions about their core composition and metallurgical techniques.

Herewith a non-invasive study methodology is proposed for underwater numismatic collections that can be applied in museums and cultural institutions consisting of a first phase of XRF analyses to inspect the main features of the surface coins followed by a second phase of punctual measurements with LIBS. The data obtained from the application of the two techniques, together with those generated by historical studies provide a better understanding of the metallurgical techniques employed, the availability of raw materials and the political and social situation, thus achieving a complete reading of the coins as historical documents.

ACKNOWLEDGMENTS

This research has been funded by the Spanish State Research Agency (AEI) through project PID2019-104124RB-I00/AEI/1013039/501100011033, by the H2020 European project IPERION HS (Integrated Platform for the European Research Infrastructure ON Heritage Science, GA 871034). Professional support of the CSIC Interdisciplinary Thematic Platform (PTI) Open Heritage: Research and Society (PTI-PAIS) is acknowledged.

REFERENCES

Alberghina, M.F., Barraco, R., Brai, M., Schillaci, T. & Tranchina, L. 2011. Comparison of LIBS and m-XRF measurements on bronze alloys for monitoring plasma effects. *Journal of Physics: Conference Series* 275: 012017.

Angelini, E., Grassini, S. & Tusa, S. 2013. 12- Underwater corrosion of metallic heritage artefacts. In Philippe Dillmann, David Watkinson, Emma Angelini & Annemie Adriaens (eds.), *European Federation of Corrosion (EFC) Series, Corrosion and conservation of cultural heritage metallic artefacts*: 236–259. Woodhead Publishing.

Anglos, D., Couris, S. & Fotakis, C. 1997. Laser diagnostics of painted artworks: laser-induced breakdown spectroscopy in pigment identification. *Applied Spectroscopy* 51(7): 1025–1030.

Anglos, D. 2019. Laser-induced breakdown spectroscopy in heritage science. *Physical Sciences Reviews* 4 (7): 20180005.

Arafat, A., Na'es, M., Kantarelou, V., Haddad, N., Giakoumaki, A., Argyropoulos, V., Anglos, D. & Karydas, A. G. 2013. Combined in situ micro-XRF, LIBS and SEM-EDS analysis of base metal and corrosion products for Islamic copper alloyed artefacts from Umm Qais Museum, Jordan. *Journal of cultural heritage* 14(3): 261–269.

Awasthi, S., Kumar, R., Rai, G. K. & Rai, A.K. 2016. Study of archaeological coins of different dynasties using libs coupled with multivariate analysis. *Optics and Lasers in Engineering* 79: 29–38.

Bakewell, P. (ed.). 2020. *Mines of Silver and Gold in the Americas*. Abingdon: Routledge.

Baldassarri, M; Cavalcanti, G.D.H., Ferreti, M., Gorghinian, A., Grifoni, E., Legnaioli, S., Lorenzetti, G., Pagnotta, S., Marras, L., Violano, E., Lezzerini, M & Palleschi, V. 2014. X-Ray Fluorescence Analysis of XII–XIV Century Italian Gold Coins. *Journal of Archaeology*, ID 519218.

Bartoli, L., Agresti, J., Mascalchi, M., Mencaglia, A., Cacciari, I. & Siano, S. 2011. Combined elemental and microstructural analysis of genuine and fake copper-alloy coins. *Quantum Electronics*, 41(7): 663–668.

Beck, L., Bosonnet, S., Réveillon, S., Eliot, D., & Pilon, F. 2004. Silver surface enrichment of silver–copper alloys: a limitation for the analysis of ancient silver coins by surface techniques. *Nuclear Instruments and Methods in Physics Research Section B: Beam Interactions with Materials and Atoms* 226 (1-2): 153–162.

Bethencourt, M., Fernández-Montblanc, T., Izquierdo, A., González-Duarte, M. M. & Muñoz-Mas, C. 2018. Study of the influence of physical, chemical and biological conditions that influence the deterioration and protection of Underwater Cultural Heritage. *Science of the Total Environment* 613: 98–114.

Borges, R., Silva, R.J.C., Alves, L.C., Araújo, M.F., Candeias, A., Corregidor, V. & Vieira, J. 2018. European silver sources from the 15th to the 17th century: the influx of "new world" silver in Portuguese currency. *Heritage* 1(2): 453–467.

Botto, A., Campanella, B., Legnaioli, S., Lezzerini, M., Lorenzetti, G., Pagnotta, S., Poggialini, F. & Palleschi, V. 2019. Applications of laser-induced breakdown spectroscopy in cultural heritage and archaeology: a critical review. *Journal of Analytical Atomic Spectrometry* 34: 81–103.

Carl, M. & Young, M. L. 2016. Complementary analytical methods for analysis of Ag-plated cultural heritage objects. *Microchemical Journal* 126: 307–315.

Castillejo, M., Oujja, M. & Sanz, M. 2020. Espectroscopias y microscopias láser al servicio del patrimonio cultural. In Secretaría General Técnica de la Subdirección General de Atención al Ciudadano, Documentación y Publicaciones (ed.), *La Ciencia y el Arte VII: Ciencias experimentales y conservación del patrimonio*: 231–250. Ministerio de Educación, Cultura y Deporte, Spain.

Ciarlo, N. C., De Rosa, H. M., Lorusso, H. N., Vázquez, C., Elkin, D. & Custo, G. 2015. Veritas Temporis Filia: Non-Destructive Analysis of Counterfeit and Regal Copper Coins from the Sloop-of-War HMS Swift (1770) by Means Of SEM-EDAX AND WDXRF. *The Numismatic Chronicle (1966-)* 175: 227–242.

Corsi, M., Cristoforetti, G., Giuffrida M., Hidalgo, M., Legnaiolo, S., Masotti, L., Palleschi, V., Salvetti, A., Tognoni, E., Vallebona, C. & Zanini, A. 2005. Archaeometric analysis of ancient copper artefacts by laser-induced breakdown spectroscopy technique. *Microchimica Acta*, 152: 105–111.

Cremers, D.A. & Radziemski, L.J. 1983. Detection of chlorine and fluorine in air by laser-induced breakdown spectrometry. *Analytical Chemistry* 55(8): 1252–1256.

Di Fazio, M., Felici, A. C., Catalli, F. & De Vito, C. 2019. Microstructure and chemical composition of Roman orichalcum coins emitted after the monetary reform of Augustus (23 BC). *Scientific Reports* 9, 12668.

Felix, V.S., Pereira, M.O., Freitas, R.P., Aranha, P. J., Heringer, P.C., Anjos, M.J. & Lopes, R.T. 2020. Analysis of silver coins from colonial Brazil by hand held XRF and micro-XRF. *Applied Radiation and Isotopes* 166: 109409.

Ferreti, M. 2000. X-ray fluorescence applications for the study and conservation of cultural heritage. In Dudley C. Creagh & David A. Bradley (eds.), *Radiation in Art and Archaeometry*: 285–296. The Netherlands: Elsevier Science.

Ferretti, M., Cristoforetti, G., Legnaioli, S., Palleschi, V., Salvetti, A., Tognoni, E., Console, E., & Palaia, P. 2007. In situ study of the Porticello Bronzes by portable X-ray Fuorescence and laser-induced breakdown spectroscopy. *Spectrochimica Acta Part B: Atomic Spectroscopy* 62(12): 1512–1518.

Fortes, F.J., Cortes, M., Simon, M.D., Cabalín, L.M. & Laserna, J.J. 2005. Chronocultural sorting of archaeological bronze objects using laser-induced breakdown spectrometry. *Analytica Chimica Acta* 554(1-2): 136–143.

Frahm, E. 2014. Scanning electron microscopy (SEM): Applications in archaeology. In Claire Smith (ed.), *Encyclopedia of global archaeology*: 6487–6495. Springer Reference.

García, S. 2020. Los protagonistas del último viaje de la fragata Nuestra Señora de las Mercedes: fuentes y representaciones. In M.D. González & D. Igual (eds.), *El mar vivido. Perfiles sociales de las gentes de mar en la larga duración (siglos XV-XXI)* 170: 193–214. Ediciones de la Universidad de Castilla La Mancha.

Gaudiuso, R., Uhlir, K. & Griesser, M. 2019. Micro-invasive depth profile analysis by laser-induced breakdown spectroscopy (LIBS): the case of mercury layers on Sasanian coins. *Journal of Analytical Atomic Spectrometry* 34: 2261–2272.

Genc Oztoprak, B., Sinmaz, M. A. & Tülek, F. 2016. Composition analysis of medieval ceramics by laser-induced breakdown spectroscopy (LIBS). *Applied Physics A* 122: 557.

Giakoumaki, A., Melessanaki, K. & Anglos, D. 2007. Laser-induced breakdown spectroscopy (LIBS) in archaeological science—applications and prospects. *Analytical and bioanalytical chemistry* 387: 749–760.

Gorghinian, A., Esposito, A., Ferreti, M. & Catalli, F. 2013. XRF analysis of Roman Imperial coins. *Nuclear Instruments and Methods in Physics Research Section B: Beam Interactions with Materials and Atoms* 309: 268–271.

Gregory, D. 2009. In situ preservation of marine archaeological sites: out of sight but not out of mind. In Vicki Richards & Jennifer McKinnon (eds.), *In Situ Conservation of Cultural Heritage: Public, Professionals and Preservation*: 1–16. Adelaide: Flinders University.

Hall, E.T. 1960. X-ray fluorescent analysis applied to archaeology. *Archaeometry* 3(1): 29–35.

Huang, J. 2012. Odyssey's Treasure Ship: Salvor, Owner, or Sovereign Immunity. *Ocean Development & International Law* 44(2): 170–184.

Inberg, A., Ashkenazi, D., Cohen, M., Iddan, N. & Cvikel, D. 2018. Corrosion products and microstructure of copper alloy coins from the Byzantine-period Ma'agan Mikhael B shipwreck, Israel. *Microchemical Journal* 143: 400–409.

Karydas, A. G. 2007. Application of a portable XRF spectrometer for the non-invasive analysis of museum metal artefacts. *Annali di Chimica: Journal of Analytical, Environmental and Cultural Heritage Chemistry* 97(7), 419–432.

Kotoula, E. & Kyranoudi, M. 2013. Study of ancient Greek and Roman coins using reflectance transformation imaging. *E-conservation magazine* 25: 74–88.

Lazic, V., Vadrucci, M., Fantoni, R., Chiari, M., Mazzinghi, A. & Gorghinian, A. 2018. Applications of laser induced breakdown spectroscopy for cultural heritage: A comparison with X-ray fluorescence and particle induced X-ray emission techniques. *Spectrochimica Acta Part B: Atomic Spectroscopy* 149: 1–14.

Lei, J., Zeng, L., Tong, H., Yu, X., Liu, J. & Hu, J. 2003. Characterization of Kangxi Coins of Tsing Empire by SEM-EDS. *Microchimica Acta* 142(1/2): 123–127.

Lorenzetti, G., Grifoni, E., Legnaioli, S., Pagnotta, S. & Palleschi, V. 2018. Analysis of the alloy of the Pisa Griffin. In A. Contadini (ed.), *The Pisa Griffin and the Mari-Cha Lion*: 105–112. Pisa: Pacini Editore.

Marcos, C. 2014. El último viaje de la fragata Mercedes. Un tesoro cultural recuperado. In *El último viaje de la fragata Mercedes. La razón frente al expolio. Un tesoro cultural recuperado (exhibition catalog)*: 53–67. Madrid: Museo Naval & Museo Arqueológico Nacional.

Mata, A. L., Carneiro, A., Neto, M. M., Proença, L. A., Salta, M. M. L., Mendonça, M. H. & Fonseca, I. T. E. 2010. Characterisation of five coins from the archaeological heritage of Portugal. *Journal of Solid State Electrochemistry* 14: 495–503.
Melessanaki, K., Mateo, M., Ferrence, S.C., Betancourt, P.P. & Anglos, D. 2002. The Application of LIBS for the Analysis of Archaeological Ceramic and Metal Artifacts. *Applied Surface Science* 197:156–163.
Memet, J. B. 2008. Conservation of Underwater Cultural Heritage: characteristics and new technologies. *Museum International* 60(4): 42–49.
Mousser, H., Amri, R., Madani, A., Darchen, A. & Mousser, A. 2011. Microchemical surface analysis of two Numidian coins. *Applied Surface Science*, 257(14): 5961–5965.
Müller, K. & Stege, H. 2003. Evaluation of the analytical potential of laser-induced breakdown spectrometry (LIBS) for the analysis of historical glasses. *Archaeometry* 45(3): 421–433.
Nieto, X. 2014. Arqueólogos y cazadores de tesoros subacuáticos. In *El último viaje de la fragata Mercedes. La razón frente al expolio. Un tesoro cultural recuperado (exhibition catalog)*: 497–506. Madrid: Museo Naval & Museo Arqueológico Nacional.
Oujja, M., Palomar-Sanz, T., Martínez-Weinbaum, M., Martínez-Ramirez, S. & Castillejo M. 2021. Characterization of medieval-like glass alteration layers by laser spectroscopy and nonlinear optical microscopy. *The European Physical Journal Plus* 136: 859.
Pardini, L., El Hassan, A., Ferretti, M., Foresta, A., Legnaioli, S., Lorenzetti, G., Nebbia, E., Catalli, F., Harith, M.A., Díaz Pace, D. Anabitarte García, F., Scuotto, M. & Palleschi, V. 2012. X-ray fluorescence and laser-induced breakdown spectroscopy analysis of Roman silver denari. *Spectrochimica Acta Part B: Atomic Spectroscopy* 74:156–161.
Rezk, R. A., Abdel Ghany, N. A., & Mostafa, A. M. 2022. Laser-Assisted Method for Cleaning and Analysis of Archaeological Metallic Coins. *Coatings*, 12(10), 1548.
Scott, D. A. 2002. *Copper and bronze in art: corrosion, colorants, conservation*. Los Angeles: Getty publications.
Selwyn, L. 2004. *Metals and corrosion: a handbook for the conservation professional*. Ottawa: Canadian Conservation Institute.
Shackley, M.S. 2011. An introduction to X-ray fluorescence (XRF) analysis in archaeology. In M. S. Shackley (ed.), *X-ray fluorescence spectrometry (XRF) in geoarchaeology*: 7–44. New York: Springer.
Siano, S. & Agresti, J. 2015. Archaeometallurgical characterisation of Donatello's Florentine copper alloy masterpieces using portable laser-induced plasma spectroscopy and traditional techniques. *Studies in Conservation* 60(Suppl. 1): S106–S119.
Valera, E. & Carrasco, R. 2014. El cargamento de la fragata Nuestra Señora de las Mercedes. Un desafio para la gestión del patrimonio cultural. In *El último viaje de la fragata Mercedes. La razón frente al expolio. Un tesoro cultural recuperado (exhibition catalog)*: 445–453. Madrid: Museo Naval & Museo Arqueológico Nacional.
Yatsuk, O., Ferreti, M., Gorghinian, A., Fiocco, G., Malagodi, M., Agostino, A. & Gulmini, M. 2022. Data from Multiple Portable XRF Units and Their Significance for Ancient Glass Studies. *Molecules* 27(18): 6068.

Lasers in the Conservation of Artworks XIII – Siano & Ciofini (eds)

LIBS to classify different shades of gildings

F. Surma, P. Schloegel, V. Aguilar & L. Rosenbaum
Epitopos, Strasbourg, France

M. Labouré
Mescla, Strasbourg, France

L. Gential
EOST, Strasbourg, France

ABSTRACT: Gilding is an important object of study in the context of heritage restoration and conservation. Gold leaves used for gilding have colors or shades according to the content of the various metals such as gold, silver, copper and others. However, time and possible alterations can modify the original shades, which makes it difficult to find a gold leaf equivalent to the original one. In order to build a model allowing us to classify gildings according to their composition, Laser Induced Breakdown Spectroscopy (LIBS) has been used. Particular gold (Au), silver (Ag), copper (Cu) and zinc (Zn) rays were studied. A Principal Component Analysis (PCA) has been performed to develop a classification model from gilding standards with known content. The resulted model has been applied on samples to identify original and restored gildings. Thereby, this approach using LIBS, could be performed on site to help restorers and gilders in their work.

1 INTRODUCTION

A gold object in jewelry is classified according to its color and composition. The color of a gold alloy depends on the other metals that constitute it, either copper to give a reddish or orange tint, or silver and zinc to make it more greenish or white (Manas 2020). Gilding is a technique that consists of applying a gold sheet to a prepared surface. The preparation is done by a grounding (layer of plaster), a bole (layer of red clay) and finally a gold leaf (Sandu et al. 2010. Different types of alloys are used, resulting in a gilding with a specific color. Nevertheless, over time, the original color could be modified as a consequence of alteration, which makes difficult for restorers to find the original shade (Ionardis et al. 2013, Kono et al. 2015). Before, the gilding identification was only visual and this can lead to inaccuracies if the surface is quite degraded. The aim of this study is to build a classification model based on the composition of known gildings and forward to use it to identify the type of gildings on site. For this purpose, LIBS technology (Laser induced breakdown spectroscopy) is the ideal instrument to collect data (Siano & Agresti 2018). Gold (Au), silver (Ag), copper (Cu) and zinc (Zn) rays were the principal studied elements and a statistical approach (PCA - Principal Component Analysis) was performed to build a classification model (Holand 2008). The model has been developed from the LIBS acquisition of five gildings with different types of gold sheets. The classification could then allow us to recognize types of gildings during measurement campaigns on site with better accuracy.

DOI: 10.1201/9781003386872-7

2 METHODOLOGY

2.1 *LIBS analysis*

The LIBS (*Laser-Induced Breakdown Spectroscopy*) is a measurement technique used to detect the chemical elements in a material. The functioning is simple: a 1064 nm pulsed laser excites the surface of the analyzed sample. The power density at the material surface induces a laser ablation. Moreover, the high-power density generates the formation of a short-lived and highly luminous plasma at the material surface. When plasma gets cooler, exited ions and atoms emit characteristic optical radiation when coming back to a lower energy state. This optical radiation is detected by optical fiber and analyzed by spectrographs (Maker et al. 1964, Runge et al. 1966). The spectra interpretation permits to determine elements composing the material. The advantage of being only micro-destructive (spot of 200 µm) (Jurado-López & Luque de Castro, M. 2003) implies that this technique is appreciated in the heritage domain (Botto et al. 2019, Perez-Rodriguez et al. 2013). Moreover, the LIBS is a portable instrument, which can be used directly on site. Finally, with the LIBS system, it is possible to repeat measurements with the same conditions (in particular temperature: here 37.5 °C) to make pseudo-quantitative assessments.

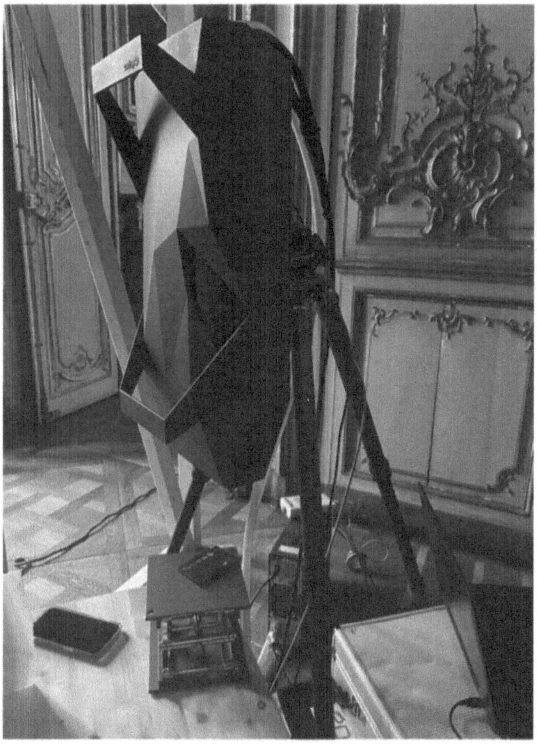

Figure 1. Use of LIBS on site («Palais Rohan, Strasbourg»).

The portable system used for this study (Figure 1) is composed of a Quantel Nd:YAG laser at 1064 nm and the focal distance is 12 cm. The laser spot diameter is very small (200 µm) and the repetition frequency of the analysis can be modified from 5Hz to 10Hz. The collected spectrum is between 230 and 900 nm and the spectral resolution is 0.1nm.

2.2 *Sampling*

Five samples of water-based gilding were analyzed for this study. The standard gildings (Figure 2) were prepared by Mescla as follows: (a) a bonding layer (skin glue), (b) a gesso layer (plaster), (c) a bole layer (red clay layer) and finally a gold leaf. The gold leaves used are from Eytzinger (Germany): red gold (Rot gold 23 carat), orange gold (Orange double gold 22 carat), white gold (WeiB Gold 12 carat), lemon gold (Citron) and green gold. A heterogeneous appearance was noticed on the leaf of lemon gold.

For each gilding, 5 laser shots have been released on 50 different zones except for the lemon gold where 25 analysis zones were performed on light areas and 15 on dark areas. The first and second shots reach the gold leaf whereas the three others reach the preparation layers. For this reason, we principally focused on the first shot. An example of spectrum is presented in Figure 3.

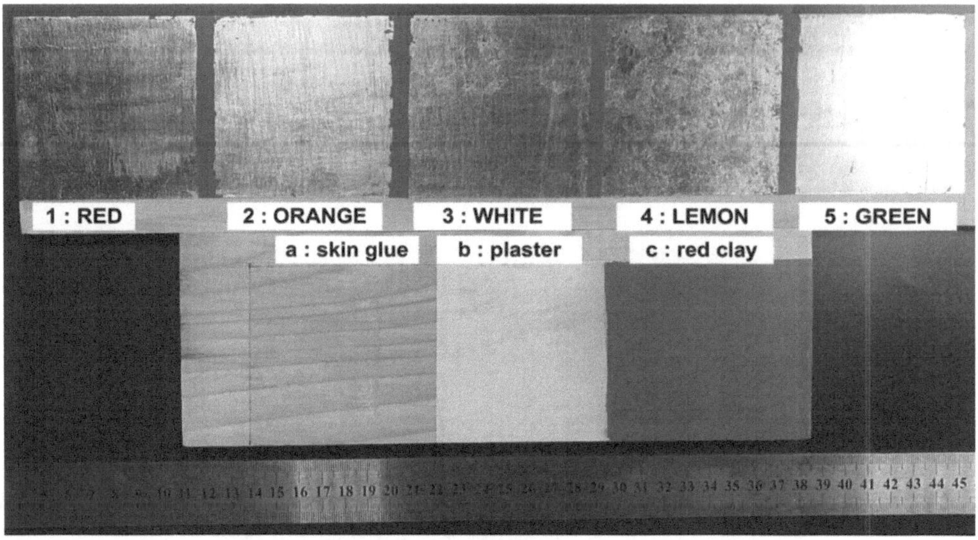

Figure 2. Different types of gildings (1 to 5) and preparation layer (a to c).

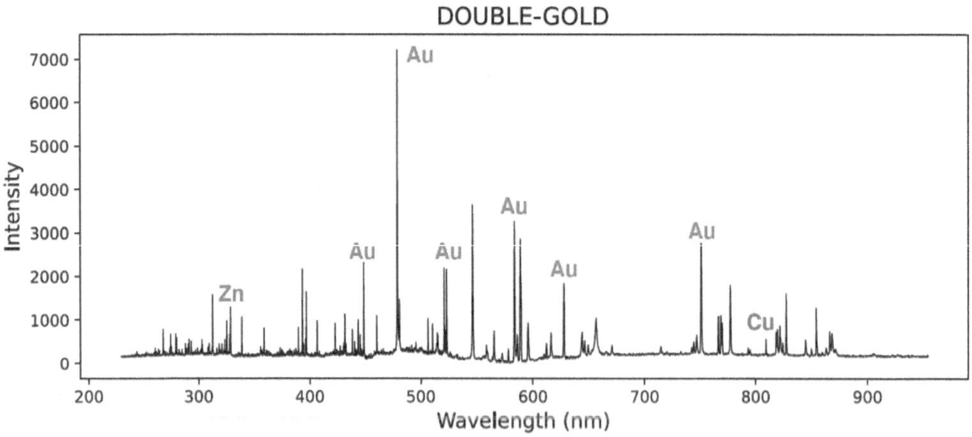

Figure 3. LIBS spectrum of Orange double gold.

2.3 *Statistical analysis (PCA)*

A great number of spectra must be treated, therefore we used a statistical method. PCA is a dimensionality-reduction method which permits the reduction of number of variables while losing as less information as possible (Cretu & Lingen 1999, Díaz et al. 2017). The objective of making PCA on LIBS spectra is to classify the different types of gilding depending on their metal element proportions: Gold (Au), Silver (Ag), Copper (Cu) and Zinc (Zn). For each element, specific wavelengths are selected (Pochon et al. 2020, Rifai & Constantin 2017) from the first shot spectra observation. We have developed a computing program to realize the PCA with the chosen rays as the variables. A 2D dimension projection has been chosen to plot the new data set.

The PCA program uses the following process (Díaz et al. 2018, Grupce et al. 2010). The range of the continuous initial variables is standardized. It means that each value is subtracted by its variable mean and then divided by its variable standard deviation. The purpose is to make the variables contribute equally to the analysis. A covariance matrix is then calculated to assess the correlation between all the variables and to identify the redundant data. The eigenvectors and the eigenvalues are calculated from the covariance matrix to determine the principal components. Those principal components are classified in order of significance. In this way, data is turned from the original axes to new axes calculated from the principal components. We attempted a minimum of 80% for the cumulated variance to accept the PCA precision (Kaiser criteria).

3 RESULTS

3.1 *Characterization of gold sheets*

PCA analysis has been realized with the five standard gildings prepared. The wavelengths used as variables are listed in Table 1. The statistical approach gives us a cumulated variance of 92.58%, which means the information loss is acceptable. The new variables are plotted on a two axed graph where each item corresponds to a laser shot. This representation shows trends on chemical composition. We can see that the different gildings are grouped in different zones (Figure 4). They are all discernible except the lemon gold which a part covers the green gold area. We can notice that the red gold and orange gold are both enriched in gold (Au) compared to other sheets. Besides, red gold seems to be enriched in copper (Cu) and white gold in zinc (Zn). In lemon and green gold sheets, there are more silver (Ag) and some zinc (Zn) too. Thereby, we obtain a clean and relevant classification of gilding with a global PCA, except for lemon gold which also is the more heterogenous gilding.

Table 1. Wavelengths used for the first PCA.

Elements	Au	Au	Ag	Ag	Cu	Zn	Zn
Wavelength (nm)	267.586	627.823	338.342	546.749	324.784	328.094	472.156

Now, in order to dissociate lemon gold from green gold, a new PCA with new wavelengths has been performed (Table 2). The wavelength has been chosen among the most differentiating rays of lemon gold and green gold spectra: calcium (Ca) and iron (Fe) have been added to the PCA. The cumulated variance of this new model is 89.41%. To understand the difference of color inside of the lemon gold, the clear zone spectra and the dark zone spectra have been separately captioned (Figure 5). We see that dark zones of lemon gold are always very spread while the light zones are very concentrated in the same area as green gold (area containing less copper (Cu), zinc (Zn) and calcium (Ca)). Thus, we can assume that the lemon gold is a mix of different gildings including green gold. This hypothesis also is supported by macroscopic observation since colors are very close.

3.2 *Identification of gildings from the 'Palais des Rohan'*

LIBS analysis has been applied on gildings from the "Palais des Rohan" of Strasbourg with the aim to determine original or restauration campaigns. Sometimes, original colors are modified by alteration or aging and it is not easy to find the right shades for restorers. Hence our classification model was tested on libs data obtained from the Palais Rohan gildings.

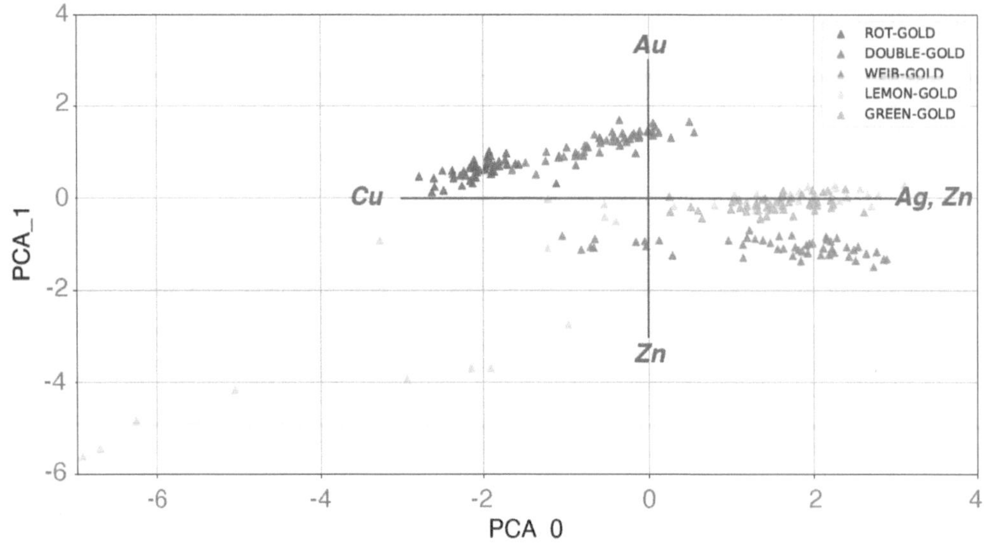

Figure 4. PCA results from LIBS shots on standard gildings.

Table 2. Wavelengths used for the second PCA.

Elements	Cu	Zn	Ca	Fe	Fe	Fe
Wavelength (nm)	809.210	636.231	458.644	344.089	358.143	274.930

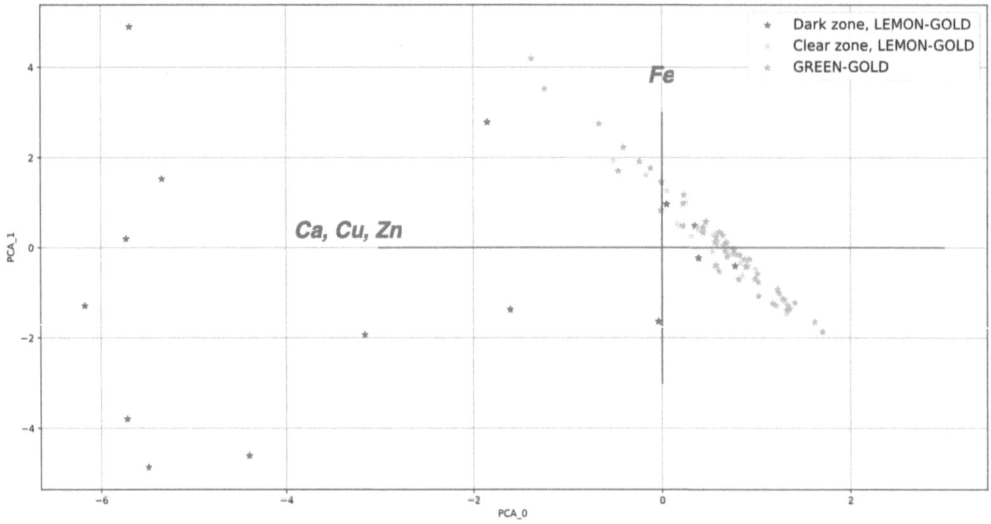

Figure 5. PCA results from LIBS shots on standard gildings.

The cumulated variance obtained by the addition of these new data is 92.81%. The wavelengths studied from the first PCA model had beed used (Table 1). Results (Figure 6) show that Rohan gildings contain more gold (Au) and copper (Cu). Original gildings are close to the red gold leaves and the restored one close to orange gold.

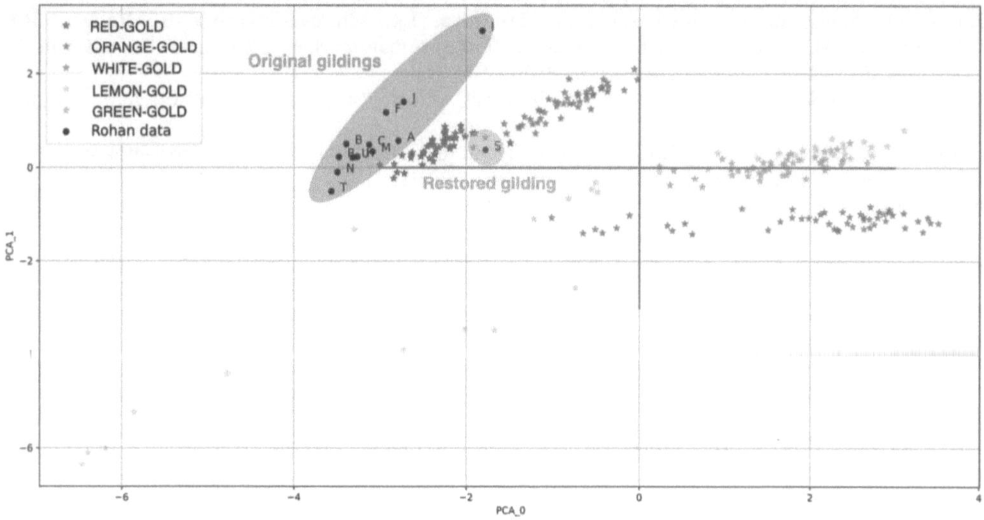

Figure 6. Classification model applied on Palais Rohan's gildings.

4 CONCLUSIONS

LIBS combined to statistical analysis (PCA) is a powerful tool to classify gildings into distinct groups according to their metal element proportions (Au, Ag, Cu, Zn, Fe). Indeed, different shades of gildings can be separated (red, orange, and white) using LIBS. Only green gold and lemon gold cannot be systematically distinguished. However, the results showed that lemon gold layer could be a mix of different types of gold, including green. Those results could be used to classify old and unknown gilding directly on site.

Gildings studied in the "Palais des Rohan" of Strasbourg appear close to the red gold leaf type (Rot Gold 23 carat). The classification model applied to the data collected from the "Palais Rohan" allowed us to identify the type of gold with precision. In heritage and restoration domain, being able to identify original or restored gildings is essential not only for restorers but also to gather information that can be useful for the reconstruction of the history (Sandu et al. 2015, Ottenwelter, et al. 2020). Thanks to the presented method, we can provide a help directly on site, using a portable technology such as LIBS. Now, we can study a gilding composition in situ and apply our model to deliver a gold type proposition.

The next step of this study is to build a bigger data base using new standard gildings which would improve the accuracy of the method.

REFERENCES

Maker, P.D., Terhune, R.W. & Savage, C.M. 1964. Optical third harmonic generation. In P. Grivet and N. Bloembergen (eds) *3rd International Conference on Quantum Electronics (Paris, France)*: 1559. Paris: Dunod Éditeur.

Runge, E.F., Bonfiglio, S. & Bryan, F.R. 1966. Spectrochemical analysis of molten using a pulsed laser source. *Spectrochimica Acta* 22: 1678–1680.

Botto, A. et al. 2019. Applications of laser-induced breakdown spectroscopy in cultural heritage and archaeology: a critical review. *Journal of Analytical Atomic Spectrometry* 34: 34–81.

Cretu, C. & Lingen, E. V. D. 1999. Different coloured gold alloys. *Gold Bulletin* 32: 115–126.

Díaz, D. et al. 2017. Quantification of gold and silver in minerals by laser-induced breakdown spectroscopy. *Spectrochimica Acta Part B: Atomic Spectroscopy* 136: 106–115.

Díaz, D., Molina, A. & Hahn, D. 2018. Effect of laser irradiance and wavelength on the analysis of gold-and-silver-bearing minerals with laser-induced breakdown spectroscopy. *Spectrochimica Acta Part B: Atomic Spectroscopy* 145: 86–95.

Grupce, O., Simileanu, M., Raškovska, A., Dzidrova, Lj., Minčeva-Šukarova B., Radvan, R., Šurbanovska, M. & Striber J. 2010. Characterization of roman marble bust using LIBS and Raman spectroscopy. A case study. *Revistapontica* 43: 563–572.
Holand, S. M. 2008. Principal Components Analysis (PCA). http://strata.uga.edu/software/pdf/pcaTutorial.pdf.
Ionardis, A. & Garcia-Guinea, J. 2013. Gold gilding and pigment identification on a post-byzantine icon from kastoria, northern greece. *Analytical Letters* 46: 936–945.
Jurado-López, A. & Luque de Castro, M. 2003. Chemometric approach to laser-induced breakdown analysis of gold alloys. *Applied spectroscopy* 57: 349–352.
Kono, M, Baldwin K.G.H., Wain A., & Rode A.V. 2015. Treating the untreatable in art and heritage materials: ultrafast laser cleaning of 'cloth-of-gold'. *Langmuir* 31(4): 1596–1604.
Manas, A. 2020. 'All glisters is not gold', a colorimetric assessment of the touchstone for gold ternary alloys. *ArcheoSciences* 44: 51–62.
Ottenwelter, E. et al. 2020. Technological characterisation of early Medieval gilded copper hollow pendants (*gombiky*), from Mikulčice (Moravia) and Prague Castle (Bohemia). *Archaeological and Anthropological Sciences* 12.
Perez-Rodriguez, M. et al. 2013. Non-invasive analytical techniques applied to characterize the components of ancient golden medallions. *Heritage Science* 4: 4.
Pochon, A. et al. 2020. Handheld laser-induced breakdown spectroscopy (LIBS) as a fast and easy method to trace gold. *Journal of Analytical Atomic Spectrometry* 35: 254–264.
Rifai, K., Constantin, M. 2017. Analysis of gold in rock samples using laser-induced breakdown spectroscopy: Matrix and heterogeneity effects. *Spectrochimica Acta Part B: Atomic Spectroscopy* 134: 33–41.
Sandu, I. et al. 2015. A comparative multi-technique investigation on materiel identification of holding layers and the conservation state of 7 portuguese mannerist altarpieces. *International Journal of Conservation science* 6: 439–454.
Sandu, I. et al. 2010. Gilding techniques in religious art between and west, 14th-18th centuries. *International Journal of Conservation science* 6: 47–62.
Siano, S. & Agresti, J. 2018. Laser-Induced Breakdown Spectroscopy (LIBS). In S. L. López Varela (ed.) *Encyclopedia of Archaeological Sciences*, Wiley. DOI: 10.1002/9781119188230.

Lasers in the Conservation of Artworks XIII – Siano & Ciofini (eds)

Use of laser additive technologies for restoration and reconstruction of artworks

V.A. Parfenov
St. Petersburg Electrotechnical University, Saint Petersburg, Russia

S.D. Igoshin
Peter the Great Saint-Petersburg Polytechnic University, Saint Petersburg, Russia

D. Kuliashou
Center of prototyping, Saint Petersburg, Russia

ABSTRACT: In recent decades, one of the most serious challenges in the preservation of cultural and historical heritage (CH) has been the rapid decay of exterior monuments due to environmental deterioration, especially in large industrial megapolises. The question on the need to re-construct damaged CH objects or to gradually replace them with copies. According to the world practice, it is the only opportunity to preserve the most important monuments of the past for descendants, at least their most valuable ones. Innovative 3D laser scanning technology combined with milling machines with numerical software control is established approach of replicating CH objects. The paper discusses the state-of-art of various 3D-printing techniques in terms of the feasibility of their use in restoration, as well as for creating replicas. In addition, we describe the case study of replication and reconstruction of the 19th century cast-iron artifact realized using combination of 3D laser scanning and 3D-printing

1 INTRODUCTION

Nowadays, the application of laser techniques in Cultural Heritage preservation has not only developed into a separate scientific and technical field, but has also become a field of wide practical application of lasers. The experience of use of lasers in this field goes back more than 50 years. The first applications are connected with the works of an American physicist, now Professor Emeritus of the University of California, John Asmus, who first applied the technique of optical holography and technology of laser cleaning for documentation and restoration of sculptural monuments in 1972 in Venice (Italy) (Lazzarini et al. 1972). Today, half a century after the pioneering works of John Asmus, there are 3 main directions in the application of lasers in CH 1. Restoration; 2. Chemical composition analysis, structural diagnosis and documentation of works of art; 3. Monitoring of monuments and the environment (i. e. the environment in which monuments exist).

In the field of restoration, the main usage of lasers is removal of natural and anthropogenic contaminations from the surface of CH objects. Laser cleaning has become very widespread in practice and is now used routinely in many museums around the world. It finds application in restoration of cultural heritage objects made of different materials: stone, metal, ceramics, fabric, paper, etc. (Cooper, M. 1998, Paraskevi et al 2016) and has been used in restoration of a number of the world-famous historical objects, including friezes of the ancient Greek temple Parthenon and monuments from Athens Acropolis in Greece (Paraskevi et al 2016) as well as a Renaissance masterpiece – Golden Gate at the Baptistery of the Cathedral of Santa Maria del Fiore in Florence (Italy) (Matteini et al 2003). Other laser technologies used in restoration include laser micro-welding (Innocenti et al 2003) and, more importantly in the context of this article, laser additive technologies.

DOI: 10.1201/9781003386872-8

As it is known, additive technologies, also known as 3D printing, are based on the layer-by-layer "growing" of objects using their 3D computer models. Therefore, 3D printing is very often used in combination with laser and optical 3D scanning, which are served to digitize CH objects. Meanwhile, most of all the works known from scientific literature on the application of 3D printing are devoted to its use for creating physical copies of monuments to replace originals either for exhibition purposes or for reconstruction of damaged and even completely destroyed objects (Hjalgrim et al. 2019).

Various additive technologies can be used for the reproduction of monuments, which use different energy sources and methods of curing the added material: laser beam, electric beam, infrared radiation, induction heating, addition of an adhesive, etc. Among the most commonly used 3D printing technologies in museum work are laser stereolithography (SLA), Selective Laser Sintering (SLS) and Fused Deposition Modeling (FDM). SLA technology ensures the highest accuracy of micro-relief reproduction of CH objects during their replication. For example, when working with small-sized objects, the accuracy of reproduction of their surface (according to the standard deviation criterion) can be no worse than 40 μm (Galushkin et al. 2019).

The additive technologies mentioned above have been used in museum work for more than 20 years. Their advantages are ease of use and the relatively low cost of the objects to be created. But they have a serious disadvantage is so that their target material group are polymers. For this reason, in those tasks where one need to make objects out of metals, these 3D-printers can used only for conducting auxiliary operations, and the final result may be achieved by means of use of other technologies.

However, there are additive technologies that allow printing objects directly (i.e. without intermediate stages) from metallic materials, such as Direct Metal Laser Sintering (DMLS or SLM) and Direct Metal Deposition (DMD or LENS) (Shahrubudin et al. 2019). The first one – DMLS – is interesting for solving tasks of metal object reconstruction from 3D-models with a quite high accuracy (of ±100 μm), but with following surface refinement, and the second one – DMD – is attractive since it allows one to carry out automated reconstruction of lost elements and surface restoration on objects with size up to several meters.

Both DMLS and DMD these technologies were used by authors of this article for reconstruction of a fragment of cast-iron fence of one of 19th century tombstones at the Alexander Nevsky Monastery in St. Petersburg. The results of this project are described in Section 2.1 and demonstrate sufficient benefits and perspectives of their usage. However, despite the main advantage of these technologies, which is the use of metal powders and, consequently, the possibility of creating copies or reconstruction of damaged objects in metal, printing directly using authentic materials in many cases is impossible, as the choice of available materials is too limited. In addition, the DMLS and DMD technologies are currently very expensive, primarily due to the high cost of metal powders.

For the reader's convenience, in the Table 1 we collected data giving the comparative information on the best known and most common 3D technologies as applied to the tasks of creating replicas and reconstructing cultural objects.

There are many other methods of 3D printing, such as LOM (Laminated Object Manufacturing), but it is not yet widespread and has been created for specific applications, although it is also a type of 3D printing. It should be noted here that the same printing technology can have several different names, as different machine manufacturers have tried to use their own unique terminology. In particular, Direct Metal Laser Sintering and Selective Laser Melting are one and the same technology, and there is also an ASTM standardised term for this technology – powder bed fusion (PBF).

As it was already above mentioned, both DMLS and DMD facilities and materials to be used for printing have very high costs. Therefore, for the tasks of museum work related to the restoration and replication of cultural objects made of metals, it may be interesting to apply such additive technologies, which are based on the use of more cost-effective technical solutions. But it should also be kept in mind that the main task of a restored object is to visually match the original material, i.e. to imitate the optical properties (color and luster) of metal objects, or to have that metal in its composition. New, metal-filled rods for FDM printing allow for just such an approach. The concept of metal-filled rods for FDM printing is similar to the composites used in industry for engineering applications, but in this case (for applications in CH preservation) the filler inside the plastic rod can be interesting in terms of performing a decorative function. This allows the plastic to be given a metal-like appearance without the use of additional coloring operations that are usually used to simulate the optical properties of metals. This possibility is due to the fact that in modern FDM bars any metal, including small fragments of material from even the cultural object being restored or copied itself, can be introduced into the plastic bar. This approach can be very special because the replica being created can inherit some of the material of the original object. Without any doubts, it is very promising approach and we demonstrated it in the frames of our work (see section 2.2).

Table 1. 3D printing technologies for replication and reconstruction of CH objects.

Technology	Material	Average field (cube side), mm	Average deviation, μm	Cost of equipment, USD	Cost of material, USD/kg
SLA[1]	polymers, ceramics	From 50 to 200	50	From 200	45
FDM[2]	polymers	From 200 to 400	100	From 150	15
SLS[3]	polymers	From 100 to 500	100	From 5 000	150
DMLS[4]	metals	From 100 to 500	100	From 15 000	80
DMD[5]	metals	From 200 to 1000	500	From 15 000	80

[1]) Stereolithography. [2]) Fused deposition modeling. [3]) Selective laser sintering. [4]) Direct metal laser sintering. [5]) Direct metal deposition.

2 CASE STUDY OF RECONSTRUCTION AND REPLICATION OF CAST-IRON HISTORICAL OBJECT

In this work we carried out experiments on application of additive technologies for reconstruction and replication of cast-iron historical object. The main goal of our work was the demonstration of principal possibility of use of laser additive technology DMLS for reconstruction of damaged metallic artifacts. One more aim was to check he feasibility of use of new additive material, such as metal-filled plastic, in conservation of artworks.

The object of our studies was a small (113.5×111.2×24.2 mm) iron star that is a decorative element of a XIX century cast-iron tombstone fence of the Alexander Nevsky monastery in St. Petersburg. This object is highly deteriorated due to corrosion and has losses of some small elements (Figure 1).

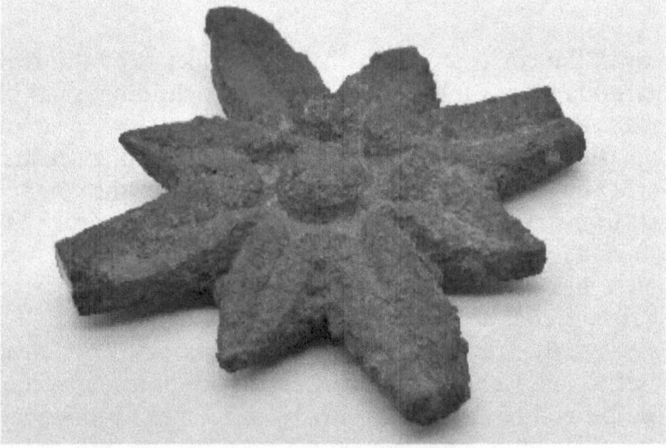

Figure 1. The object for reconstruction.

2.1 *Reconstruction of cast-iron star by means of DMLS and DMD*

The main idea of our project was to recreate the losses of individual elements of the star using a combination of 3D scanning technology, direct metal laser sintering, and laser cladding.

Firstly, we analyzed the chemical composition of the star using X-Ray Fluorescence (XRF) using XL3t-32280 equipment. The analysis results are shown in Table 2.

Table 2. Chemical composition of cast-iron star.

Element	Fe	Si	Co	Mn	P	Zn	Pb	Ti	Ag	Other
Content, wt.%	91.05	6.38	0.79	0.45	0.34	0.24	0.22	0.17	0.13	0.24

We did find that the main chemical element is Fe (its concentration is of about 91%). Additionally, there are permanent impurities typical for gray cast-iron (the total concentration is only about 7.2%).

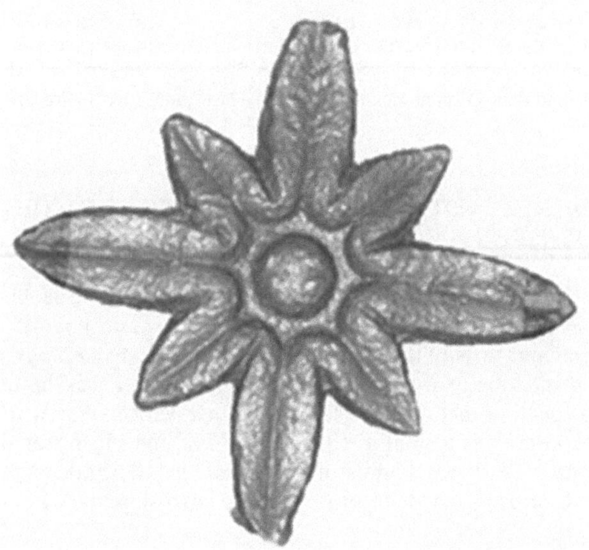

Figure 2. 3D model with added rays.

Secondly, we removed the corrosion from the star surface. For the removal of corroded layers, we tried to use different treatment techniques: laser cleaning, sand blasting and chemical treatment; the latter gave the best result.

Then, we carried out 3D scanning of the star using laser triangular scanner Konica Minolta V-910 and created its 3D model. The next stage of our work was the computer modeling. The missing ray of the cast-iron star was modeled with the Zbrush software. The main distinguishing feature of this software is the ability to "sculpt" 3D objects. The surviving cones were projected, processed, and "glued" to the main body of the star in the parts where there were losses (Figure 2), where blue ends are reconstructed lost parts of the star.

It was decided to separate those experiments into two stages: firstly, we created one of the star rays, which we proposed to join with the star; the second stage of reconstruction will be based on a reconstruction of the lost end of another star ray by direct metal deposition. It is obvious that such a method of reconstruction is more complicated since the very precise movement of a laser beam is needed, and such a task can be solved by means of the development of specialized software.

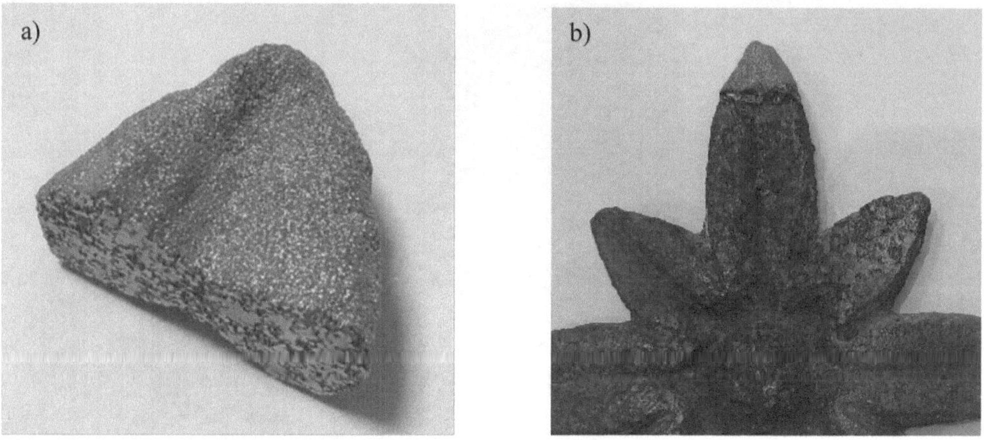

Figure 3. a) DMLS printed star ray; b) welded star ray.

A separate ray of the star was grown by the DMLS method according to the CAD model. We used a CW fiber Ytterbium laser with an output power of up to 80 W. Stainless-steel powder 316 L20 (manufacturer – Hoganas, Belgium) with a granule size of 53 μm was used for printing. The result of creating this piece can be seen in Figure 3a.

We then applied direct laser deposition (DLD) using the OKTA-Printer machine developed at the Peter the Great St. Petersburg Polytechnic University. The essence of the DLD method is to directly feed a metal powder or composite powder mixture into a molten bath formed by a laser beam. The local action of heating the laser beam, the selective effect, as well as the high accuracy of modern robotic manipulators allow using this technology as an advanced and modern method for repairing and restoring products. The main parts of the OKTA-Printer are the powder cladding head and the CAMAU robot. For joining, we used a nickel alloy powder Inconel 625 (manufacturer – Hoganas, Belgium) with a particle size distribution of 50–150 μm. It is one of the most popular alloys in laser cladding and DLD. To connect the two pieces, they were joined to each other with glue, then welded in one pass on each side. The following parameters were used for welding: laser spot width 1.5 mm, laser power 700 W, robot movement speed 1500 mm/s, and powder feed 20 g/min. The result of the star reconstruction is shown in Figure 3b.

It should be noted that as far as we know this case study is the first application of DMLS and DMD technologies for restoration of metal CH objects.

2.2 *Replication of cast-iron star by means of technology FDM using metal-filled rods*

Next stage of our work was connected with replication of the same cast-iron star by means of technology FDM using metal-filled rods. In the experiments we used 3D printer Raise3D Pro2 of FDM type made by Raise3D INC., USA. The printing was done with a locally produced metal-filled plastic bar. This filament contains 20% by volume bronze powder with a particle size distribution of approximately 100 μm. The binder is PLA plastic also in bronze color. A general view of the filament is shown in Figure 4.

The main goal of this part of our experiments was the study of feasibility of use of this new type of filaments containing metallic powder for restoration and replication of artworks, and we achieved this task by printing the tips of the cast-iron star and its replica.

Preparation of 3D model needed for printing was performed in IdeaMaker software. Printing was done at 220 °C nozzle temperature, 60 °C table temperature, 0.6 mm nozzle diameter, 0.2 mm lift height, 2 wall thickness, and 20% fill. The print time was 5 hours 3 min, and the weight of the object without supports was 52 g, which is 25% more than the print weight of pure PLA.

Since any 3D printing technology is based on layer-by-layer "growing" the created objects, 3D printing creates so-called supports that need to be removed after the printing process is complete.

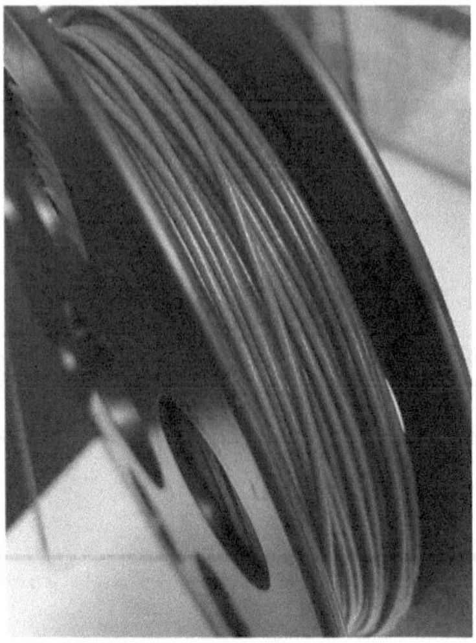

Figure 4. General view of a spool of metal-filled plastic filament with bronze powder content.

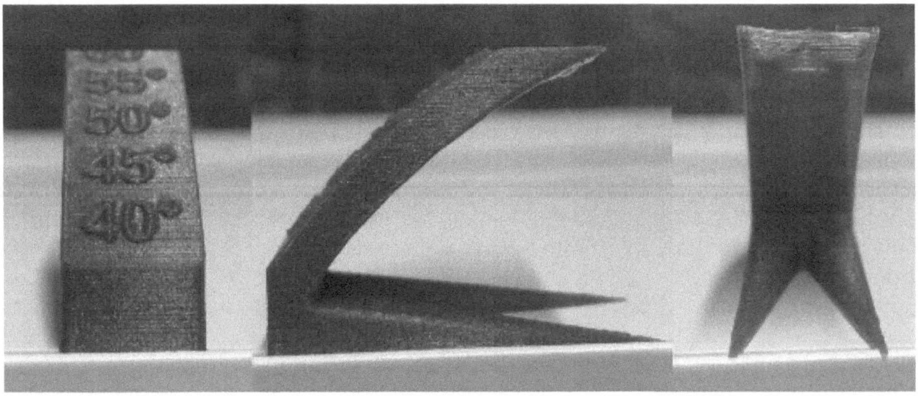

Figure 5. Arrangement of test for determination of the maximum allowable tilt angles of the metal-filled filament (for 0.6 mm nozzle and 0.2 mm lift height).

As the weight of the metal-filled filament exceeds the weight of the conventional filament (without metal filler), a special test was conducted to determine the maximum allowable angle of the wall of the star made on the printer without supports (Figure 5), which showed that at an angle of inclination greater than 55 degrees the next layers were not stable, and they bent. The larger this angle while printing, the better. Thus, the number of supporting structures is reduced. Where the main part touches the supporting structure, the surface of the part will look worse. Also, the removal of supporting structures is a time-consuming technological operation. Further, we entered of 55 degrees into the program to set the conditions for the arrangement of supporting structures. If a part feature is rotated more than 55 degrees relative to the table, the program will place a support underneath it.

Supports were placed semi-automatically with the removal of unnecessary columns. The result of printing, as well as the arrangement of supports is shown in Figure 6a. Also, since the main component of the filament is PLA plastic, immediately after printing its gloss prevails on the surface of the replica, which is an undesirable effect. To remove the shine of the plastic and activate the shine of the metal, post-processing was performed after printing, namely, the surface of the replica-ca star was machined with a boron and copper brush at low rpm to avoid causing uncontrolled melting of the plastic (Figure 6c).

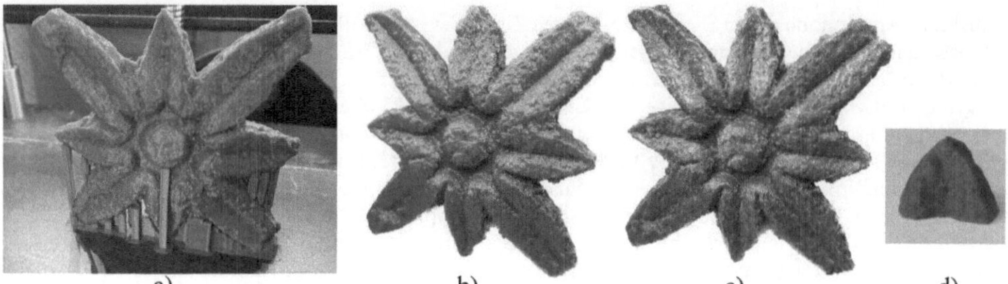

a) b) c) d)

Figure 6. General view of printed star: a – printed star replica with supports, b – replica with supports removed, c – replica after machining, d – tip created by printing with a metal-filled filament.

Similarly, the missing parts of the original star from the same material were printed using the model. The image of computing 3D model of one of the tips of the star, which was obtained by 3D modeling (for reconstruction of one of missing tips), and the restored tip are shown in Figure 6d. So, proposed approach has been shown to demonstrate the possibilities of replication and reconstruction of metallic CH objects by this new approach to 3D printing as well.

3 DISCUSSION AND CONCLUSIONS

The analysis of the scientific literature and the results of our experimental studies show that the use of additive technologies can effectively resolve the urgent tasks of modern museum work associated with the creation of physical copies, as well as the restoration and reconstruction of CH objects. Such work is possible as a result of using additive technologies in combination with laser and optical 3D-scanning technologies as well as 3D-modelling.

We demonstrated that the use of laser additive DMLS technology is appropriate in cases where it is necessary to ensure high mechanical strength and durability of created copies of monuments or reconstructed fragments of damaged objects. In all other cases for the restoration and reconstruction of metallic CH objects one can use a simpler and more budget-friendly FDM 3D printing technology, using plastics, imitating metals and containing them in their composition. This mate-rial allows to imitate shine and color of different metals, including the color of copper and bronze. It allows to carry out replication and reconstruction of sculptural monuments and objects of decorative art without application of artistic coloring.

In particular, using this approach it is possible to reconstruct in a new way the monument "Eve at the fountain", the project of which was described in Ignatiev et al. 2017. The replica of this monument was created using FDM technology from PLA plastic, which has a white color. Therefore, to imitate the color of zinc (the material from which the original sculpture was created) at the final stage of the project it was necessary to perform artificial patination. Using the approach described in this article, imitating the desired color of the copies of the monuments is unnecessary. This allows the copies to be displayed not only indoors, but even outdoors.

In conclusion, we hope that our proposed new approaches, related to the use of additive technologies in museum work, will be further developed and widely used since they open new perspectives in the use of laser techniques in CH preservation.

REFERENCES

Bertasa, M. & Korenberg, C. 2022. Successes and challenges in laser cleaning metal artefacts, A review, *Journal of Cultural Heritage* 53: 100–117.

Cooper, M. 1998. Laser Cleaning in Conservation: An Intfilamentuction, Oxford: Butterworth-Heinemann.

Galushkin, A., Gonobobleva, S., Parfenov, V. & Zhuravlev, A. 2019. Application of 3D Scanning for Documentation and Creation of Physical Copies of Estampages, *Restaurator, International Journal for the Preservation of Library and Archival Material* 40(1): 1–14.

Hjalgrim, H., Lynnerup, N., Liversage, M. & Rosenklint, A. 1995. Stereolithography: Potential Applications in Anthropological Studies, *American Journal of Physical Anthropology* 97 (3): 329–333.

Ignatiev P. P., Osipov D. V., Parfenov V. A. & Tishkin V. O. 2017. Reconstruction of the sculpture "Eve at the spring" from the estate "Sergievka" using 3D laser scanning, computer modeling and additive technologies. Obstchestvo. Sreda. Razvitie. 2017. No. 2. pp. 69–74.{in Russian}.

Innocenti, C., Pieri, G., & Yanagishita, M.R., Pini, R., Siano, S. & Zanini, A. 2003. Application of laser welding to the restoration of the ostensory of the martyr St. Ignatius from Palermo, *Journal of Cultural Heritage* 4(Suppl. 1): 362–366.

Lazzarini, L., Asmus, J & Marchesini, M.L. 1972. Laser for cleaning of statuary, initial results and potentialities, *1st Int. Symp. on the Deterioration of Building Stone*: 89–94. La Rochelle.

Matteini, M., Lalli, C., Tosini, I., Giusti, A. & Siano, S. 2003. Laser and chemical cleaning tests for the conservation of the Porta del Paradiso by Lorenzo Ghiberti, *Journal of Cultural Heritage* 4(Suppl. 1): 147–151.

Paraskevi, P., Papakonstantinou, E., Frantzikinaki, K. & Panou, A. 2016. The two-wavelength laser cleaning methodology. Theoretical background and examples from its application on CH objects and monuments with emphasis to the Athens Acropolis sculptures, *Heritage Science* 4: 9.

Shahrubudin, N., Lee T.C. & Ramlan R. 2019. An Overview on 3D Printing Technology: Technological, Materials, and Applications, *Procedia Manufacturing* 35: 1286–1296.

LASER TREATMENTS

Painted surfaces

Lasers in the Conservation of Artworks XIII – Siano & Ciofini (eds)

Laser cleaning in the conservation of archaeological artifacts: Polychrome wooden objects from ancient Egypt

F. Zenucchini, C. Ricci & A. Piccirillo
Centro per la Conservazione ed il Restauro dei Beni Culturali «La Venaria Reale», Venaria Reale, Italy

T. Cavaleri
Centro per la Conservazione ed il Restauro dei Beni Culturali «La Venaria Reale», Venaria Reale Italy and Department of Economics, Engineering, Society and Business Organization (DEIM), University of Tuscia, Viterbo , Italy

I. Cacciari
Institute of Applied Physics "Nello Carrara", Consiglio Nazionale delle Ricerche, Florence, Italy

M. Borla
Soprintendenza Archeologia, Belle Arti e Paesaggio per la città Metropolitana di Torino, Italy

S. Aicardi
Museo Egizio di Torino, Italy

P. Buscaglia
Centro per la Conservazione ed il Restauro dei Beni Culturali «La Venaria Reale», Venaria Reale Italy and Department of Applied Science and Technology (DISAT), Politecnico di Torino, Torino, Italy

ABSTRACT: This work describes some of the main results obtained over the years in the application of different lasers to the cleaning of ancient Egyptian painted wooden objects, with the primary goal of removing superficial blackening and altered synthetic adhesives applied within previous conservation treatments. The project's following phases entailed a systematic comparison of laser sources commonly used in the cultural heritage field, including Nd:YAG 1064-nm and 532-nm lasers in Q-Switching (QS) and Long Q-Switching (LQS) regimes, both alone and in combination with dry cleaning and solvent mixtures. A multi-analytical campaign, aimed to characterize the objects' materials and techniques, but also to assess their specific state of preservation and to monitor the cleaning process. Results demonstrate how the target materials, layered on the surfaces, could be removed with high spatial control, well preserving the underlying substrate and without mechanical stress to the treated surfaces.

1 INTRODUCTION

The Centro Conservazione e Restauro "La Venaria Reale" (CCR) works in partnership with "Light for Art" of El.En. group S.p.A. since 2007 with the aim of testing different laser systems in the context of scientific research, academic activity and conservation projects. Over the years, particularly satisfactory and significant results have been obtained on numbers of objects and materials.

In 2014, the Museo Egizio of Turin (ME) provided CCR staff with the opportunity to study and conserve several painted wooden coffins belonging to its collections for the renovation of its permanent exhibition, and to become partner of a wider international research project, called Vatican Coffin Project-VCP (Amenta, 2014). In this context, laser cleaning was performed on the lid of Nesimendjem's (Buscaglia, 2018) with successful results.

DOI: 10.1201/9781003386872-9

Since 2014 the activity in the frame of VCP has focused on the investigation on original artistic materials and techniques, on the conservation history of the artifacts, on specific conservation issues and deterioration processes. CCR scientific department constantly supported the study and the monitoring of treatments performing a multi-analytical campaign; besides this, for each case study developed collaborations aimed to integrate investigation techniques available to respond effectively to topics of specific interest or problems observed.

The long-standing collaboration with ME concerns the team to expand this line of research and identify a new conservation approach on painted objects from ancient Egypt. Some examples of wooden coffins as well as objects belonging to funerary equipment are further mentioned.

Each case study considered in this contribution presents common features for what concerns materials of interventions applied in their relatively recent history, i.e. since the second half of XX century, and, mainly, they all come from documented archaeological excavations, carried out by the Egyptologist E. Schiaparelli and his Italian team during his direction in the Museo Egizio.

Especially concerning the finds coming from Schiaparelli's excavations in Queen's Valley it seems important to briefly introduce the historical context, due to some peculiarities of the discovered tombs, which must necessarily be considered for a correct understanding of the layering of materials on some treated objects.

The Italian archaeological mission M.A.I. (Missione Archeologica Italiana), directed by E. Schiaparelli, excavated in 1903-04 the 20th Dynasty tombs built for two sons of Ramses III (Setherkhepeshef, QV 43, and Khaemuaset, QV 44; Guzzon, 2017).

The two tombs, already looted during the centuries, presented a very messy situation at the time of the discovery, as documented today by archive photos. Coffins were heaped up together with fragments of human mummies, few surviving grave goods and a few statuettes and boxes for shabti. Moreover, the tomb of Setherkhepeshef was found blackened by a fire, as written in some notes "the walls are blackened by the smoke and identical mess of blacking and burnt is also on coffins" (Schiaparelli, 1923).

When all the archeological finds were taken outside the two tombs to prepare them for transport to Turin, no precise provenance of the coffins was registered. Archival photos confirm the state of preservation of some of the coffins and testify a superficial black layer, apparently powdered at that time (Buscaglia et al, in press). Far along the years in the Museum, those objects, as others belonging to the collection, went through undocumented treatments and maintenance interventions.

1.1 *Case studies*

From numbers of case studies treated over the time, some cases, representative of different issues, were selected. All the objects mentioned are made of wood with ground layers and painted surface; the Egyptian palette is mainly composed by pigments such as: Egyptian blue, red ochre, yellow ochre, orpiment, carbon based black, copper based green and Egyptian green; from literature is known that natural gum was mainly used as pigment medium. (Lee, 2000; Newman and Serpico, 2000). The conservation histories of the items are related each other; indeed, issues are similar.

The Nesimendjem's Coffin lid (S. 05227), Third Intermediate Period/21st-25th Dynasty (1070-656 BCE), presented a natural resin as original finish covered by a thick black layer probably due to the fired tomb, and acrylic resin spread on both the black surface and on previous cleaning tests.

The statuette of bearer (S. 08795), Middle Kingdom, 12nd Dynasty (1976- 1872 BCE), the model of craft activity (S. 14375/03), Old Kingdom/6th Dynasty (2330-2190 BCE) and the coffin box of Padiamenemipet (Cat. 2235/2), Third Intermediate Period/22nd-23rd Dynasty (946-712 BCE) presented a very sensitive calcium carbonate based preparation layer and colors without any original resinous finish, altered in a gray hue by the presence of dust incorporated in the acrylic resin (Paraloid B72).

The coffin lid of Padiamenemipet (Cat. 2235/1), Third Intermediate Period, 22nd-23rd Dynasty (946-712 BCE) and the outer coffin lid of Besenmut (S. 5217), Late Period/24th-25th Dynasty (740-655 BCE), had very sensitive painted surfaces covered by black carbon particles due probably to the fired tomb, that had become coherent together with the acrylic resin.

1.2 *Research aims*

The aim of this research is to provide valuable insights for conservators, researchers, and institutions dealing with the preservation of cultural heritage objects.

The established practice provides for a preliminary analytical campaign for the correct characterization of both original and intervention materials to assess treatments and evaluating the effectiveness of different cleaning methods for the removal of previous treatments materials.

Starting from the first analytical campaign on ME coffins in 2014, the main data that emerged was the extended presence on the surfaces of an acrylic resin identified as Paraloid B72. This product became very common in Europe in 1983 and is to date used in the field of conservation as an adhesive and finishing coating (Horie, 1987; Koob, 1986). The investigation campaign also deepened the understanding of some important witness areas, which are the uncleaned test window left on the painted surfaces in previous interventions (Figure 1), useful to understand the original layering of materials and the eventual presence of layers taken off by the surface previously. The research also includes laser tests on these areas.

The results of the study demonstrate that the target materials overlapping the original painted decoration are confined to superficial layers. Physical methods, such as laser cleaning provide high spatial control, with no interference or mechanical stress for the treatment of extremely depolymerized surfaces. During the last twenty years the application of Nd:YAG lasers (solid-state lasers) on painted surfaces has been described in many papers (Zenucchini, 2022; Buscaglia, 2018; Brunetto, 2017; Ferrarato, 2017; Mansi, 2017; Gottardo, 2014; Graue, 2011; Siano, 2010; Castillejo, 2003).

However, the use of highly retentive gels has also shown positive results on extremely sensitive artworks (Manfredda et al., 2021). Therefore, we focused the study on the comparison of cleaning methods such as water solutions and/or solvents confined in high retentive gels, dry cleaning and lasers, all recently widely used in the field of conservation.

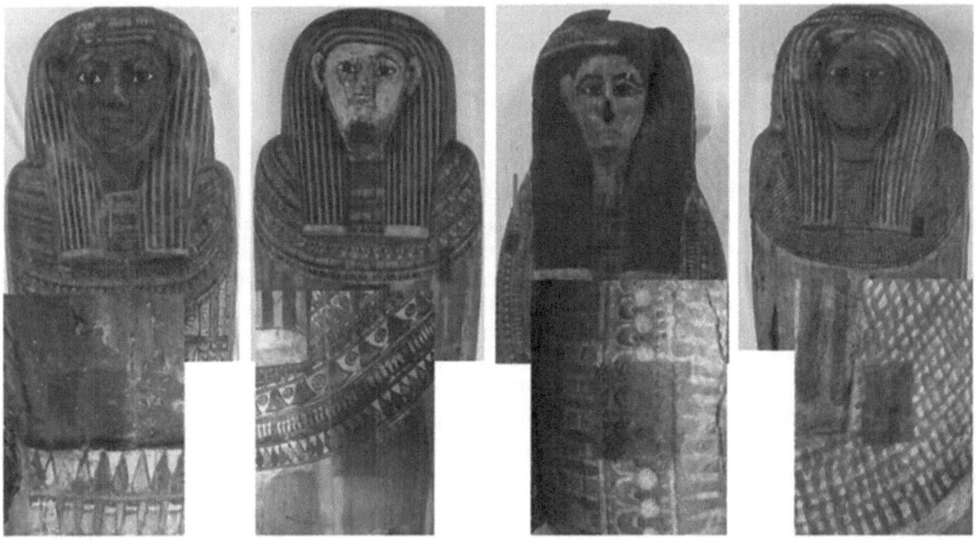

Figure 1. Uncleaned test window on a series of coffins lid studied, evidence of original layers and burn marks to be investigated.

2 MATERIALS AND METHODS

2.1 *Analytical campaign*

The multi-analytical campaign aimed not only to characterize the objects' materials and techniques, but also to obtain a comprehensive understanding of their conservation issues and to assess the cleaning process. After a preliminary investigation of the artifacts through multi-spectral imaging techniques, both microscopy and spectroscopy techniques were applied. In particular, the analytical study included: the use of a USB digital microscope and an optical microscope (OM) for a visual inspection of the surface and the examination of the cross-sections of a few samples; X-ray fluorescence (XRF) spectroscopy to determine the elemental composition of the constituent materials; Fourier-transform infrared (FTIR) spectroscopy for the characterization of the organic compounds present; Scanning electron microscopy coupled with energy-dispersive X-ray spectroscopy (SEM/EDS) for the analysis of the stratigraphy and elemental composition of the samples. Furthermore, a 3D digital microscope prototype (MICRO3D), developed by IFAC, allowed for non-destructive and contact-less surface measurements (Cacciari, 2012; Cacciari, 2013; Siano, 2017).

2.1.1 *Multispectral imaging*

For the acquisition of the ultraviolet-induced visible fluorescence (UVF), the lighting of the object was achieved by means of two 5-W Madatec LED UV lamps with emission peak at 368 nm. Images were acquired in the 380-780-nm spectral range with a Nikon D810 DSLR camera equipped with a complementary metal oxide semiconductor (CMOS) silicon sensor as well as a Hoya UV-IR Cut high-pass filter, providing a resolution of 7360 x 4912 pixels.

For Infrared reflectography (IRR), the lighting of the object was achieved by means of two 800-W Ianiro Varibeam Halogen lamps. Images were acquired in the 850-1000-nm spectral range (near infrared, NIR) with a Nikon D810 DSLR Full Spectrum camera, modified to extend its spectral sensitivity in the 350-1000-nm range and providing a resolution of 7360 x 4912 pixels, equipped with a complementary metal oxide semiconductor (CMOS) silicon sensor as well as an 850-nm B+W 093 high-pass filter. A 24-color X-Rite ColorChecker Classic reference was used for NIR images and their processing to obtain infrared false color images.

Infrared false color images were obtained in the RGB color space of Adobe Photoshop by using two reflection images acquired in the visible and NIR spectral ranges. In particular, the green (G) and red (R) components of the visible image are transferred into the blue (B) and green (G) channels, while the red (R) component is replaced with the NIR image. This methodology yields false color images of the NIR-R-G (RGB) type.

The Visible-induced infrared luminescence (VIL) was also used. Lighting of the object was achieved with a Nikon SB-910 portable flash with multiple Hoya UV-IR Cut filters to remove the infrared light component. Images were acquired in the 850-1000-nm spectral range with a Nikon D810 DSLR camera equipped with a complementary metal oxide semiconductor (CMOS) silicon sensor as well as an 850-nm B+W 093 high-pass filter, carried out by means of Adobe Lightroom and Adobe Photoshop software, included a color correction conducted by inserting a 24-color X-Rite ColorChecker Classic reference for the evaluation and reduction of the reflected light component, and an Egyptian blue reference (Kremer Pigmente n. 10060) in the field of view.

2.1.2 *USB Digital Microscope*

Microphotographs were collected using an RS PRO USB portable digital video microscope, model number 196-4075, equipped with a 5-megapixel image sensor as well as LED lighting with visible illumination. Images obtained with this system have a maximum resolution of 2592 x 1944 pixels, while the ranges of manual focusing and magnification are 0-150 mm e 20-200x, respectively.

2.1.3 *Cross section preparation and Optical Microscopy*

Cross sections were prepared by embedding each sample within a double layer of Struers EpoFix resin. After removal of the excess resin, the sample surface was finely polished using Struers abrasive cloths of progressively finer grits to expose the paint stratigraphy, thus enabling observation at high magnifications and scientific analysis with various instrumental techniques.

Multi-layered samples were then observed and photographed under visible and ultraviolet light using an Olympus BX51 minero-petrographic microscope equipped with an Olympus DP71 digital camera. In both cases, image acquisition and processing were performed by means of analySIS FIVE proprietary software.

2.1.4 *X-ray fluorescence spectroscopy (XRF)*
XRF analysis was performed using a Micro-EDXRF Bruker Artax 200 spectrometer equipped with a fine focus X-ray source including a molybdenum anode and a Si(Li) silicon drift detector (SDD) with an 8-μm beryllium window, providing an average resolution of approximately 144 eV for the full width at half maximum of the manganese Kα line. Measurements were carried out using 30-kV voltage, 1300-μA current, 60-s acquisition time, 1.5-mm collimator, with no filter, by fluxing helium gas onto the measurement area to improve the technique's detection limits (corresponding, with a helium flux, to Z=11, sodium).

2.1.5 *Fourier-transform infrared (FTIR) spectroscopy*
FTIR analysis was performed with a Bruker Vertex 70 FTIR spectrometer coupled with a Bruker Hyperion 3000 infrared microscope and equipped with a mercury cadmium telluride (MCT) detector. Scrapings were analyzed as a bulk in transmission mode through a 15x objective, upon compression in a diamond cell. Data were collected in the 4000-650 cm-1 spectral range, at a spectral resolution of 4 cm−1, as the sum of 64 scans. Spectra were interpreted by comparison with published literature and spectral libraries available at the CCR scientific laboratories.

2.1.6 *Scanning electron microscopy with energy-dispersive X-ray spectroscopy (SEM/EDS)*
Cross sections were observed and analyzed with a Zeiss EVO60 scanning electron microscope equipped with a lanthanum hexaboride (LaB6) cathode and a silicon drift detector (SDD), and coupled with a 40 mm^2 Oxford Ultim Max EDS microprobe for semi-quantitative elemental analysis. Samples were analyzed without any pretreatment in variable pressure mode, using an accelerating voltage of 20 kV and a pressure of 20 Pa.

2.1.7 *The portable MICRO 3D, developed by IFAC*
This prototype represents an implementation of "shape from shading" technique. The device includes an optoelectronic imaging group (CCD digital camera and a fixed focus objective) and a calibrated stage for translating this group along the optical axis. In particular, given the reduced thickness of the layer which is generally removed with the cleaning methods used, it was decided to use an optical system with an adequate depth of focus (of the order of tens of microns). This choice, for purely optical reasons, results in a field of view of approximately 1.9×1.4 mm. During the scan, images with a resolution of 400×300 pixels (pixel size of 4.77 μm) were acquired. In order to improve the signal-to-noise ratio, each saved image was obtained from the average of 5 images.

The software for data acquisition and processing was developed by IFAC: it allows the user to create a map of the heights of the surface under inspection either using a false color scale or with original color textures. It can also calculate both linear and areal roughness according to the UNI standard.

2.2 *Cleaning methods*

2.2.1 *Laser systems*
The methodological approach adopted includes a comparison of different laser systems commonly used in the field of Cultural Heritage. In particular, it is intended to compare ablation channels such as the photomechanical effect (Siano, 2008) using Nd:YAG lasers: the Q-Switched (Thunder Art, Quanta System) that emits at 1064 nm, 532 nm, and 355 nm, with a pulse duration of 8 ns (QS), the laser beam is transported via a 7-mirror articulated arm and it features an interchangeable fixed-focal handle, which can be replaced with a homogenizer handle; the Long-Q-Switched (EOS 1000 LQS, El.En. Group) emits at 1064 nm with a pulse duration of 100 ns and an energy of 130 mJ per pulse (LQS). It features a fiber optic bundle with a variable focal handle that can adjust the spot size from 1 to 6 mm. Specific filters can be applied to the handle to decrease the energy of the outgoing beam, for instance, the T50 filter can halve the minimum energy of 130 mJ, while the T25 filter allows only 25% of the bundle to pass.

Irradiation with short pulses ranging from 8 to 100 nanoseconds theoretically should not cause any thermal effects, but instead produce a vaporization effect with low fluence values and spallation effects with higher fluence values. In our experiments, the LQS laser was used with a fixed energy setting of 130 mJ (LQS-I). Subsequent energy values (250, 380 mJ) increased the pulse duration up to 200 microseconds. The frequency was also an important parameter to control in order to prevent surface heating.

Laser cleaning has been applied either alone or in combination with other cleaning methods described below.

2.2.2 *Dry cleaning and chemical methods*

Dry cleaning was performed with polyurethane makeup sponges (Deffner&Johann®), technologically upgraded sponges for obtaining a very soft mechanical action with no residues released (Daudin- Schotte, 2013).

In most cases, original materials have proved to be extremely reactive to wetting solutions, therefore their application on the surfaces, even though in the confined form with agarose to measure pH and conductivity, resulted not to be safe. In order to control and supply the chemical action, various tests were carried out using high-retentive gel systems: water based solutions confined in polysaccharide natural gels (Gellan and Agar Agar), polar solvents (acetone) confined in polyvinyl alcohol-borax type, and water-in-oil nanostructured fluids (Nanorestore Cleaning®, Baglioni, 2014; Domingues, 2013) confined in PVA-based gels (Peggy®) and pHEMA/PVP-based gels (Nanorestore Gels®).

2.3 *Mock-ups*

An experimental set up, consisting of two different series of mock-ups, was designed in order to reproduce the layering of materials, as found in many case studies, as well as reported for other similar objects in scientific literature (Pages-Camagna 2010; Serpico and White, 2000; Scott, 2014).

In particular, the first series of mock-up (Set 1) shows a calcium carbonate based ground layer covered by a natural resin layer (pistacia varnish), while the second one (Set 2) has no resinous finish. A detailed description of the two sets of mock-ups is reported in Figure 2.

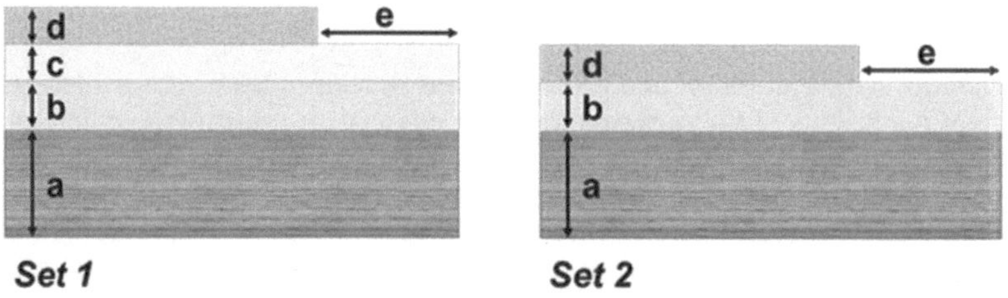

Figure 2. Schematic representation of the mock-ups, respectively with (Set 1, on the left) and without (Set 2, on the right) a resinous finish applied on the ground layer. Layers composition is: (a) wooden support; (b) ground layer composed of micronized calcium carbonate (particle size < 50 µm, Bresciani) binded with Arabic gum (powder, Bresciani) dissolved in water (15%); (c) natural resin (being Pistacia Resin highly documented in literature, a Mastic dissolved in turpentine essence, provided by AN.T.A. RES srl, was selected); (d) soiling obtained with Paraloid B72 dissolved in Acetone at 5%, with subsequent application of dirt. All mock-ups have an unsoiled area (e), that has been left to verify potential changes in color, thickness and roughness of the layer to preserve after treatments.

More specifically, Set 1 reproduces the Coffin lid of Nesimendjem (S. 05227) and the coffin lid of Padiamenemipet (Cat. 2235/1), while Set 2 reproduces the superficial layers of some of the cases listed in this paper, such as the statuette of bearer (S. 08795), the model of craft activity (S. 14375/03) and the coffin box of Padiamenemipet (Cat. 2235/2). In both cases, the surface of the mock-ups was later soiled with acrylic resin and dust (Figure 3).

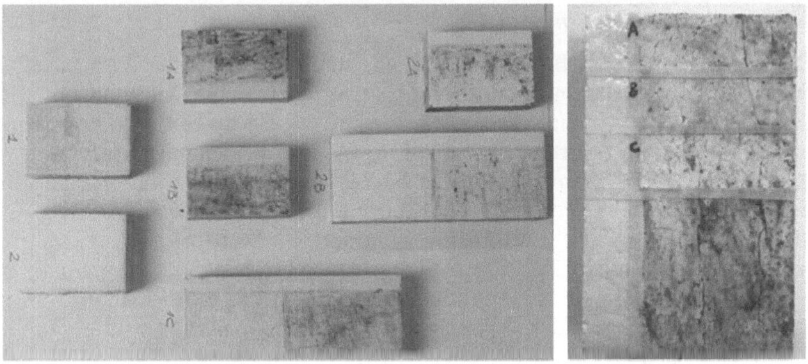

Figure 3. The two different series of mock-ups. On the right one of the mock-ups from Set 1 showing laser tests: A and B LQS and C QS.

Although aware of not reaching an aging of materials comparable to the original and intervention ones, both sets were artificially aged in a UV Solar box (UV Solar box Hereus Suntest CPS equipped with a filtered- coated quartz glass simulating a 3 mm window glass, cutting λ<300 nm- Xenon lamp and with an average irradiation of 750 W/m^2 and an internal temperature of about 50 C°), following two different steps: the first, after the application of the materials mimicking the original ones (1500 hours), and the second after soiling the surfaces (1000 hours).

3 RESULTS

3.1 *Preliminary cleaning tests on mock-ups*

The goal was to explore and understand the potential outcomes of laser irradiation with these specific pulse durations (8 to 100 ns). Moreover, a selection of gels has been tested in comparison, allowing us to understand which method could better preserve the original layers under the acrylic resin.

Cleaning tests have been performed on both series of mock-ups.

The results demonstrated that gels with the ability to partially remove the acrylic resin needed an additional swabbing action, risky for the original layers in both Set 1 and Set 2. The LQS laser (100 ns) was the most respectful in presence of the natural varnish, while the QS (8 ns) produced a spallation effect. Both lasers had a good result on set 2 respecting the white ground layer.

The evaluation of the effectiveness and safeness was observed by using a portable video microscope; in addition, a 3D microscopy technique, applied on selected areas for a more in-depth analysis of the cleaning efficiency, provided valuable insights into the best practices for the preservation (Figure 4).

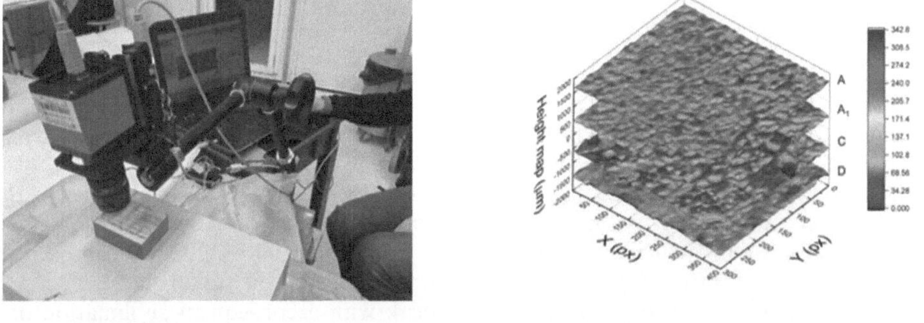

Figure 4. MICRO 3D, developed by IFAC during the evaluations on mock-up. The 3D reconstructions showed micromorphology of the four zones obtained with a false color scale.

3.2 *Nesimendjem's Coffin lid (inv. S. 05227)*

In this case already published (Buscaglia 2018) the main issue was the removal of the black layer covering most of the painted surfaces. Extensive investigations made by FTIR and SEM/EDS confirmed the presence of a natural resin covered by black carbons components and acrylic resin all over the surface (Figures 5–6). After a preliminary assessment, the LQS laser (LQS VARIO prototype model by El.En.) was selected. Ablation channels are determined from laser parameters setting and pulse duration together with dirt layer and object surface (Siano 2008); in this case it was also very important the effect of the beam profile that provides a homogenous action. The range of fluences between 0.30 – 1.16 J/cm^2, frequency from 1 to 5 Hz made it possible for a confined laser action to absorb mainly from the black dirt on top. The natural resin underneath allows a secondary spallation with high reflection of energy as well as a partially spallation effect. In this condition a self-limiting process takes place with a high spatial control and low interference with the original materials. After the irradiation the partially detached residues were removed with PU sponges. USB video microscopy and UV fluorescence were used to constantly monitor the effectiveness and extremely respectfulness of the laser action compared to the not documented previous cleaning tests that removed the original varnish layer (Figure 5).

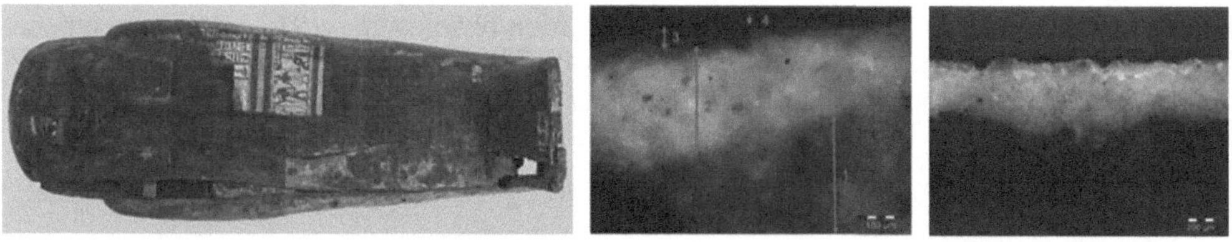

Figure 5. Nesimendjem's Coffin lid before treatment. On the right the cross sections in visible and UV Fluorescence show the layers covering most of the painted surface: 3 natural resin, 4 acrylic resin and black carbons components.

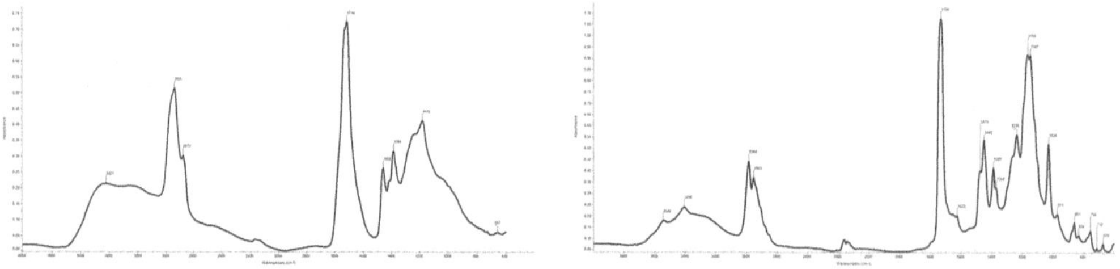

Figure 6. The FTIR spectrum of samples in Figure 5, in layer 3 a natural resin was identified; on the right the FTIR spectrum of the layer 4, Paraloid B72 was identified.

3.3 *Statuette of an offering bearer (S. 08795)*

In the case of the painted wooden statuette, the main issue was the presence of a transparent glossy layer of acrylic resin (identified by means of FTIR) which over time incorporated atmospheric particulates becoming gray coloured. The UV fluorescence highlighted the non-homogeneous presence of this overlapped material (Figure 8).

Following previous and successful results obtained by cleaning with a physical method, lasers have been considered as the first choice for the removal of the altered non-original materials. The Long-Q-Switched at 1064 nm with a pulse of 100 ns (LQS-I) has been used with a range of fluences from 0.30 to 0.40 J/cm^2 with a frequency of 1 to 3 Hz.

The action performed produces a rarefaction peak with a consequent sublimation of the material by a primary spallation effect, also in this case the beam delivery throws the optical fiber concern a homogenous effect on the surfaces (Figure 9).

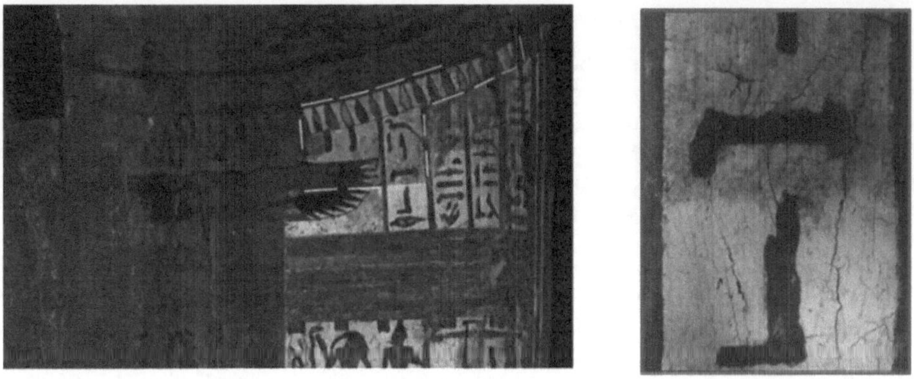

Figure 7. Detail of the laser test in UV fluorescence, adjacent to an existing test. On the right, the magnification (100×) shows that the laser cleaning respected the original resin beneath the black layer.

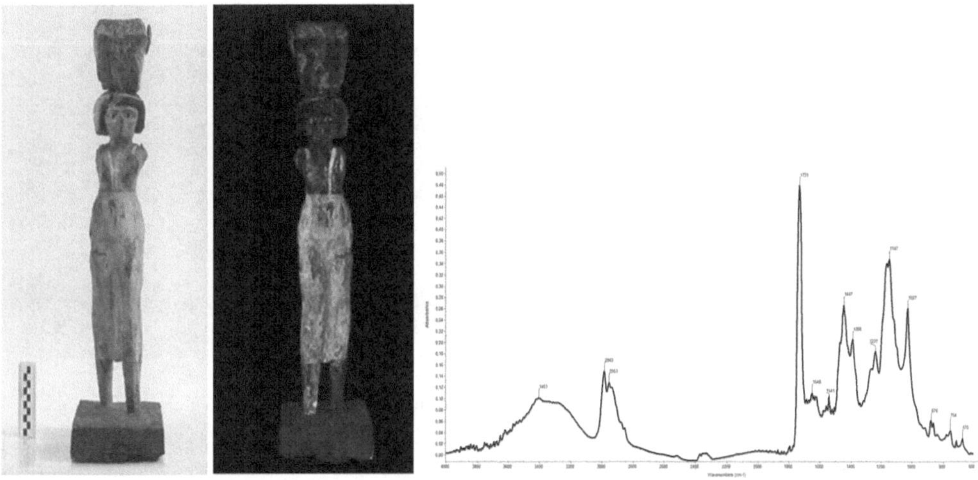

Figure 8. Statuette of the offering bearer. The UV fluorescence in the middle shows the non-homogeneous presence of overlapped materials; in the FTIR spectrum Paraloid B72 was identified.

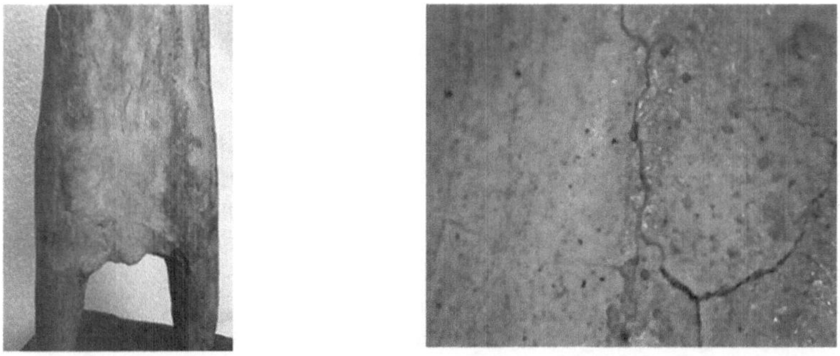

Figure 9. Detail of laser cleaning with LQS and a magnification by USB video microscope (100x).

3.4 *Model of craft activity (S. 14375/03)*

The extremely reactive calcium carbonate ground layer of this artifact led us to believe it was not suitable at a cleaning with chemical methods. Indeed it underwent just a comparison of physical methods in dry conditions (Figure 10). The main conservation issue was the graying induced by dust adhered to the substrate where highly diluted solutions of Paraloid B72 have been applied, as documented in a published report on the intervention (Marocchetti, 2008): «cohesion problems were solved by subsequent applications of acrylic resin in a 1% to 5% solution».

A Q-Switched laser (with pulse duration of 8 ns) was tested using 1064 and 532 nm harmonics. The macro observation of the results led to the choice of operating with the wavelength of 532 nm, in a range of fluence 0,20 – 0,50 J/cm² and, ablation rate of 5 Hz for a vaporization-mediated removal of the graying. The second harmonic concerns in certain conditions better optical effects of light reflection in addition to a more superficial action. Laser irradiation was combined with micro-vacuuming to prevent re-deposition of debris as droplets on the surface after laser treatment.

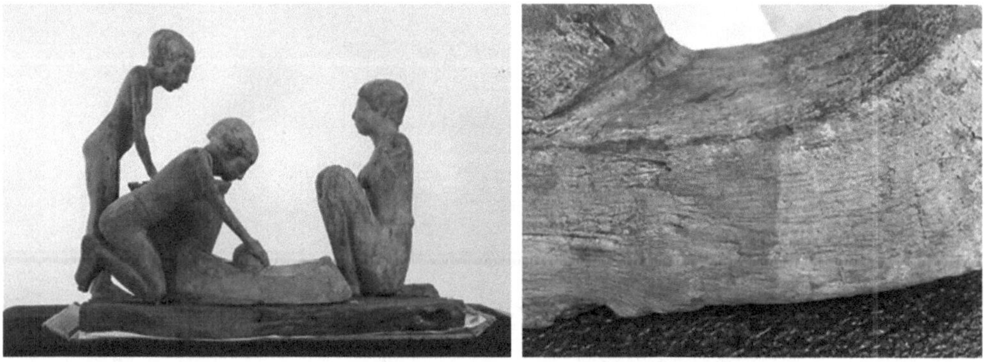

Figure 10. Model of craft activity and detail of laser cleaning with a QS at 532 nm.

3.5 *Coffin box of Padiamenemipet (Cat. 2235/2)*

The focus of this case was the removal of an inhomogeneous grayed material, from the flat edge of the coffin box (Figure 11), where an uncleaned test window witnessed the complete stratigraphy current before the previous cleaning intervention. The uncleaned test window analyzed by OM, and SEM/EDS revealed the presence of an organic layer (it is supposed to be a natural resin) applied on the white ground layer (calcium carbonate with an organic binder). Today traces of oxalates remaining on the surface are probably evidence of this resinous finish left in traces in previous cleaning intervention; while the acrylic resin, spread all over the surface, incorporated part of residues that probably got adhered during the time (Figure 12).

The aim was to respect those traces of oxalate through the selective removal of the grayed acrylic resin. The comparison between different cleaning methods was carried out to identify the appropriate approach. Free and confined solvent solutions in high retentive gels were tested: no convincing timing for application was found, even in the case of use of confining systems, resulting not completely effective with short time applications and a little invasive with longer applications, getting in this case partially solubilized the calcium carbonate-based ground layer and observing changes in the surface roughness (see Section 3.1).

The lasers selected for trials were LQS and QS lasers at 1064 nm and 532 nm both in dry conditions.

The better results were obtained at 1064 nm QS mode, with a pulse duration of 8 ns; the range of fluency was 0.25-0.70 J/cm² at 4 Hz of frequency; the handpiece output used

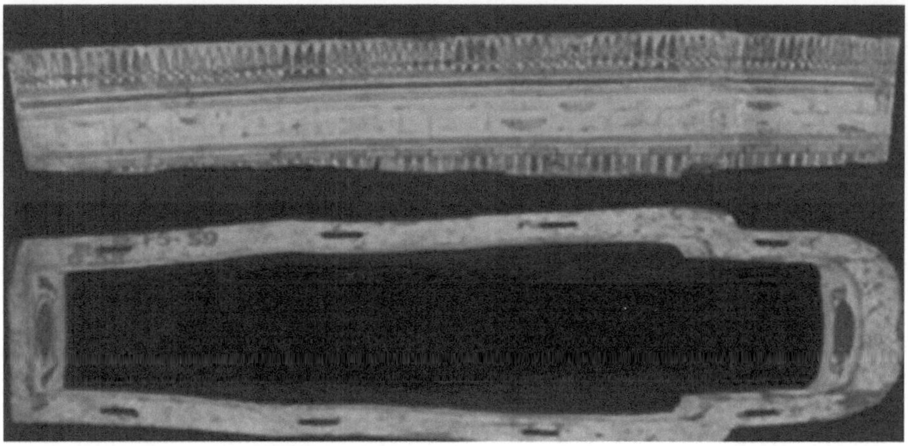

Figure 11. Coffin box of Padiamenemipet, flat edge and left side photos of the coffin.

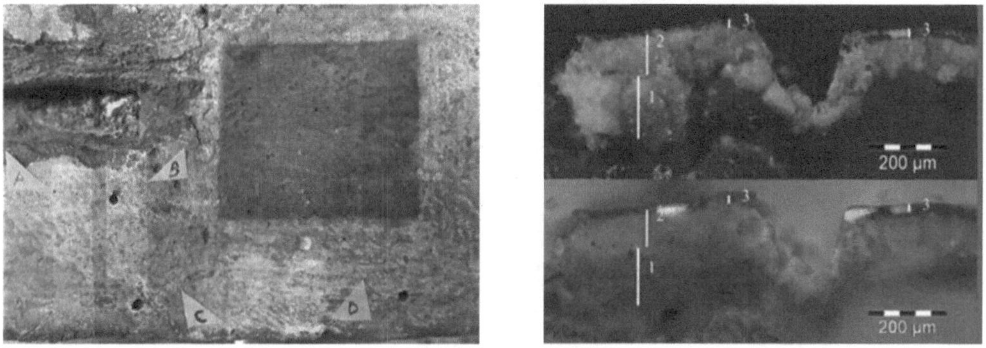

Figure 12. On the left, non-treated witness area and preliminary laser tests; on the right, cross-sections of a sample taken from the witness area, imaged with OM under visible and UV light. Layers 1 and 2 show the ground layers, while layer 3 shows organic compounds and pigment grains.

a homogenizer lens. The short pulses at low fluency concern the removal of graying through a rapid vaporization effect below any threshold.

The results were investigated using a USB video microscope in visible light. In order to assess the selectivity of laser cleaning, a test was performed irradiating a glass slide located immediately over the area to be treated, in order to collect the materials sublimated by laser. Subsequently analyzed by FTIR, the results confirm only the presence of Paraloid B72 on the glass slide. The cleaning was also inspected by a digital 3D microscope (MICRO 3D), here it is reported the result obtained in a transition area (laser cleaning - graying acrylic resin). The micromorphology 3D measurements evidenced a very thin transition step between layers (approximately 15-19 µm), thus the laser cleaning has not heavily modified the roughness values of the surface (Figure 13).

3.6 *Coffin lid of Padiamenemipet (Cat. 2235/1)*

The coffin lid as well as its box (3.5 Coffin box of Padiamenemipet) underwent previous conservation interventions (Figure 14). Probably within the same intervention the painted surface was cleaned selectively and unevenly obtained different results, strongly influencing the various thickness and porosity of colors and their state of preservation.

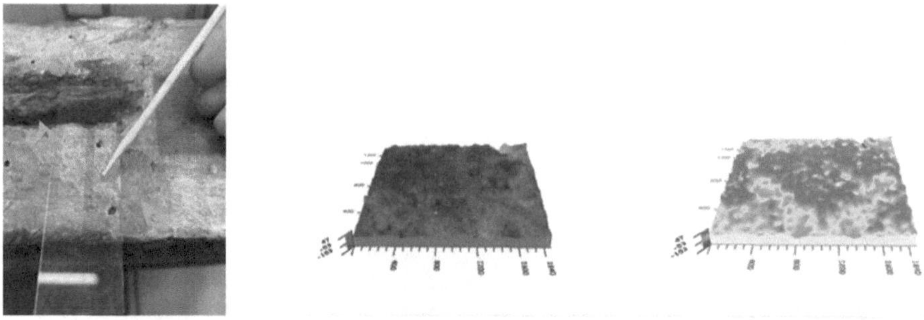

Figure 13. Glass slide used for collecting the materials sublimated by laser. On the right side the micromorphology 3D measurements in visible light and in false color scale.

The first problem concerns the orpiment colored areas (identified on the basis of presence of As, detected with XRF). These areas, where the colored layer was thin and quite porous, appeared laid out as a background finish on specific parts of the decorative registers, perhaps to create a contrast with the white background of the others, being distributed alternately; FT-IR analysis provided a spectrum suggesting the presence of an organic medium (nonetheless the specific binder for the orpiment layer was not clearly identified due to the very low amount of material). This layer apparently absorbed the acrylic resin detected (Paraloid B72) and underwent an optical alteration becoming gray (probably due to a lower percentage of pigment in the original mixture combined to a deterioration of the organic material during time). The issue in this case was to remove the altered acrylic resin while distinguishing the original color.

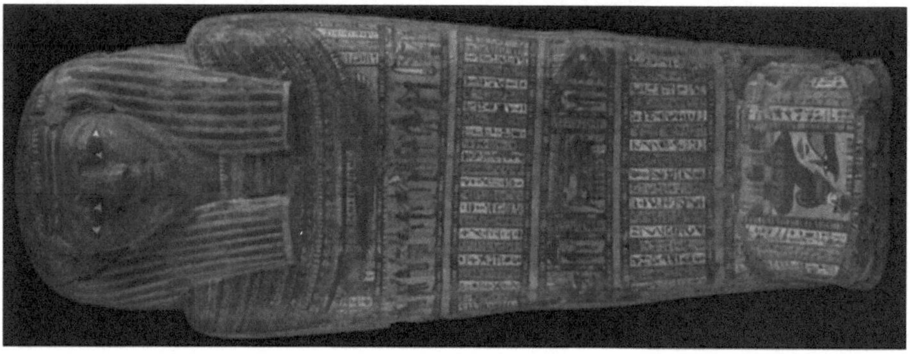

Figure 14. Coffin lid of Padiamenemipet, before treatment.

The preliminary tests performed were essential. The surface proved to be extremely reactive to the wetting solution. Despite this, in consideration of the good results obtained with a superficial cleaning using highly retentive hydrogels (Manfredda et al. 2021), we decided to test nanostructured fluids (Nanorestore cleaning®) confined in a twin-chain polymer hydrogels based on poly (vinyl alcohol), developed within the H2020 European project NANORESTART (grant agreement 646063), today marketed by Center for Colloid and Surface Science of Florence with the name of Peggy gels®. In particular we decided to confine Nanorestore Cleaning G®, suitable for synthetic materials, confined in Peggy gel 5®, that has the more retentive structure among the flexible ones available, testing different timing in the view of understanding the efficacy and failure point. A constant support has been employed using a USB video microscope: the macro observation revealed that the resin and color were linked together; in this condition gels failed to be selective, solubilizing the acrylic resin together with the orpiment pigment.

Two Nd:YAG lasers at 1064 nm were also tested in LQS and QS regimes. The physical action was carried out in dry conditions. The better result was obtained using the Long-

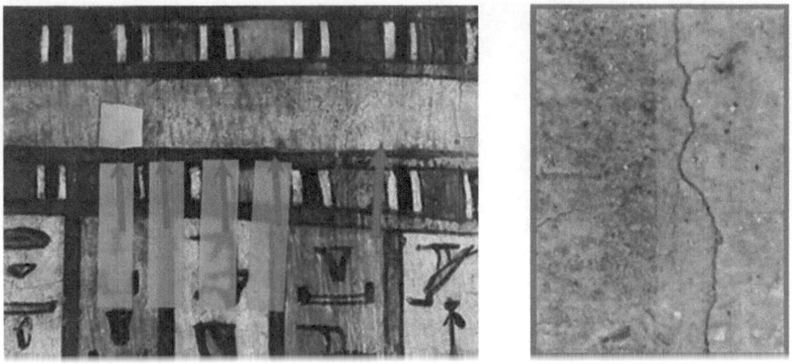

Figure 15. Preliminary tests performed on orpiment; arrows indicate the application of Nanorestore Cleaning G® confined in Peggy gel 5®. On the right side the LQS laser cleaning by USB video microscope (100×).

Q-Switched laser with a pulse of 100 ns; the fluence ranged from 0.29 to 0.50 J/cm2 with a frequency of 3 Hz producing a rarefaction peak, we observed a primary spallation effect. Optical fiber beam delivery concerned an homogenous effect of cleaning (Figure 15).

The laser has proven to be very selective and hence to distinguish the grayed material from the orpiment since the radiation was mainly absorbed by the gray.

Visual outcomes of the cleaning methods tested were observed in transition areas (dirt-gels vs dirt-laser) using a USB video microscope and 3D digital microscope (MICRO3D).

The specific insight evidenced that both the areas cleaned with laser and the ones cleaned with gel show a similar morphology to those not cleaned; this data should confirm the absence of a peeling action in both cases and also demonstrates the efficacy of the laser treatment in dry conditions.

The second issue concerned the red ochre color of the face (characterized by means of XRF and multispectral imaging). The red layer demonstrated to be extremely compact compared to the orpiment one, probably due to the pigments' different morphological characteristics and, perhaps, to a different concentration of pigment in the layer. Moreover, the reactivity of the earth-based pigment to variations of humidity over the time has probably induced a strong flaking of the layer and cracking with plastic deformations. In addition, in this case, Paraloid B72 seemed not to penetrate in the layer underneath, forming a thick glossy layer, over an uneven layer of black carbonaceous particles.

Given the delicate nature of the surface, it was considered very risky to attempt any mechanical cleaning action. Therefore, the team decided to focus the intervention on the physical action of the laser. A Long-Q-Switched laser using optical fiber delivery was selected; tests in dry conditions were considered too close to the damage threshold because of the necessary raising of fluency, while tests in wet conditions allowed the use a minor fluence with best results; a liquid plays an important role in photomechanical and pressure wave propagation increasing the absorption of radiation by the material to be removed (Siano 2008).

The irradiation through the solvent solution (Isooctane 80%, Acetone 20%) was performed with a range of fluences between 0.50 - 1.17 J/cm^2 with 1 to 3 Hz of frequency.

Laser cleaning has been constantly monitored with a USB video microscope connected to a computer display (Figure 16).

3.7 *Outer coffin lid of Besenmut (Cat. S.5217)*

The Besenmut' coffin lid has been one of the most challenging and peculiar case studies recently faced by the CCR team, not only for the relevant size of the wooden structure (257 cm in length and 115 cm in width) (Figure 17). The characterization of materials and technique (OM), in fact, highlighted different stratigraphies: all over the body only a brown ground layer was present (mixture of calcium carbonate and dolomite, with in iron-titanium and aluminum silicate inclusions); the upper part of the lid, including the collar, face and wig,

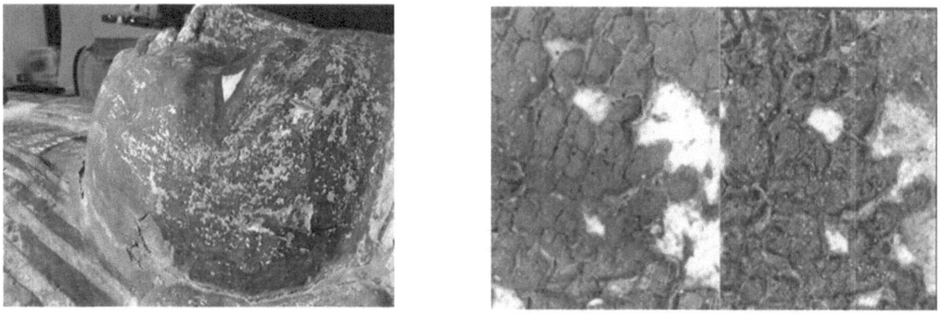

Figure 16. Detail of laser cleaning with LQS and a magnification before and after cleaning by USB video microscope (100×).

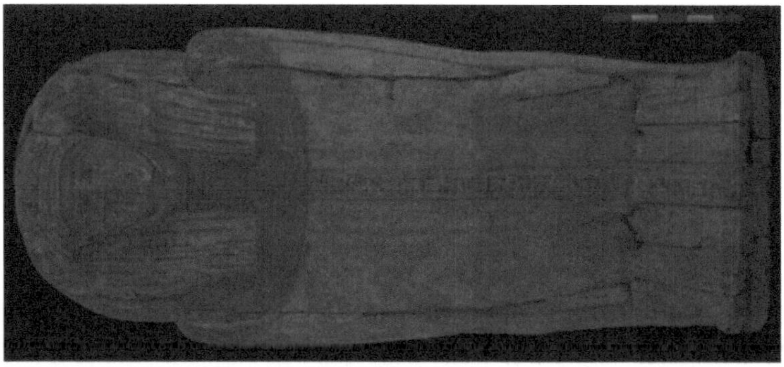

Figure 17. Outer coffin lid of Besenmut, before treatment.

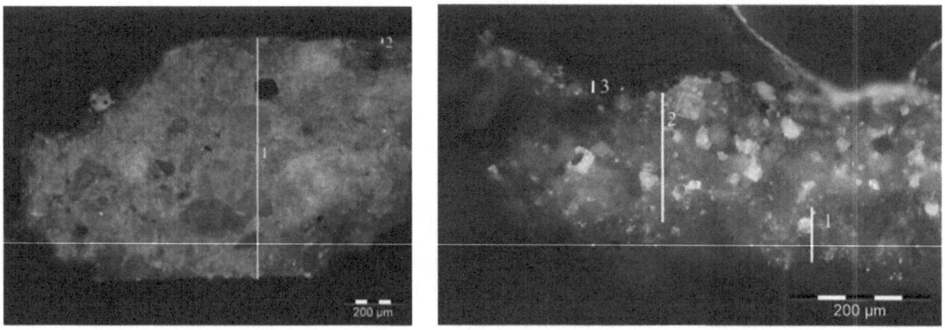

Figure 18. Cross sections by OM. On the left side the brown ground layer of the body; on the right side the sample taken from the uncleaned witness area on the wig, showing the white ground at layer 1. Layer 2 shows yellow and red pigments sequentially and layer 3 carbon particles.

showed instead the presence of a white ground layer (calcium carbonate, magnesium, quartz and gypsum) under the painting materials (the Egyptian traditional palette included in this case copper based green, yellow ochre, red ochre, Egyptian blue, and lines of a carbon based black) (Figure 18).

As already described in some of the previous case studies, the presence of acrylic resin and uneven residues of black carbon particles could cause long-term damage to the objects and also interfere with the correct reading of the decoration. The removal of this layer was performed by testing a selection of cleaning methods confining pH and conductivity calibrated water solutions and nanostructured fluids in high retentive gels: polyvinyl alcohol-borax type, rigid gels of gellan and agar agar, PVA-based gels (Peggy®) and pHEMA/PVP-based gels (Nanorestore Gel Dry®). None of the systems listed were fully effective as they either caused surface whitening (possibly a scattering due to a limited solution of the material to remove) or affected the pigments (the action at the interface Paraloid B72-colors and Paraloid B72+black residues on colors is not completely manageable). Laser tests were performed using short pulses: LQS-I (100 ns, 130 mJ) and QS laser (8 ns) at 1064 nm and 532 nm.

Trials evaluation has been carried out with different monitoring methods: 4K micro camera connected with a laptop, USB video microscope, OM and Multispectral imaging.

The QS in IR wavelength (1064 nm) demonstrated to be the most effective for the cleaning of all decorative surfaces (range of fluency was 0.25 - 0.50 J/cm² at 1 Hz of frequency) using the homogenizer lens to avoid Gaussian beam marks. The physic effect obtained was a primary spallation effect below the vaporization threshold, it was necessary to combine a micro vacuuming to avoid a re-deposited of plasma during the cleaning process. A micro-fragment observed in top view by OM showing a transition area between dirty and laser cleaning (Figure 19). The multispectral images helps in checking the homogeneity during the laser cleaning: UV fluorescence highlights the green copper (altered in black due to aging) after the remove of black opaque carbon residues, while VIL (Verri, 2013) luminescence clearly increase the response of the Egyptian blue following after the removal of the carbon black (Figure 20).

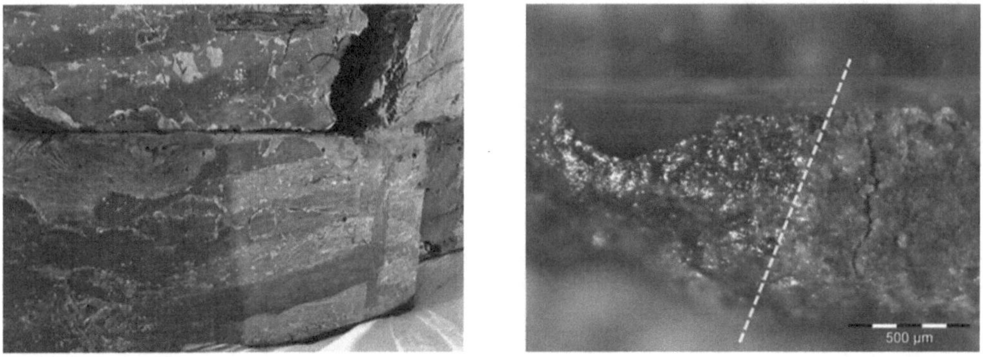

Figure 19. Laser cleaning of the wig with QS in IR 1064 nm. On the right side is a micro-fragment of the brown body observed by OM during laser cleaning.

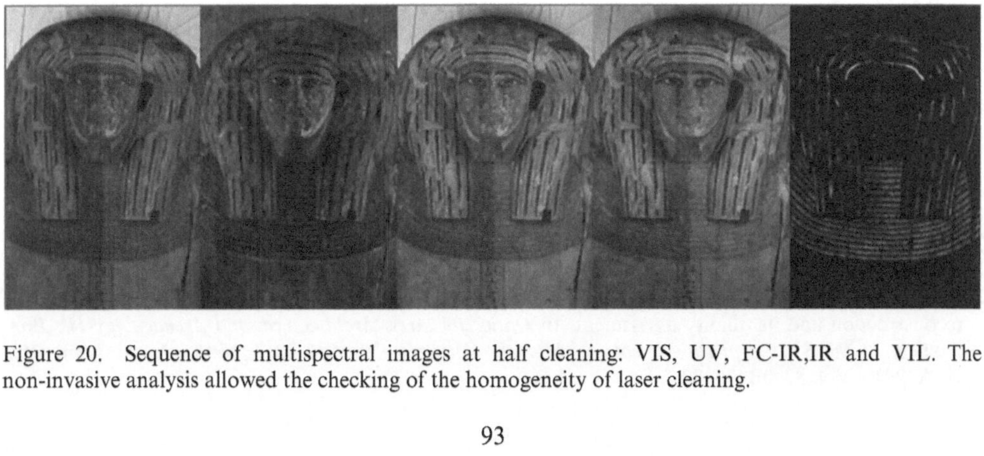

Figure 20. Sequence of multispectral images at half cleaning: VIS, UV, FC-IR,IR and VIL. The non-invasive analysis allowed the checking of the homogeneity of laser cleaning.

4 CONCLUSIONS

As is well known, every conservation process represents an important opportunity for knowledge and understanding of the treated and studied artifact, and cleaning can produce results that either aid or hinder the correct perception of the painted decoration. Moreover, properly preserving ancient artifacts means ensuring their accessibility for future generations.

This study focused on the conservation of several wooden Egyptian finds, on the cleaning issues strongly linked both to the characteristics of the original artistic technique and to the presence of previous intervention materials, and on the importance of a comprehensive preliminary analytical campaign to understand the original technique and to support the conservation project design.

Different cleaning methods were tested, always comparing their specific performances to assess their effectiveness and safety. Chemical and physical methods should always be compared to tailor cleaning treatments for each specific case study and scientific insights should always support the evaluation of their effectiveness and safeness.

The results demonstrated that laser cleaning (using Nd:YAG short pulses LQS and QS systems at 1064 nm and 532 nm) was found to be an effective and safe method for removing black carbon particles and acrylic resin from Egyptian painted surfaces, both with and without original resinous finishes. To date, no comparable results have been achieved with chemical cleaning methods.

This study achieved important goals, such as providing data for Egyptologists, eliminating long-term damage risks, and providing reference cases for laser cleaning on archaeological painting materials.

ACKNOWLEDGEMENTS

The authors would like to acknowledge the institutions that supported the present work. First of all, we want to thank al the Vatican Coffin Project research group (Musei Vaticani, Vatican City State, Musée du Louvre, Paris, Rijksmuseum van Oudheden, Leiden, Museo Egizio, Turin, Centre de Recherche et de Restauration des Musées de France, Paris) for sharing their knowledge and approach for research and conservation of Egyptian coffins. Moreover, we thank the National Institute of Nuclear Physic (INFN) and its CHNet and the Department of Chemistry and Industrial Chemistry of Pisa University that provided analytical insights for interpreting the execution techniques and assessing the conservation treatments. Finally, a special thank to the Center for Colloid and Surface Science (CSGI) for the availability and support.

REFERENCES

Amenta, A. 2014. The Vatican Coffin Project. In E. Pischikova, J. Budka and K. Griffin, *Thebes in the First Millennium BC*: 483–499. Newcastle upon Tyne: Cambridge Scholars Publishing.

Baglioni, P., Berti, D., Bonini, M., Carretti, E., Dei, L., Fratini, E. & Giorgi, R. 2014. Micelle, microemulsions, and gels for the conservation of cultural heritage. Advances in Colloid and Interface Science (205): 361–371.

Buscaglia, P., Cardinali, M., Cavaleri, T. & Ferraris, E. Study and conservation of some Late Period coffins coming from Queens' Valley, the *Second Vatican Coffin Conference*. Città del Vaticano, 6-9/06/2017 (in press).

Buscaglia, P., Cardinali, M., Cavaleri, T., Croveri, P., Ferraris di Celle, G., Piccirillo, A. & Zenucchini, F. 2018. Nesimenjem and the Valley of the Queens' Coffins. In H. Strudwick and J. Dawson (eds). *Ancient Egyptian Coffins: Past Present Future*: 83–95. Oxford: Oxbow Books.

Brunetto A., Giovagnoli A., Laurenti M.C., Mano M. & Zanini A. 2017. Laser cleaning on the Mut Temple wall paintings, Gebel Barkal (Sudan), APLAR 6.

Cacciari, I., Ciofini, D., Mascalchi, M., Mencaglia, A. & Siano, S. 2012. Novel approach to the microscopic inspection during laser cleaning treatments of artworks. *Anal. Bioanal. Chem.* 402: 1585–1591.

Cacciari, I., Mencaglia, A.A. & Siano, S. Micromorphology of gold jewels: A novel algorithm for 3D reconstruction and its quality assessment. In *Optics for Arts, Architecture, and Archaeology IV. Proceedings of the SPIE Optical Metrology, Munich, Germany*, 30 May 2013; SPIE: Washington, DC, USA, 2013; Vol. 8790B: 70–80.

Castillejo M., Martín M., Oujja M., Rebollar E., Domingo C., García-Ramos J. & Sánchez-Cortés S. 2003, Effect of wavelength on the laser cleaning of polychromes on wood. *Journal of Cultural Heritage* 4: 243–249

Daudin- Schotte M., Bisschoff, M., Joosten I., Keulen H. & Van den Berg K.J. 2013 Dry Cleaning Approaches for Unvarnished Paint Surfaces" 2013 publication at: https://www.researchgate.net/publication/281066098

Domingues, J., Bonelli, N., Giorgi, R., Frattini, E., Gorel, F. & Baglioni, P. 2013. Innovative hydrogels based on semi-interpenetrating p(HEMA)/PVP networks for the cleaning of water-sensitive cultural heritage artifacts. *Langmuir* 29 (8): 2746–2755.

Ferrarato B., Zenucchini F., Ferrari Di Celle G., Piccirillo A. & Gulmini M., 2017. Pulitura laser del blu egizio su superfici archeologiche. In A. Brunetto (ed) *APLAR 6*.

Gottardo M. 2014. Coperchio di sarcofago egizio in arenaria dipinta: considerazioni sugli esiti e sulla messa a confronto di diverse modalità applicative dell'ablazione laser utilizzata per la pulitura. In A. Brunetto (ed) *APLAR 5*.

Graue B., Brinkmann S. & Verbeek C. 2011. Laser Cleaning of ancient Egyptian wall paintings and painted surfaces. In D. Saunders (ed), *LACONA IX Lasers in Conservation of Artworks*, London: Archetype.

Guzzon, E. 2017. The wooden coffins of the late Third Intermediate Period and Late period found by Schiaparelli in the Valley of the Queens (QV 43 and QV 44). In Amenta, A. & Guichard, H. (eds) *Proceedings of first Vatican coffin conference*: 191–198. 19–22 June 2013, Città del Vaticano: Ed. Musei Vaticani.

Horie, C.V. 1987. Materials for conservation. Organic consolidants, adhesives and coatings. Oxford: Butterworth-Heinemann.

Koob S. 1986. The Use of Paraloid B-72 as an Adhesive: Its Application for Archaeological Ceramics and Other Materials. *Studies in Conservation* 31(1):7–14. DOI:10.1179/sic.1986.31.1.7

Lee, L. & Quirke, S. 2000. Painting materials. In Nicholson, P.T., Shaw, I. (eds) *Ancient Egyptian Materials and Technology*. Cambridge: Cambridge University Press.

Manfredda, N., Buscaglia, P., Gallo, P., Borla, M., Aicardi, S., Poggi, G., Baglioni, P., Nervo, M., Scalarone, D., Borghi, A., Re A., Guidorzi L. & Lo Giudice A. 2021. An Ancient Egyptian Multilayered Polychrome Wooden Sculpture Belonging to the Museo Egizio of Torino: Characterization of Painting Materials and Design of Cleaning Processes by Means of Highly Retentive Hydrogels. *Coatings* 11: 1335. https://doi.org/10.3390/coatings11111335.

Mansi, S., Zenucchini, F., Croveri, P. & Spagnoli, F. 2017. Pulitura laser della cassetta porta-ushabti (Cat. 2441) del Museo Egizio di Torino. In A. Brunetto (ed) *APLAR 6*.

Marocchetti E.F. 2008. La scultura in legno al Museo Egizio di Torino. Problemi di conservazione e restauro In *Materiali e strutture. Problemi di conservazione sulla scultura. Nuova serie anno VI* (11-12): 9–31.

Newman, R. & Serpico, M., 2000. Adhesives and binders. In Nicholson, P. T. and Saw, I. (eds) *Ancient Egyptian Materials and Technology*: 475–494, Cambridge: Cambridge University press.

Pagès-Camagna, S. 2010. A Review of Pigment Research at the Louvre Egyptian Collections. In C. Rozeik, J. Dawson & M. Wright (eds) *Decorated Surfaces on Ancient Egyptian Objects: Technology, Deterioration and Conservation*. Cambridge: Fitzwilliam Museum: University of Cambridge.

Serpico, M. & White, R. 2000. Resins, Amber and Bitumen. In Nicholson, Paul T. and Saw, I. (eds) *Ancient Egyptian Materials and Technology*: 430–474. Cambridge: Cambridge University press.

Scott, D. 2014. A review of ancient Egyptian pigments and cosmetics. *Studies in Conservation* 61(4): 185–202. DOI 10.1179/2047058414Y.0000000162

Schiaparelli, E. 1923. Relazione sui lavori della Missione Archeologica Italiana in Egitto (1903–1920). Volume primo, *Esplorazione della "Valle delle Regine" nella necropoli di T*ebe: 185–206, Torino.

Siano, S. 2008, Principles of Laser Cleaning in Conservation. In M. Schreiner, M. Strlic, R. Salimbeni (eds) Handbook on the Use of Lasers in Conservation and Conservation Science. Bruxelles: Cost Action G7.

Siano. S. & Salimbeni, R. 2010. Advances in Laser Cleaning of Artwork and Objects of Historical Interest: The Optimized Pulse Duration Approach. *Accounts of Chemical Research* 43(6): 739–750. https://doi.org/10.1021/ar900190f

Siano, S., Mencaglia, A.A. & Cacciari, I. 2017. Microscopy Optoelectric Device with Focus Scanning. U.S. Patent 9,612,427 B, 4 April 2017. https://patents.google.com/patent/WO2013018030A1/un

Verri, G., Saunders, D., Ambers, J. & Sweek, T. 2013. Digital mapping of Egyptian Blue: conservation implications. *Studies in Conservation* 55 (Suppl. 2): 220–224.

Vigorelli, L., Re, A., Guidorzi, L., Cavaleri, T., Buscaglia, P., Nervo, M., Del Vesco, P., Borla, M., Grassini, S. & Lo Giudice, A. 2022. Multi-analytical approach for the study of an ancient Egyptian wooden statuette from the collection of Museo Egizio of Torino. *ACTA IMEKO* 11(1):1–10.

Zenucchini F., Tasso T., Croveri, P., Iannaccone R., Cardinali M. & Manchinu P. 2022. Pannelli cinesi Coromandel (tecnica Kuan Cai): approcci di pulitura specifici basati sullo studio delle diverse campiture di colore. In A. Brunetto (ed) *Atti del 7° Convegno APLAR*.

Lasers in the Conservation of Artworks XIII – Siano & Ciofini (eds)

You can clean but cannot touch. Graffiti removal from prehistoric pictographs at Hueco Tanks State Park & Historic Site using laser ablation process

A. Dajnowski
The Conservation of Sculpture and Objects Studio, Inc., Forest Park, IL, USA

T.J. Tague Jr.
Bruker Scientific, LLC, Billerica, MA, USA

T. Roberts
Texas Parks and Wildlife Department, Fort Davis, TX, USA

M. Strutt
Texas Parks and Wildlife Department, Austin, TX, USA

B.A. Price
Philadelphia Museum of Art, Philadelphia, PA, USA

N.M. Kelly
Bruker Nano Analytics, Denver, CO, USA

D.W. Tague
University of Texas at Dallas, Richardson, TX, USA

K.R. Sutherland
The Art Institute of Chicago, Chicago, IL, USA

B.A. Dajnowski
G.C. Laser Systems Inc., Forest Park, IL, USA

ABSTRACT: The Hueco Tanks State Park & Historic Site, near El Paso, Texas, USA, is culturally and spiritually significant for several Native American Tribes. It has the largest concentration of prehistoric mask pictographs (rock paintings) in North America. This National Historic Landmark, also a popular recreational destination, was vandalized numerous times over the past century. The pictographs and rock formations were disfigured by extensive graffiti paint. Laser ablation was used to remove the graffiti without affecting the original pictographs and rock formations. This process has restored the legibility and beauty of the pictographs while respecting their sacred legacy.

To carefully assess the laser cleaning treatment, it was necessary to analyze the pictograph and graffiti. The site is considered sacred, so physical sampling of the pictographs was not permitted. A mobile lab was set up at the cave sites to perform in situ, non-invasive Raman (Bravo and Sentinel, Bruker Scientific, LLC), infrared (Alpha II, Bruker Scientific, LLC), XRF (Tracer 5i, Bruker Nano), and microscopic analysis of the pictographs. Goethite, hematite, quartz, gypsum, feldspar, and calcium oxalate were detected. The graffiti paints were sampled and analyzed using FTIR, Py-GCMS, and SEM-EDS. Cellulose nitrate and alkyd-based paints with chrome yellow, Prussian blue, iron oxide, burnt umber, zinc oxide, calcium carbonate, and barium sulfate pigments/fillers were detected.

DOI: 10.1201/9781003386872-10

Based on this information, Conservation of Sculpture and Objects Studio Inc. prepared mockups on local stone substrates to simulate the pictographs and graffiti paint environment and optimize the laser parameters. After the optimal parameters were determined, they were applied in the field using 1064nm pulsed lasers to remove the graffiti safely and effectively without affecting the original works. This paper describes the research, preparation, and graffiti removal from the Hueco Tanks pictographs and rock formations using selected laser systems.

1 INTRODUCTION

Hueco Tanks State Park and Historic Site is located near El Paso, Texas, USA. The site centers on four rocky foothills rising as high as 400 feet above the surrounding desert floor. These foothills were formed millions of years ago after igneous masses intruded into the overlying sedimentary rock and were exposed as that softer overlying rock eroded (Wise 1977). The rock hills are composed of porphyritic syenite and dotted with vertical and horizontal hollows, or *huecos* in Spanish (Bourke and Viles 2007, Richardson 1909). Some horizontal *huecos* can hold thousands of gallons of water for months after rainstorms, providing important sources of water in the arid environment. (Greenwood 1944, Turney 1983)

Based on the earliest temporally diagnostic artifacts discovered at Hueco Tanks, the cultural history began with the arrival of Paleoindians as early as 11,000 years ago, and extended to the Historic Ranching Era (Howard, et al. 2010). Both archeological deposits and Native American rock imagery are present, but the site is most renowned for its rock imagery. 304 imagery panels are known, containing an estimated 3,000 to 6,000 individual figures. Several different rock art styles and time periods are represented. The rock art motifs include over 200 masks or face-like pictographs, the largest concentration in North America (Schaafsma 1980). The masks and many other figures were created by the Jornada Mogollon people, early agriculturalists in the region. Hueco Tanks was a focal point in their spiritual landscape during the Formative Period 1,800 to 550 years ago and remains a sacred place for several Native American tribes (Sutherland 1995).

The Texas Parks and Wildlife Department (TPWD) acquired Hueco Tanks in 1969. Prior to that date and until the establishment of a Public Use Plan by TPWD in the late 1990s, the rock imagery at Hueco Tanks was the frequent target of graffiti vandals. Beginning in 1993, TPWD began to work with professional conservators to treat non-Historic graffiti at the site, but it was not until 2009 that the lasers were used to remove graffiti from the site. This paper discusses two of the nine pictographs that were laser cleaned between 2009 and 2021. These are located at sites E01F and N19C and named "HC" and "CASTRO," respectively, based on their graffiti. (Figure 1). The HC covered a yellow dancing figure and CASTRO, an elaborate red animal composition.

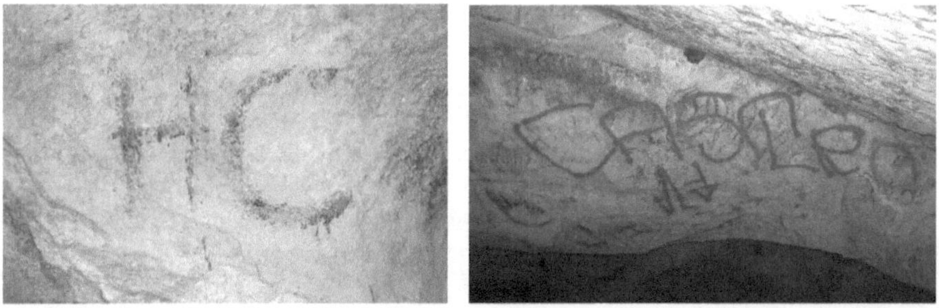

Figure 1. Left: Site E01F "HC" blue graffiti over Dancing Figure pictograph. Right: Site N19C "CASTRO" grey graffiti paint over red animal pictograph.

2 RESEARCH AND TESTING PHASE

The site had been vandalized with graffiti sometime between 1950 and 1960. Before the on-site laser project began, a pilot cleaning test of graffiti was performed at the Clate and Donna Cave at Mescalero Canyon site #52, which did not have pictographs. During the test, it was determined that by use of laser, the graffiti could be removed safely without affecting the stone. Afterward, a contract was issued by TPWD; however, the final approval belonged to representatives of the Native American Tribes, which was given in 2011.

Characterization of the materials of the pictographs and graffiti paint was essential to inform the development of the laser treatment protocols. Because Hueco Tanks and the pictographs are sacred, sampling the rock paintings was not permitted. In 2010 and 2011, the pictographs were studied non-invasively in the field using portable Raman, X-Ray Fluorescence (XRF) and Fourier transform infrared spectroscopy (FTIR). Instrumental conditions were as described by Tague (2011). The iron-based pigments, goethite (FeOOH) and hematite (Fe_2O_3), were detected at sites E01F and N19C, respectively, and along with the minerals quartz (SiO_2), gypsum ($CaSO_4 \cdot 2H_2O$), and feldspar. An example Raman spectrum collected in situ from the Dancing Figure pictograph at site E01F is shown in Figure 2. The Raman bands (243, 296, 385, 548 cm^{-1}) and corresponding XRF data (not shown) indicate the yellow pigment, goethite, was used to create the image.

Also shown in Figure 2 is a Raman spectrum of red pigment collected from the animal pictograph at site N19C. The bands suggest hematite (411 cm^{-1}), gypsum (1007 cm^{-1}), feldspar (510 cm^{-1}), and calcium oxalate (1468, 1482 cm^{-1}), which was detected as encrustation. Rock art has been found to be encapsulated by calcium oxalate that is thought to result from the action of oxalic acid from lichen on calcareous substrates. The oxalate layer at Hueco Tanks was fortuitous for our purposes because it served as a protective layer during the laser cleaning of the graffiti.

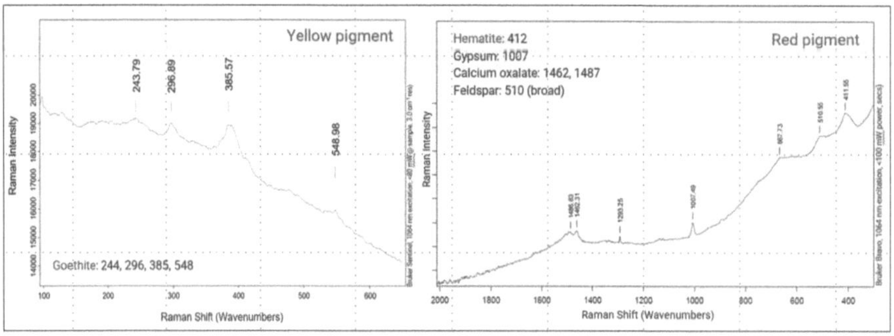

Figure 2. Raman spectra acquired in situ. Left: Goethite yellow pigment found in the Dancing Figure pictograph at Site E01F. Right: Red pigment of the animal composition pictograph at Site N19C suggesting hematite, gypsum and calcium oxalate.

The graffiti paints were sampled and analyzed in the lab using scanning electron microscopy with energy dispersive spectrometry (SEM-EDS), FTIR microspectroscopy (μFTIR), and pyrolysis-gas chromatography mass spectrometry (Py-GCMS). The dark blue "HC" graffiti at site E01F was determined to be a cellulose nitrate-based (CN) paint containing natural resin (probably pine resin), drying oil, phthalates, and Prussian (iron blue) pigment ($Fe_4[Fe(CN)_6]_3$). In contrast, the dark grey "CASTRO" graffiti at N19C was alkyd-based with a composition that included pentaerythritol, ortho-phthalate, as well as titanium dioxide (TiO_2) and zinc oxide (ZnO) pigments. Example FTIR spectra acquired from the graffiti paints with pertinent references are shown in Figure 3.

Further details of the graffiti analysis results and instrumental conditions have been discussed by Lins and Price (2011).

Numerous pieces of loose stone were collected from Hueco Tanks for the substrates. To imitate the graffiti, modern alkyd and CN paints were applied over iron-based pigments. Laser tests were then performed using EOS 1000 LQS and EOS 1000 SFR (El.En. S.p.A., Calenzano, IT) and

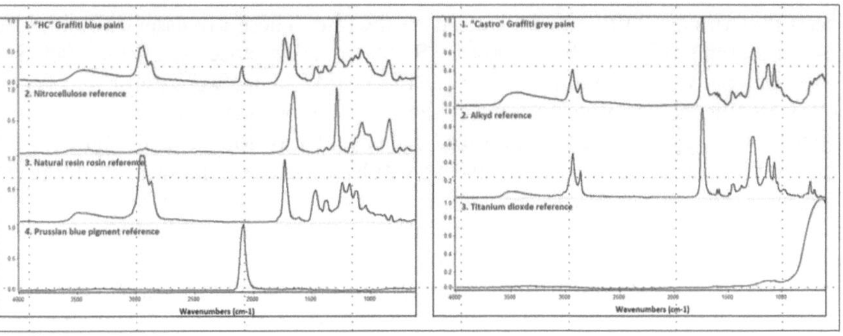

Figure 3. μFTIR spectra. Left: (1) "HC" cellulose nitrate-based graffiti paint from site E01F compared with (2) nitrocellulose, (3) natural resin, and (4) Prussian blue pigment references. Right: (1) "CASTRO" alkyd-based graffiti paint from site N19C compared with (2) alkyd and (3) titanium dioxide references.

several laser settings to determine the optimal/safest operating parameters. The testing started using single pulses to carefully monitor and control the precision of graffiti removal, as we could observe the results after each laser pulse. The best results in removing graffiti and uncovering the pigments were achieved using 100 ns pulses (EOS 1000 LQS) with the laser set to 130 mJ with a 50% filter. This gave us an energy level distributed over the cleaning spot of 65 mJ. Depending on the spot size chosen (1.5–4 mm), the fluence used ranged between 0.5–3.7 J/cm². Using this system, we selectively removed the graffiti from the pigments on the mock-ups (Figure 4).

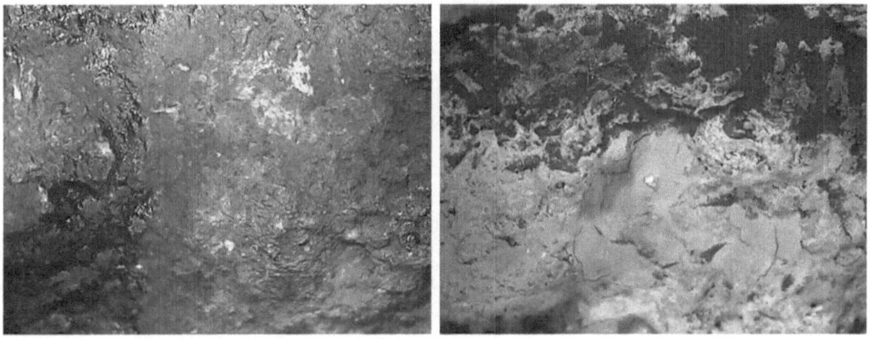

Figure 4. Left: Mockup of black alkyd-based paint over red iron oxide pigment. Alkyd paint was successfully removed from red pigment using out of focus EOS 1000 SFR 250 mJ spot size 5 mm. Right: Mockup of blue cellulose nitrate-based paint over yellow pigment). The paint was successfully removed using EOS 1000 LQS wet out of focus 130 mJ, spot size 5 mm.

After the laser protocol was established, the test results were presented to representatives of TPWD and local Native American Tribes. Before the treatment commenced at the site, the laser cleaning process was demonstrated at their request. The demonstration was done in two phases: offsite and observation of graffiti removal at the CASTRO site. The representatives of the tribes were pleased and impressed that the laser light source could be used in such a selective way that the sacred artworks were not damaged or altered. The use of lasers for graffiti removal was therefore approved.

3 TREATMENTS

3.1 *Phase 1*

The treatment portion of the project proceeded in two phases. During the first phase, two water-cooled El En lasers EOS 1000 SFR (Short Free Running) and EOS 1000 LQS (long

Q-switched) that run on 240V, were brought to the site. These are essentially low-frequency laser cleaning systems that fire: EOS 1000 LQS 100 to 200ns pulses and EOS 1000 SFR 60 to 130μs. The laser parameters optimized during the testing phase were employed (vide supra). Only the lowest fluence was used to remove the graffiti from the areas with pictographs. The higher fluence was applied to the vandalized areas that did not have pictographs under the graffiti paint. To ensure that the pigments were fully protected, the laser cleaning was done wet by spraying the surfaces of the pictographs with a mist of distilled water. The water provided a cooling effect, which was important since we were cleaning iron-based pigments. Creating too strong energy flux on the surface could have resulted in phase and color changes of the iron pigment. The cleaning results were monitored with a Zarbeco field microscope (Figure 5). Images were taken with the same magnification before and after treatment. The treated areas were photographed using daylight, UV, and IR illumination. Fans were used to blow air over the cleaned area to prevent condensation of the ablated graffiti vapor.

The EOS 1000 LQS laser effectively removed graffiti from the yellow, and EOS 1000 SFR was better for removing graffiti from the red pigment of the pictographs covered by black paint.

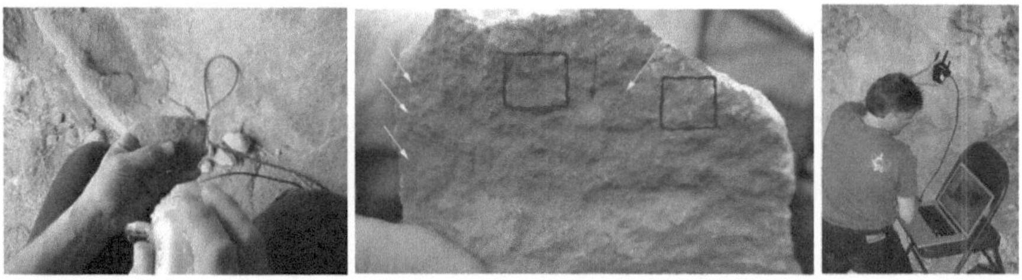

Figure 5. Images illustrating monitoring process of the laser cleaning. Image top left – testing settings on the mock-up sample. Right – constant monitoring with a microscope. Middle image- mock-up sample that illustrates: 1. correct settings that did not affect dry pigment – green boxes, 2. settings rejected – white arrows. The red dot in the middle – blue arrow indicates the spot of the aiming beam of the laser.

The first phase lasted two months, during which approximately 50% of the graffiti at sites N19C and E01F was removed (Figures 7 and 9). At this point, the project was paused due to funding limitations. Additional testing was then conducted at the studio to refine the process further. At this time, a new 1064nm tunable high-frequency GC-1 conservation laser system from G.C. Laser Systems had become available, providing new treatment possibilities. We found that this new system offered extreme precision and effectiveness in laser graffiti removal. Two significant logistical advantages of the system were that it used 110V and had a hot spot-free high-speed circular scan. The variable level of fluence, variable pulse duration, variable pulse repetition, and variable rotation of the cleaning spot while being air-cooled made the system most suitable for this project. Before and after images of the test samples cleaned with the GC-1 laser are shown in Figure 6.

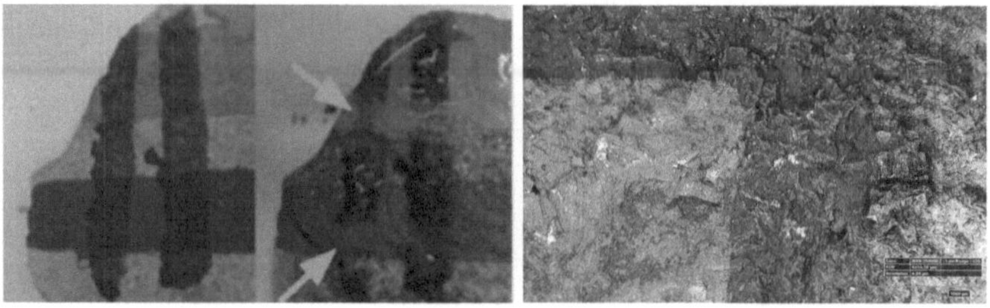

Figure 6. Left: Before and after ablation. Right: Hirox image: 35× - red pigment after ablation.

With over 30 different pulse durations to choose from in one machine ranging from 10 ns-500 ns, the GC-1 laser provided us with unmatched precision and tunability with a much broader range of cleaning parameter options than the lasers previously used. After determining with the mock-up samples that safe cleaning of the pigments could be achieved below a pulse energy of 0.14 mJ with a 50 ns pulse, we ensured that this setting was never higher. The pulse duration was set to 50ns, pulse frequency was set to 70kHz and scan speed RPM was set to 12,000. The maximum fluence determined to be safe was 0.55 J/cm^2. As before, the cleanings were performed wet to further protect the pigments from the risk of phase change from laser plasma and fans were used to move the ablated particles.

3.2 *Phase 2*

During the second phase of the treatment, the N19C site was revisited, where the GC-1 laser proved very efficient in removing graffiti. It was expected that at least two or more weeks of cleaning would be required, yet the site's graffiti was removed in a few days. Even more remarkable was the cleaning of site E01F. To our pleasant surprise the remaining 50% of the graffiti was cleaned there in less than an hour, while it was anticipated that more than a week would be needed based on our previous experience with the LQS and SFR lasers!

After the laser treatment, Raman spectra were acquired from the surfaces to verify the graffiti removal. Figure 8 shows an example Raman spectrum obtained from site E01F after clearing the Prussian blue "HC" paint. Removal of the Prussian blue is noted by the absence of a band at 2157 cm^{-1}.

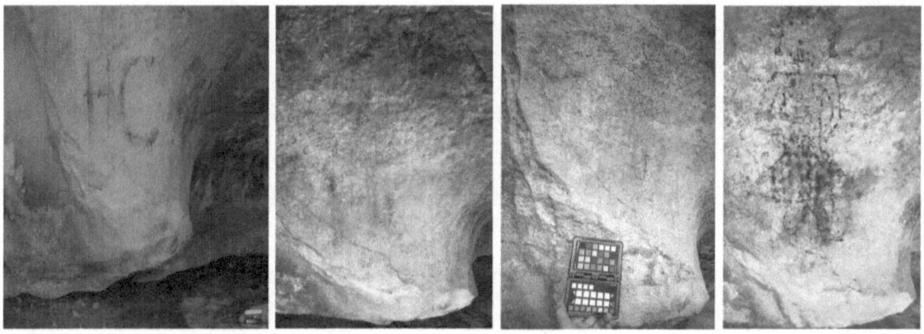

Figure 7. Site E01F: from left BT, After phase 1, After treatment, AT DStretched YRD filter.

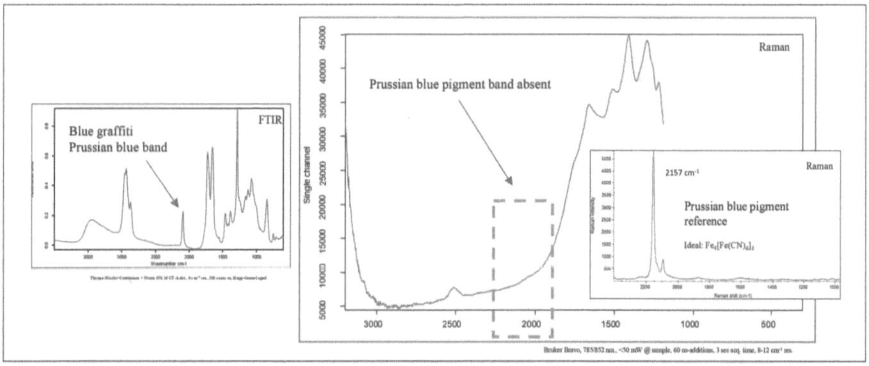

Figure 8. Left: FTIR spectrum of the blue "HC" graffiti alkyd paint containing Prussian blue pigment from site E01F. Right: Raman spectrum collected from the surface after laser cleaning the graffiti paint. The blue dashed box shows the absence of a Prussian blue band in the region where it would be expected, corroborating the graffiti removal. Inset: Reference Raman spectrum of Prussian blue pigment.

Another tool we used to evaluate the condition of the surfaces and treatment results was a DStretch processing of the visible images. DStretch, or decorrelation stretch, is a digital process used to enhance (stretch) the color differences in a color image. The method includes the removal of the inter-channel correlation found in the input pixels. The Jet Propulsion Laboratory developed the process for remote sensing to enhance multispectral images, and NASA used it to enhance Mars Rover images. Silicon Valley programmer, Jon Harman, modified and implemented the technique for rock art research (Harmon 2008).

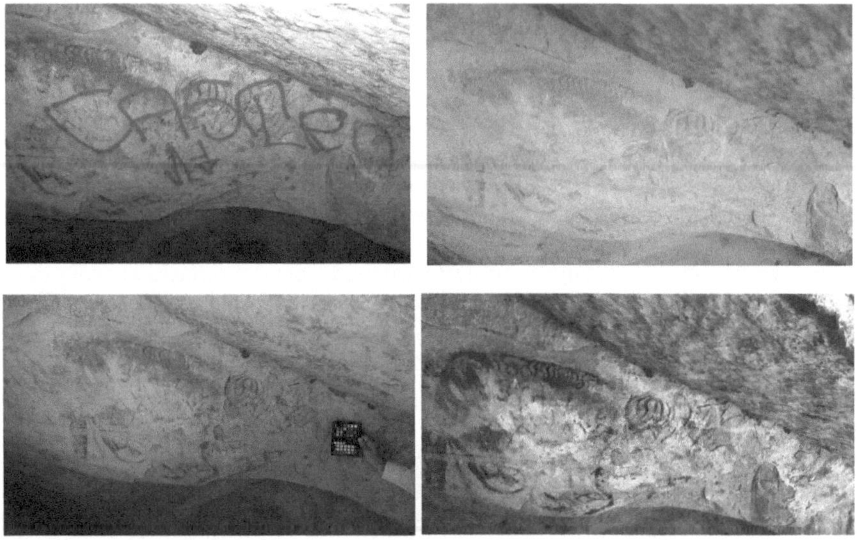

Figure 9. Site N19C. Top row: Before treatment (left), same area after phase 1 cleaning (right). Bottom: After treatment – phase 2 – daylight (left), after treatment DStretch YRD Filter (right).

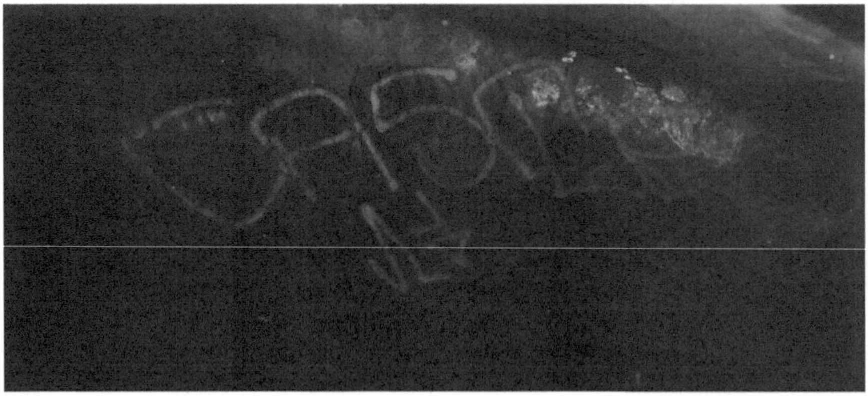

Figure 10. After treatment image of the site N19C showing residue of the graffiti that is visible only under UV illumination.

The UV image in Figure 10 shows that there is residue of the graffiti on the surface of the stone. Our treatment goal was to remove all graffiti visible to the naked eye from the surface of the pictograph without disturbing the surfaces of the pictograph and stone. The goal was successfully achieved (Figure 9). Trying to remove all residue could have resulted in damaging the original pictograph. Although there is residue of the graffiti paint in the stone crevices, the esthetic value of the original pictograph was uncovered and preserved. The pictograph will be viewed by most people who will look at it during the day and see a pictograph that is visually intact and free of graffiti. In cases like this, stopping treatment at the right point is critical. It is a perfect example of "less is more" (Figure 10).

4 CONCLUSIONS

Our treatment goal for this project, to remove all visible graffiti from both sites, was achieved. Systematic analysis, testing, and preparation allowed us to identify safe laser parameters for this complex cleaning problem. Two different laser cleaning technologies were used during this project and the lessons learned from the initial work with low pulse repetition frequency lasers were directly translated to a much more productive and highly tunable high frequency GC-1 laser system. This project demonstrated that a strong collaboration between several dedicated organizations can make it is possible to clean something without touching it, and to effectively remove modern graffiti paint from prehistoric pictographs.

ACKNOWLEDGEMENTS

We would like to thank Texas Parks and Wildlife Dep., Mescalero Apache and Tigua representatives for selecting our team for this project. This publication is dedicated to the memory of Andrew Lins who greatly contributed to the success of this project.

REFERENCES

Bourke, Mary C. & Heather A. Viles 2007. *A Photographic Atlas of Rock Breakdown Features in Geomorphic Environments*, Planetary Science Institute, Tucson, AZ.

Greenwood, C. L. 1944. Opening Routes to El Paso, 1849, *Southwestern Historical Quarterly* Vol. 48(2), pp. 262–272.

Harman, Jon 2008. *Using Decorrelation Stretch to Enhance Rock Art Images*. http://www.DStretch.com/AlgorithmDescription.html

Howard, Margaret, Logan McNatt, Terri Myers, Tim Roberts, and Amy Ringstaff 2010. *10,000 Years at Hueco Tanks State Park and Historic Site, El Paso County, Texas*. Report prepared by the Cultural Resources Program, Texas Parks and Wildlife Department, Austin, Texas.

Lins, P. Andrew and Beth Price 2011. *Final Report on the Non-Invasive Analysis of Pictographs and on Analysis of Graffiti at The Hueco Tanks Site, Texas Parks and Wildlife Department, El Paso County, Texas*. RFP#802-10-44458. Philadelphia Museum of Art, Philadelphia, Pennsylvania.

Richardson, George B. 1909. *El Paso Folio, Texas*. Geologic Atlas of the United States, Folio 166. U. S. Geological Survey, Department of the Interior, Washington, D. C.

Schaafsma, Polly 1980. *Indian Rock Art of the Southwest*. University of New Mexico Press, Albuquerque.

Sutherland, Kay 1995. *Rock Paintings at Hueco Tanks State Historical Park*. Austin: Texas Parks and Wildlife Press.

Tague, Tom 2011. *Report on Hueco Tanks Pictographs*. Not published.

Turney, W.F. 1983. Prehistoric Water Reservoirs—The Southwest. In C.H. Lange (ed) *Southwestern Culture History: Collected Papers in Honor of Albert H. Schroeder*: 43–57, Archaeological Society of New Mexico Paper 10. Santa Fe, New Mexico: Ancient City Press.

Wise, Henry M. 1977. *Geology and Petrography of Igneous Intrusions of Northern Hueco Mountains, El Paso and Hudspeth Counties, Texas*, Unpublished Master's thesis, The University of Texas at El Paso.

Lasers in the Conservation of Artworks XIII – Siano & Ciofini (eds)
 ISBN 978-1-032-47995-8

Preliminary study and implementation of nanosecond NIR fibre laser in conservation of polychrome heritage objects

A. Faron
Doctoral School Academia Artium Humaniorum, Nicolaus Copernicus University in Toruń, Poland

M. Iwanicka
Faculty of Fine Arts, Nicolaus Copernicus University in Toruń, Poland

ABSTRACT: The aim of this work is to present a preliminary study in the field of laser cleaning of painted and gilded heritage objects. A novelty of this attempt consists in the new type of laser and its parameters, which were not yet reported extensively in the literature. The equipment applied to this work was based on RedEnergy G4 EP-Z Pulsed Fibre Laser (SPI Lasers, UK) which is characterized by an emission wavelength of 1060 nm, average power of 100 W and a very high pulse repetition frequency (1 kHz – 4 MHz), with pulse time adjustable in the range of 4 ns – 2 μs and a very small spot diameter of ca 200 μm. As examples of implementation of such laser in conservation of polychrome and gilded works of art were not satisfactorily reported, there was a need to perform a series of basic tests and trials that led us to defining a safe range of parameters for certain materials and painting supports. Dedicated software for fluence estimation was created, enabling comparison with the experimental conditions reported in the literature.

1 INTRODUCTION

Applications of laser cleaning in the conservation of art have already been described many times over the course of the last thirty years (Siano et al. 2003, Brunetto 2004, Bordalo et al. 2006, Marczak et al. 2008, Chiantore & Rava 2012, Domasłowski 2011, Koss & Marczak 2015, Matteini et al. 2016, Moretti et al. 2019). In spite of such a long time since the initial application, new ways to develop efficient procedures to control laser cleaning are still being sought. Usage of laser cleaning is widely spread in conservation of unpainted objects, such as stone and other mineral substrates (Pouli et al. 2012). Polychrome surfaces are probably the most complex kind of materials for the laser method, seriously prone to damages (Pouli et al. 2000, de Cruz et al. 2000, Castillejo et al. 2002, 2003, Bordalo et al. 2006, Pouli et al. 2012, Siano et al. 2015).

During last decades, when laser cleaning have been developed intensively, the most common instrument employed to laser cleaning of cultural heritage objects was Nd:YAG system. Following a long testing period also Er:YAG laser started to be used. Recently, on the market of cleaning services there has been a significant growth in the number of fibre laser systems for cleaning of various surfaces (Hildenhagen et al. 2011, Dajnowski A. et al. 2017). Nevertheless, in comparison to Nd:YAG and Er:YAG systems, there is a considerable shortage of literature reporting results of fibre laser cleaning attempts, which limits the possibility to compare parameters and access to up-to-date knowledge for laser operators. The exception are results associated with industrial applications such as shipbuilding and metal surfaces processing (Li et al. 2020, Vu et al. 2021), but the needs of these fields are far from conservation of artworks. Such a phenomenon proves that there is a lot to develop in the subject of safe

DOI: 10.1201/9781003386872-11

cleaning with this kind of novel tool in cultural heritage conservation and still there is a need for fundamental studies. Research works reported in this topic focus on wide range of materials. Removal of similar types of unwanted layers has been reported in the literature albeit by means of Nd:YAG and Er:YAG lasers, particularly cleaning heritage objects from dirt (de Cruz et al. 2000), overpaints (Siano et al. 2015) and aged coatings (Taarnskov et al. 2011, Dajnowski A. et al. 2011, Brunetto et al. 2020, Hellen 2020).

The aim of the present study is to present cleaning protocols relatively safe for most common materials and problems encountered in art conservation practice while using IR nanosecond fibre laser with very high pulse repetition frequency. The experimental part has been supported by lab-developed software which allows the laser operator to better identify parameters of the cleaning process, namely the fluence of the laser beam.

2 METHODOLOGY

2.1 *Laser equipment and software*

All experiments described in this work were conducted by means of an industrial galvo marking system built around a RedEnergy G4 EP-Z Pulsed Fiber Laser (SPI Lasers, UK), which was applied to the conservation of cultural heritage probably for the first time. Previously reported applications of this laser concerned the removal of layers from industrial objects (Dondieu et al. 2020, Dudek et al. 2022). The laser operates at wavelength of 1060 nm and it is characterized by the adjustable pulse duration of 5 ns – 2 μs, high pulse repetition frequency (20 kHz – 4 MHz), and maximum average power of 100 W. It is equipped with focusing optics and two perpendicular galvo scanners providing scanning speed in the range of 6 – 17 m/s and a spot of ca 200 μm in diameter.

The bespoke software allows for a programmable pattern generation and modification of scanning speed, average power, pulse duration and frequency. For the aim of this paper, a pattern composed of parallel lines covering a small (up to 1×1 cm^2) area was used. Unless stated otherwise in the text, in this contribution test areas were scanned horizontally, vertically, as well as at 45 and 135 degree angles (4 passes altogether).

The laser head was mounted on a stable stand to precisely control distance to the object and the scanned area position. It is worthwhile to note that the overall homogeneity of laser ablation depends both on the pulse parameters and the scanning pattern. To integrate these factors, the additional, lab-developed software was prepared with LabView programming environment essentially for calculation of the fluence and visualisation of the coverage of the cleaned surface with the laser pulses. The latter was used for prediction of the homogeneity of the cleaning process.

2.2 *Objects of examination*

Experiments were carried out on three groups of objects. The first group of unaged samples was prepared for the preliminary tests that mainly enabled to estimate the range of feasible laser parameters. The majority of samples in this group were made by applying a few layers of animal glue and chalk priming on cardboard support. The priming was subsequently polished, and some of the samples additionally painted with oil colours (Rembrandt, Talens). Moreover, a ready-made, primed linen canvas on stretcher (Talens) covered with tempera paint (raw sienna, Maimeri) with artificially made craquelure was used. The samples were artificially homogenously soiled with charcoal powder. The last type of sample from the first group was a sample of sandstone with a thick, opaque layer of grey, mineral paint.

The second group consisted of naturally aged samples and model objects on wood (brass/gold gilding with overpaints – indigo oil colour with dammar varnish) and intonaco (wall painting executed using the *fresco* technique soiled with soot from burning candle).

In the third group, three heritage objects of no historical or artistic value, donated to science were cleaned: an oil painting on canvas (20th cent.) covered with a layer of yellowed varnish, a part of a wooden polychrome sculpture (lime, unknown provenience and time of execution), and a fragment of historical wooden construction (oak, 18th cent.), two of the latter overpainted with industrial oil paints.

2.3 *Monitoring of the treatment*

Along with the removal of undesired layers, a monitoring protocol of assessment of the treated surface has been implemented. Analyses were conducted before and after the cleaning process to evaluate changes in the particular layers. Examination techniques chosen for the purpose of this study were mostly non-destructive: evaluation in visible and ultraviolet light (Pinna et al. 2012) as well as optical microscopy (Nikon SMZ745T, DeltaPix CMOS camera).

Optical coherence tomography was used in order to assess the possible changes in the surface topography during the laser cleaning. It is a non-invasive optical technique developed for medical imaging of subsurface semi-transparent layers with micrometric precision (Drexler, Fujimoto et al. 2015), which has been used for many years in heritage science (Targowski et al. 2015) also for monitoring cleaning procedures (Iwanicka et al. 2018, Moretti et al. 2019, Faron 2021, Iwanicka et al. 2022).

The laboratory-made instrument used in this study was working in the spectral band 750–960 nm with 12 μm lateral and 3.2 μm axial resolutions (Targowski et al. 2020). Here, OCT was used mostly as a profilometric tool to generate topography maps showing the change in the painting's surface before and after the laser varnish removal test. These maps were generated by extraction of surface profiles from 3D OCT data cubes of 150 B-scans collected before and after the laser treatment. Then the surface profiles obtained before cleaning were subtracted from the ones acquired after in order to create differential maps (coded in false colour scale) showing the amount of material removed.

3 RESULTS

3.1 *Experiment no. 1 – laser cleaning of painting layers with surface dirt*

The experiment was carried out on model samples in two stages. In the first stage two types of mock-ups were used – from the first (chalk ground – Figure 1a and oil paint – Figure 1e) and second (*fresco* wall painting – Figure 1h) groups described (Section 2.2).

The objects were soiled with charcoal dust rubbed dry onto the surface (Figures 1b, f). Trials using different sets of parameters were performed (Table 1) and for every object a satisfactory cleaning effect was achieved (Figures 1c, d, g, h).

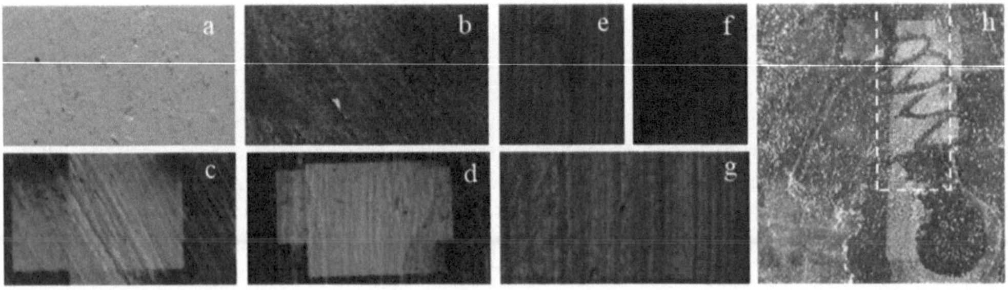

Figure 1. a) White chalk ground before application of artificial dirt. b) White chalk ground covered with a layer of grounded charcoal. c) White chalk ground after laser cleaning (Table 1, set 1). d) White chalk ground after laser cleaning (Table 1, set 2). e) Green umber (oil paint, Rembrandt, Talens) on white chalk ground before application of artificial dirt. f) Green umber on white chalk ground covered with a layer of grounded charcoal. g) Green umber on white chalk ground after laser cleaning (Table 1, set 2). h) Fragment of *fresco* wall painting covered with a layer of artificial dirt (soot from burning candle), white rectangle marks laser cleaning test area (Table 1, parameter set 2).

Table 1. Sets of parameters for cleaning different substrates from artificial surface dirt.

	1	2	3	4	5
Parameter set number	Satisfactory results		Unsatisfactory results		
Frequency [kHz]	500	80	20	30	30
Fluence per one pass [J/cm2]	0.671	0.403	0.574	0.502	0.561
Scanning speed [m/s]	15	8.5	15	15	15
Pulse duration [ns]	13-16	46-53	35-261	35-108	37-88
Number of passes	4	4	4	4	8

Conclusions derived from this stage were a base for solving a more complex problem – laser cleaning of tempera paint layer extensively damaged by artificially caused craquelure and delaminations (sample on factory-made primed canvas) – Figure 2a. In this case the paint layer was not only water-sensitive but also extremely fragile to mechanical action. Successful laser cleaning (Figure 2c) may be considered a promising alternative to traditional conservation approaches.

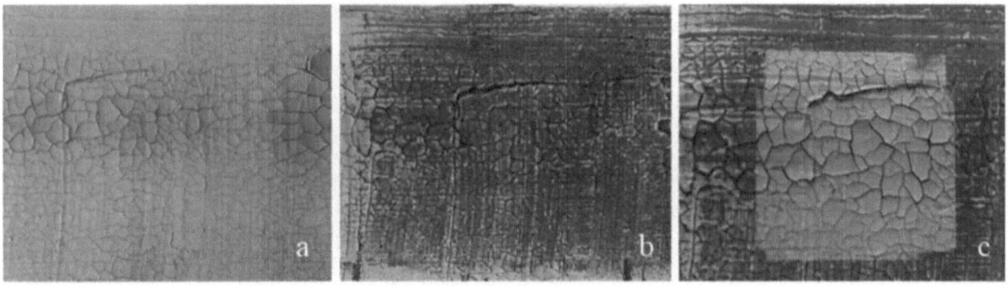

Figure 2. a – layer of raw sienna (tempera paint, Maimeri) on white primed canvas before application of dirt, craquelures were caused artificially; b – layer of raw sienna on white primed canvas covered with a layer of grounded charcoal; c – layer of raw sienna on white primed canvas after laser cleaning (Table 1, parameter set no 2).

3.2 *Experiment no. 2 - laser removal of overpaints from different supports*

For the purpose of these trials, objects from all three groups were used (Figure 3). The first group was represented by mock-up of overpainted sandstone. The mock-up of brass gilding covered with semi-transparent oil layer representing the second and a part of a wooden polychrome sculpture and a fragment of historical wooden construction from the third group were used, as they better imitate the situation of removal of old secondary layers.

Table 2. Sets of parameters removal of overpaints from different supports.

	1a (step 1)	1b (step 2)	2a (step 1)	2b (step 2)	3a (step 1)	3b (step 2)	4 (step 1)	5
Parameter set number	Satisfactory results							Unsatisfactory result
Frequency [kHz]	30	30	30	80	30	30	30	40
Fluence per one pass [J/cm^2]	2.46	1.94	2.46	2.05	2.46	1.62	0.707	1.51
Scanning speed [m/s]	10	10	10	8,5	10	8,5	8,5	10
Pulse duration [ns]	35-261	35-261	35-261	46-53	35-261	46-53	47-330	30-197
Number of passes	8	8	32	40	10	20	10	40

The tests were performed as described in the Section 2.1. Additionally, the surface was prewetted before laser cleaning. Semi-transparent oil layer from brass gilding was wetted with isopropanol, in other cases water was used. To achieve an acceptable effect of cleaning that leads to removal of the thick overpaint and to leave the original substrate undamaged, in case of objects shown in Figure 3c–f, there was a need to perform the treatment in two steps. In the first step parameters with higher energy were chosen in order to remove the bulk of the secondary material (step 1). A thin layer of overpaint was left intentionally, in order to be able to perform the final cleaning (step 2) in a more controlled manner.

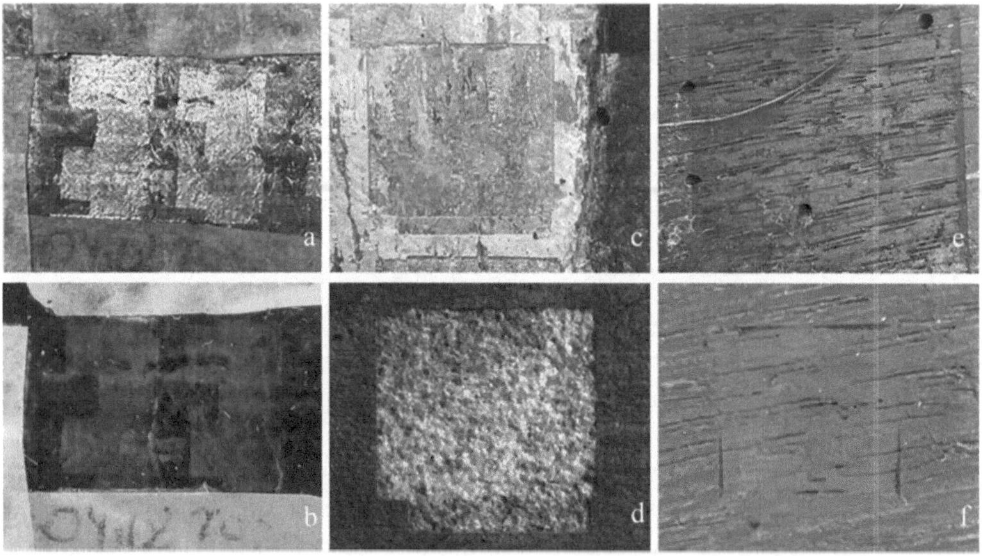

Figure 3. a – semi-transparent layer (oil paint (Talens) with dammar varnish) on brass gilding after laser cleaning, VIS light (Table 2., parameter set no 4); b – semi-transparent layer (oil paint (Talens) with dammar varnish) on brass gilding after laser cleaning, UV light (Table 2., parameter set no 4) (photo: A. Adamski); c – fragment of wooden sculpture (18th cent.?, lime) with a traces of yellowish ground overpainted with two thick layers of pink and grey industrial oil paint, after laser cleaning, VIS light (Table 2., parameter set no 1a, 1b); d – opaque layer of mineral paint on sandstone after laser cleaning (Table 2., parameter set no 3a, 3b); e – fragment of historical wooden construction (18th cent. ?, oak) overpainted with thick, red brown, industrial oil paint after laser cleaning (Table 2., parameter set no 2a, 2b); f – fragment of historical wooden construction (18th cent. ?, oak) overpainted with thick, red brown, industrial oil paint after unsatisfactory laser cleaning (Table 2., parameter set no 5).

3.3 *Experiment no. 3 – laser cleaning of paintings from aged and yellowed transparent layers*

This experiment was conducted on oil painting on canvas which appearance was disturbed by semi-transparent, yellowed aged varnish layer (Figure 4a–c) from the third group of objects (Section 2.2). The process of removing varnish was performed in three steps. Firstly, the layer of water-gel based on hydroxypropylcellulose (Klucel^{TM}) was applied for about 20 minutes. Then after the removal of the gel, the same area was laser irradiated. Contrary to the experiments described above, here the laser passed only twice in horizontal lines. A promising result in this experiment was achieved with two sets of parameters. In the first round a frequency 200 kHz, fluence 0.71 J/cm^2, scanning speed 12 m/s and pulse duration in the range of ca 20 ns were applied. In the second round the frequency was 300 kHz, fluence 0.77 J/cm^2 with the remaining parameters unchanged. As a result of laser treatment (Figure 4f, white rectangle), the surface of varnish turned opaque and cracked. Finally, in the upper part of the irradiated area the disintegrated varnish was removed mechanically in a controlled manner with a scalpel (Figure 4f, white arrow) and the original painting layer remain unaltered. OCT microprofilometric models (Figure 4d, e) show no damage to surface topography due to cleaning (specifically no traces of regular scanning pattern). In the differential map (Figure 4g) it is clearly visible that the laser irradiation did not cause direct removal of material. The removal happened only in the area of further scalpel treatment.

It is worth to note that the layer of varnish in this case was very difficult to remove homogenously with classical conservation techniques (solvent/mechanical cleaning) without laser irradiation. Combining more than one method of treatment during laser cleaning was reported many times in literature for varnish removal (Brunetto et al. 2020, Hellen 2020) but by means of different types of lasers (high energy, low frequency).

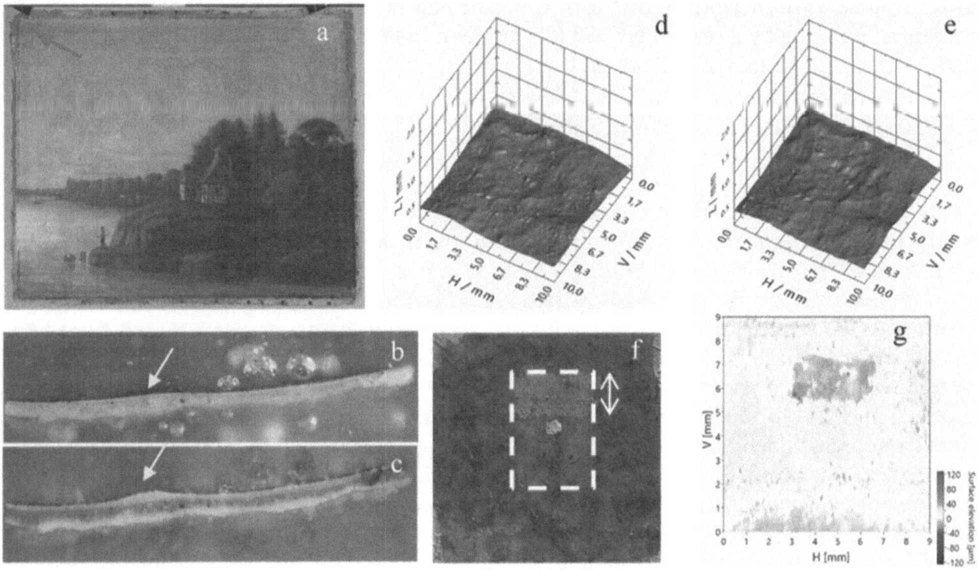

Figure 4. "Landscape with lake", oil on canvas, 20th c., painting donated to science: a – visible scattered light photography, red arrow marks sample collection spot; b, c – cross-section of the sample, visible light and UV-induced fluorescence image, respectively, yellow arrows mark the varnish layer; d – OCT surface model before laser cleaning test; e – OCT surface model after laser cleaning test; f – macrophotography after laser cleaning test; g – OCT differential map after laser cleaning test.

4 DISCUSSION

Surface cleaning was performed using fluences in the range of approx. 0.4 to 0.7 J/cm^2 with the scanning speed in the range of 8.5 to 15 m/s. Very short pulses (less than 20 ns) were used together with a very high scanning speed and higher frequencies (a few hundred kHz). Slightly longer duration of pulses (approx. 50 ns) were successfully used with a frequency of 80 kHz and a scanning speed of 8.5 m/s. Despite the use of a similar fluence as in the above cases (approx. 0.5 J/cm^2), some tests were negatively evaluated where lower frequency (approx. 20-30 kHz) was used together with pulses with a duration of 100 and 200 ns.

Removal of overpaint in the case of thick, well-aged layers (Figure 3c–f) was performed in two steps. In the first step, fluence in the range of approx. 2-2.5 J/cm^2 was used, while in the second, when removing the remaining thin layer of material, fluence of approx. 1.6-2 J/cm^2 was implemented. The scanning speed was 10 m/s in most cases. However, if the thin layer of overpaint was to be removed, the scanning speed was reduced to 8.5 m/s. Fluence values were adjusted depending on the problem. Specifically, in most cases of the removal of thick overpaint layers it was impossible to achieve the effect of satisfying cleaning using fluence values below 2 J/cm^2 (Table 2, parameter set no 5; Figure 3f).

It is interesting to discuss some implications of the results described in this contribution to conservation practice. In case of fragile, cracked, flaking water-sensitive paint layer it was possible to remove the superficial dirt layer to a satisfactory degree (Figure 2c) without mechanical action or pre-consolidation of the paint (which would be a standard treatment, albeit posing threats of its own to the heritage object). In the group of heritage objects removing two overlaid secondary layers of industrial oil paints of different colour and composition (Table 2, parameter set no 1a, 1b; Figure 3c) was performed gradually layer by layer, leaving the original layer undamaged – in this case yellowed chalk-glue ground layer on lime wood. Another outcome of conducted experiments concerns the resistence of different types of wood to laser irradiation. Lime, which is a softer wood in comparison to oak wood, was more vulnerable to mechanical damage in course of laser treatment. Finally, in the case of removing hardly soluble varnish from a 20th cent. oil painting introducing laser irradiation prior to mechanical cleaning by a scalpel proved to significantly facilitate the process and considerably decrease the risk of damaging the paint layer.

5 CONCLUSIONS

A noticeable similarity in effectiveness of certain sets of parameters applied to remove each of tested type of undesired layers on different supports was observed. According to results obtained in course of this research it is possible to define range of laser irradiation parameters as a response for a specific problem (e.g. removing of soot-based soiling from oil paint).

Although the development of the method of application of nanosecond pulsed fibre laser requires more research and experiments, the results obtained until now, allow us to hope for a wider range of applications in cleaning of cultural heritage objects. The experiments described in this contribution prove the feasibility of this type of laser for removing undesired layers from brass gilded objects, wood, and paintings on canvas in different states of preservation of the paint layer. Especially, the ability to obtain a homogenous effect of removal on large areas at once in a relatively short time is the most attractive feature from the conservator-restorer's point of view. The riskiest configuration of removal of overpaints concerns the removal of thick paint layer from the underlying original paint layer, which will be the subject of future research. In all the cases, there is always a need for a skilled and trained laser operator, who is able to evaluate to what extent the cleaning should be performed and adjust the cleaning parameters accordingly. Using laser may be nevertheless considered an interesting alternative to traditional conservation-restoration approaches in problematic cases.

ACKNOWLEDGEMENTS

The authors wish to thank professor Piotr Targowski and staff of the Department of Biophotonics and Optical Engineering at the Institute of Physics, Faculty of Physics, Astronomy and Informatics, Nicolaus Copernicus University in Toruń. The work is partially financed by the programme: Excellence Initiative – Research University of Polish Ministry of Education and Science and with the partial use of the infrastructure of the Centre for Modern Interdisciplinary Technologies NCU.

REFERENCES

Bordalo R., Morais P. J. et al. 2006. Laser cleaning of easel paintings: An Overview. *Laser Chemistry* 2006: 1–9. https://doi.org/10.1155/2006/90279

Brunetto A. 2004. *L'utilizzo della strumentazione laser per la pulitura delle superfici nei manufatti artistici.* Saonara: Il Prato.

Brunetto A., Bono G. & Frezzato F. 2020. Er:YAG laser cleaning of 'San Marziale in Gloria' by Jacopo Tintoretto in the Church of San Marziale, Venice. *Journal of the Institute of Conservation* 43(1): 44–58. DOI: 10.1080/19455224.2019.1706596.

Castillejo M., Martín M., Oujja M., Silva D., Torres R., Manousaki A., Zafiropulos Z., van den Brink O.F, Heeren R. M. A., Teule R., Silva A. & Gouveia H. 2002. Analytical study of the chemical

and physical changes induced by KrF laser cleaning of tempera paints. *Analytical Chemistry* 74(18): 4662–4671. https://doi.org/10.1021/ac025778c.
Castillejo M., Martín M., Oujja M., Rebollar E., Domingo C., García-Ramos J.V. & Sánchez-Cortes S. 2003. Effect of the wavelength on the laser cleaning of polychromes on wood. *Journal of Cultural Heritage* 4(3): 243–249. https://doi.org/10.1016/S1296-2074(03)00049-9.
Chiantore O. & Rava A. 2012. *Conserving Contemporary Art: Issues, Methods, Materials, and Research.* Los Angeles: The Getty Conservation Institute.
de Cruz Adele, Wolbarsht M.L., & Hauger S. A. 2000. Laser removal of contaminants from painted surfaces. *Journal of Cultural Heritage* 1(1): 173–180. https://doi.org/10.1016/S1296-2074(00)00182-5
Dajnowski A. & Dajnowski B.A. 2017. Using the new G.C Laser Cleaning System for cleaning and surface preparation for re-gilding of a large outdoor bronze monument of Alexander Hamilton. In Targowski P., Walczak M., Puoli P. (eds), *Laser in the Conservation of Artworks XI: Proceedings of the International Conference LACONA XI, Kraków, Poland 20-23 September 2016*: 217–288. Toruń: Wydawnictwo Naukowe Uniwersytetu Mikołaja Kopernika. DOI: 10.12775/3875-4.15
Dajnowski A. & Lins A. 2011. The practical use of lasers in removing deteriorated Inralac coatings from large bronze monuments. In Radvan R., Asmus J. F., Castillejo M., Pouli P., Nevin A. (eds), *Laser in the Conservation of Artworks VIII: Proceedings of the International Conference on Lasers in the Conservation of Artworks VIII*: 47–52. London: Taylor&Francis.
Domasłowski W. 2011. *Zabytki kamienne i metalowe, ich niszczenie i konserwacja profilaktyczna*, Toruń: Wydawnictwo Naukowe Uniwersytetu Mikołaja Kopernika.
Dondieu S. D, Wlodarczyk K. L., Harrison P., Rosowski A., Gabzdyl J., Reuben R.L & Hand D. P. 2020. Process Optimization for 100 W Nanosecond Pulsed Fiber Laser Engraving of 316L Grade Stainless Steel. *Journal of Manufacturing and Materials Processing* 4(4): 110–125. https://doi.org/10.3390/jmmp4040110
Drexler W Fujimoto J.G (eds). 2015. *Optical Coherence Tomography: Technology and Applications*, 2nd ed. Cham, Heidelberg, New York, Dordrecht, London: Springer.
Dudek M., Wawryniuk Z., Nesteruk M., Rosowski A., Cichomski M., Kozicki M. & Święcik R. 2022. Changes in the Laser-Processed Ti6A14V Titanium Alloy Surface Observed by Using Raman Spectroscopy. *Materials* 15(20): 7153–7165. https://doi.org/10.3390/ma15207153
Faron A. 2021. Instrumental techniques aiding overpaints removal – MA-XRF imaging supported with OCT in conservation practice. Fondazione Centro per la Conservazione e il Restauro dei Beni Culturali "La Veneria Reale" (ed), *Proceedings 2021: Young Professionals Forum 2021*: 59–66. Genoa: Sagep Editori. https://www.sfogliami.it/fl/244402/qjym9pz2ktxhs5qcjrgs21rr8yqckutv.
Hellen R. 2020. Preliminary study into the reduction and removal of naturally aged varnishes from painted surfaces using Er:YAG laser in a two-step cleaning process with solvents. *Journal of the Institute of Conservation* 46 (1): 79–93. https://doi.org/10.1080/19455224.2019.1706594
Hildenhagen J. & Dickman K. 2011. Compact short pulsed fiber laser offers new possibilities for laser cleaning. In Radvan R., Asmus J. F., Castillejo M., Pouli P., Nevin A. (eds), *Laser in the Conservation of Artworks VIII: Proceedings of the International Conference on Lasers in the Conservation of Artworks VIII*: 29–32. London: Taylor&Francis.
Koss A., Marczak J. 2015, *Lasery w konserwacji dzieł sztuki i zabytków: zasady – eksploatacja – bezpieczeństwo*. Warszawa-Kraków: Międzyuczelniany Instytut Konserwacji i Restauracji Dzieł Sztuki.
Iwanicka M., Moretti P., van Oudheusden S., Sylwestrzak M., Cartechini L., van den Berg K.J., Targowski P. & Miliani C. 2018. Complementary use of Optical Coherence Tomography (OCT) and Reflection FTIR spectroscopy for in-situ non-invasive monitoring of varnish removal from easel paintings. *Microchemical Journal* 138: 7–18. doi: 10.1016/j.microc.2017.12.016
Iwanicka M., Moretti P., Pilz K., Doherty B., Cartechini L., Geldof M., de Groot S., Miliani C. & Targowski P. 2022. Congregation leaving the Reformed Church in Nuenen by Vincent van Gogh: a combined multi-instrumental approach to analyse the painting's stratigraphy in support of varnish removal. *Heritage Science* 10(1): 167. doi: 10.1186/s40494-022-00789-0.
Li X., Wang D., Gao J., Zhang W., Li C., Wang N. & Lei Y. 2020. Influence of ns-laser cleaning parameters on the removal of the painted layer and selected properties of the base metal. *Materials* 13(23): 5363–5380. https://doi.org/10.3390/ma13235363
Marczak J., Koss A., Targowski P., Góra M., Strzelec M., Sarzyński A., Skrzeczenowski W., Ostrowski R. & Rycyk A. 2008. Characterization of laser cleaning of artworks. *Sensors* 8(10): 6507–6548. doi: 10.3390/s8106507
Matteini M., Mazzeo R. & Moles A. 2016. *Chemistry for Restoration: Paintings and Restoration Materials*. Florence: Nardini Editore.
Moretti P., Iwanicka M., Melassanaki K., Dimitroulaki E., Kokkinaki O., Daugherty M., Sylwestrzak M., Pouli P., Targowski P., Van den Berg K.J., Cartechini L. & Miliani C. 2019. Laser

cleaning of paintings: in situ optimization of operative parameters through non-invasive assessment by optical coherence tomography (OCT), reflection FT-IR spectroscopy and laser induced fluorescence spectroscopy (LIF). *Heritage Science* 7 (44). https://doi.org/10.1186/s40494-019-0284-8

Pinna D., Galeotti M. & Mazzeo R. 2012. *Współczesne metody badań obrazów sztalugowych: Podręcznik konserwatora-restauratora*. Toruń: Wydawnictwo Naukowe Uniwersytetu Mikołaja Kopernika.

Pouli P. & Emmony D. C. 2000. *The effect of Nd:YAG laser radiation on medieval paintings. Journal of Cultural Heritage* 1(1): 181–188. DOI:10.1016/S1296-2074(00)00143-6

Pouli P., Oujja M. & Castillejo M. 2012. Practical issues in laser cleaning of stone and painted artefacts: optimisation procedures and side effects. *Applied Phisics A* 106: 447–464. https://doi.org/10.1007/s00339-011-6696-2

Siano S., Salimbeni R., Pini R., Giusti A. & Matteini M. 2003. Laser cleaning methodology for the preservation of the *Porta del Paradiso* by Lorenzo Ghiberti. *Journal of Cultural Heritage* 4(1): 40–146. https://doi.org/10.1016/S1296-2074(02)01138-X.

Siano S., Osticioli I., Pavia A. & Ciofini D. 2015. Overpaint removal from easel paintings using LQS Nd: YAG laser: The first validation study. *Studies in Conservation* 60(1): 49–57. DOI: 10.1179/0039363015Z.000000000207

Taarnskov B., Pouli P. & Bredal-Jørgensen J. 2011. Laser cleaning studies for the removal of tarnishing from silver and gilt silver threads in silk textiles. In Radvan R., Asmus J. F., Castillejo M., Pouli P., Nevin A. (eds), *Laser in the Conservation of Artworks VIII: Proceedings of the International Conference on Lasers in the Conservation of Artworks VIII*. London: Taylor&Francis: 67–74.

Targowski P., Iwanicka M., Rouba B.J. & Frosinini C. 2015. OCT for Examination of Artwork. In Drexler W., Fujimoto G. (eds), *Optical Coherence Tomography Technology and Applications*. Cham-Heidelberg-New York-Dordrecht-London: Springer:2473–95

Targowski P., Kowalska M., Sylwestrzak M. & Iwanicka M. 2020. OCT for examination of cultural heritage objects. In Wang M.R., (ed.), *Optical Coherence Tomography and Its Non-medical Applications*: 147–164. London, UK: IntechOpen, doi:dx.doi.org/10.5772/intechopen.88215

Vu T.T. & Hoang H.H. 2021. Investigating the effect of pulsed fiber laser parameters on the roughness of aluminium alloy and steel surfaces in cleaning processes. *Lasers in Manufacturing and Materials Processing* 8: 113–124. https://doi.org/10.1007/s40516-021-00139-1

Lasers in the Conservation of Artworks XIII – Siano & Ciofini (eds)

Egyptian limestone polychrome statues: Laser cleaning in comparison with traditional methods

E.A. Furgiuele
Università degli Studi di Torino (SUSCOR), Centro per la Conservazione ed il Restauro dei Beni Culturali "La Venaria Reale", Venaria Reale, Turin, Italy

F. Zenucchini & P. Croveri
Centro per la Conservazione ed il Restauro dei Beni Culturali "La Venaria Reale", Venaria Reale, Turin, Italy

M.C. Capua
Historical Studies Department, Università degli Studi di Torino (SUSCOR), Turin, Italy

ABSTRACT: This comparative study focuses on identifying the best intervention methodology for small limestone sculptures from Heliopolis that hold traces of red ochre, part of the *Museo Egizio di Torino* collection. After a thorough characterization diagnostic campaign using non-invasive investigations (XRF and SEM-EDS), original constitutive materials have been identified and a set of preliminary tests were carried out by employing traditional methods and three different types of Nd:YAG lasers (Q-Switch, Long Q-Switch and Short Free Running). The selected method successfully uncovered painted, as well as unpainted, surfaces with minimized risk of damage achieving a gradual, controlled, and selective cleaning.

1 INTRODUCTION

The *Museo Egizio di Torino* houses approximately 1550 finds from Heliopolis, which were excavated during the campaigns led by Ernesto Schiaparelli, the museum director from 1894 to 1928 (Sbriglio & Ugliano, 2014). From this group, seven small limestone polychrome sculptures were selected for study and restoration.

The archaeological nature and historical significance of these statues were crucial in defining the appropriate cleaning methodology. As a result, restoration efforts focused on minimal intervention, maximum selectivity and careful product selection, particularly with regard to surface decohesion and sensitivity to water solutions, due to the presence of a carbonate-chalky layer.

2 MATERIALS AND METHODS

2.1 *The artefacts*

The Heliopolitan statues (Figure 1) date from the 7th to the 3rd century B.C (Late and Greek-Roman Age) and have been identified as fertility votive offerings, displaying iconographic types typical of Lower Egypt (Thomas, 2015). The artifacts include an Isis lactans (S.3371), a lying woman (S.3373), two Bastet sphynxes (S.3423, S.3424) and two ithyphallic figures, a Harpocrates (S.3412) and a knight (S.3418). Some of them feature a preparation layer and a paint layer, which may be red (S.3374, S.3412, S.3418, S.3424) and/or black (S.3371, S.3374, S.3418).

These artifacts were excavated by Schiaparelli's archaeological mission between 1903 and 1906 and have never been restored before. Therefore, the layer adhering to the surfaces was mainly composed of excavated earth material with a thickness and tenacity that differed from

DOI: 10.1201/9781003386872-12

sculpture to sculpture. The primary conservation issue, apart from the thick layer of deposit, was the surface decohesion of both the limestone and the paint layer. Some types of degradation, such as black stains (S.3424), carbonate concretions (S.3374, S.3418) and traces of previous biological attack (S.3373, S.3412, S.3424) were only found on a limited number of artifacts.

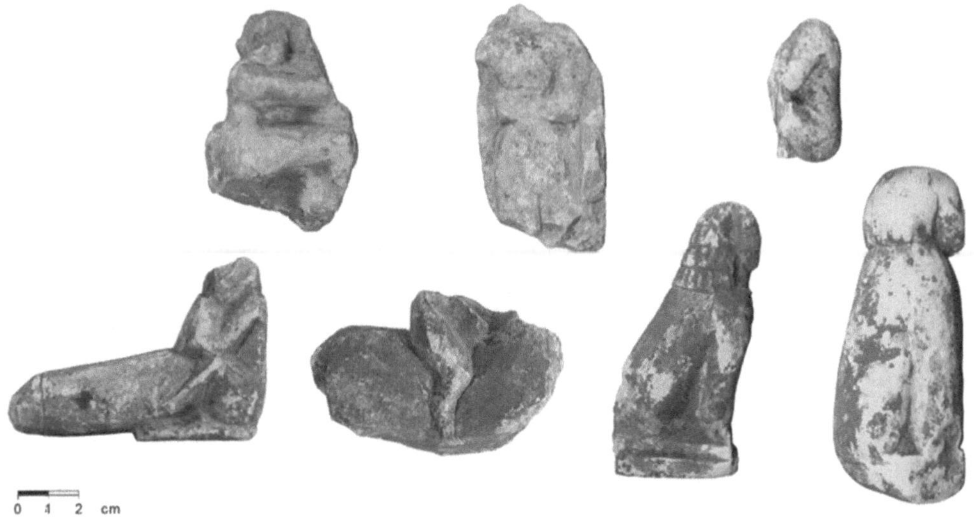

Figure 1. Heliopolitan statues before cleaning. Top (from left to right): S.3371, S.3373, S.3374. Bottom (from left to right): S.3412, S.3418, S.3423, S.3424.

2.2 *Experimental*

Constituent materials and execution techniques were investigated trough non-invasive scientific analysis, supported by careful macroscopic and microscopic observations. In agreement with the owner and the conservation authority, it was decided not to carry out para-destructive or destructive investigations, as they could have compromised such small and degraded artifacts.

The measurements were carried out on stone, preparation layer and pigments. These features were identified using a portable X-ray Fluorescence (XRF) instrument (Micro-EDXRF Bruker ARTAX 200), providing elemental composition on a variable area between about 0.65–1.50 mm, and a Scanning Electron Microscopy equipped with Energy Dispersive Spectroscopy (SEM-EDS) instrument (Zeiss EVO60 – Bruker Quantax 200). The sculptures were so small that it was possible to place them directly inside the SEM analysis chamber.

Table 1. The different kinds of cleaning methods tested on samples.

Method	Class	Type	Composition
Mechanical	Makeup sponge	Deffner&Johann	Polyurethane ether, high viscosity
	Sponge	Wishab Akapad	Styrene butadiene rubber
		Smoke Sponge	Isoprene rubber
Chemical	Physical gel	Agar Agar	Polysaccharide based on agarose and agaropectin
		Gellan Gum	Exopolysaccharide of bacterial origin
	Chemical gel	Nanorestore Gel	p(HEMA)/PVP
		PVA/borax	Polyvinylalcohol and tetraboradodecahydrate ion
Physical	QS	Thunder Art	Nd:YAG (λ 1064-532-355nm) – 8 ns
	LQS	Eos 1000	Nd:YAG (λ 1064nm) – 100 ns
	SFR	Smart Clean II	Nd:YAG (λ 1064nm) – 30÷110 μs

A set of preliminary cleaning tests were carried out on samples made of a carbonate substrate, a chalky preparation layer, an ochre pigment without any binder and a coherent deposit of silt and clay. The samples were prepared to get as close as possible to the real situation. They didn't aim to perfectly recreate the artefacts but to be functional to compare different cleaning methods on a surface without cohesion and sensitive to water solutions.

Laboratory tests were performed using different methods and then the results were evaluated through microscopy examination. Overall, three kind of sponges, four gels, and three types of lasers were tested, as shown in Table 2.

Table 2. Parameters used for each of the three lasers tested.

Laser system	Wavelenght (nm)	Pulse	Energy (mJ)	Spot (mm)	Frequency (Hz)	Fluence (J/cm^2)
QS	1064	8 ns	121	7,5	2	0.27
LQS	1064	100 ns	32.5	4	3	0.26
SFR	1064	30÷110 µs	75	3	4	1.98

In particular, laboratory irradiation tests on prepared samples and cleaning trials on the artworks were performed using fibre-coupled QS, LQS, SFR Nd:YAG (1064nm) lasers as summarized in Table 2. Furthermore, some trials using FR (280-250µs) Er:YAG were also carried out to evaluate the removal of biological traces, but with no satisfactory result.

The laboratory experimentation provided information on the ablation processes associated with the different laser parameters, which allowed qualitative interpretations of the irradiation phenomenology observed on artifacts.

3 RESULTS

3.1 *State of preservation*

After observation under a stereo microscope, all the artworks' constitutive material seemed to have the same morphological characteristics: fossiliferous carbonated sedimentary rocks, fine-grained, with a whitish colour and a warm hue.

Therefore, the stone supports were assumed to be limestones, hypothesis confirmed by SEM qualitative analysis (Figure 2). Furthermore, similarities could be noted with a limestone from Tura (Mokattam Group, Middle Eocene), a quarry near Heliopolis (Harrell, 2012) and a characteristic fossil (*Umbilicosphaera protoannulus*) of this kind of stone (Klemm & Klemm, 2009) has been identified through SEM image analysis (Figure 3).

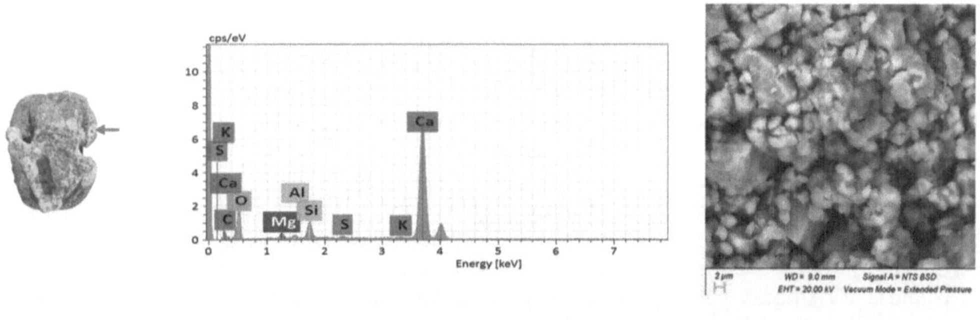

Figure 2. S.3374 limestone analysis. Left: EDS spectrum with a significant Ca peak. Right: SEM photograph of calcite with equal-sized grains, few pores and a few fossils and nanofossils.

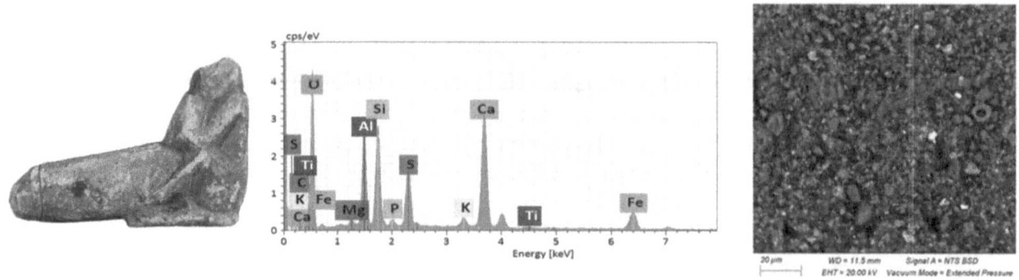

Figure 3. S.3412 surface analysis. Left: EDS spectrum in which S and Ca peaks are indicative of a chalky component in the preparation (calcium sulphate) and the Si, Al, Fe peaks show the presence of red ochre. Right: SEM photograph in which few fossils, likely coccoliths, can be seen.

Consistent with available literature (Aston *et al*, 2000), a white-coloured and very fine-grained layer of preparation could be seen in some artifacts (S.3373, S.3412, S.3418, S.3423) under optical microscope. Later, SEM and XRF data showed the presence and the carbonate-chalky nature of this layer (Figure 3).

Two colours were visible on the surfaces: red (S.3374, S.3412, S.3418, S.3424) and black (S.3371, S.3374, S.3418). XRF analyses revealed that the red pigment was an earth ochre based on ferrous minerals whereas the black pigment was generally interpreted as an organic black, having no significant elements with an atomic weight greater than 11.

3.2 *Cleaning tests on prepared samples*

As first, cleaning tests were carried out on prepared samples using different kinds of mechanical, chemical and physical methods.

3.2.1 *Mechanical methods*

A preliminary selection was made from the wide range of available materials, considering the artifacts' surface features (which cannot withstand excessive mechanical action) and data from previous studies, carefully considering composition, texture, density, residues released, mechanical action and level of abrasion (Daudin et al., 2012).

Three sponges were tested: an EU sponge (Deffner&Johann®), a SBR sponge (Wishab Akapad®) and an IR rubber (Smoke Sponge ®).

The best results among the mechanical methods were obtained with the Deffner&Johann sponge, which achieved an even cleaning degree. Thanks to its microporosity, it was able to perform a mild action, trapping within its structure dirt particles, but also few pigment ones. The other sponges were considered too aggressive, as they damaged the polychrome surface.

3.2.2 *Chemical methods*

Water could have been a risk factor given the artworks' composition and their preservation state: the chalk-based preparation and the paint layer were very sensitive to water solutions meantime the surface alterations and the increased porosity could have led to greater diffusion into the substrate and, therefore, to chemical-physical interactions (Cremonesi, 2012).

In order to control water supply on surface different gelling systems were selected evaluating several factors as chemical inertia towards the substrate, chemical and physical stability, syneresis and transparency (Baglioni *et al.*, 2014).

Four different gels were tested: Agar agar gel, a polysaccharide based on agarose and agaropectin (applied both hot and cold) (Campani *et al.*, 2007), Gellan gum gel, an exopolysaccharide of bacterial origin (applied both hot and cold), two Nanorestore gels, a polymer of p(HEMA)/PVP (Nanorestore Gel® MWR and Nanorestore Gel® Peggy 5) (Domingues *et al.*, 2013) and a PVA/borax gel, a polymer of polyvinylalcohol and a borate ion. The different support systems were left on the samples' surface for 3 minutes.

Among these chemical methods, the worst proved to be the hot-applied systems, releasing significant amounts of water and partially removing the samples' ochre paint layer. Nearly all gels could be easily removed except for Peggy 5, which seemed to exert a slightly sticky action.

The best performance has been achieved by the Gel Nanorestore MWR® and, above all, by PVA/borax, the transparent gelled system that fitted perfectly to surface while respecting its integrity. Both systems could be easily removed and didn't release any harmful residues.

However, this method, as well as the mechanical one, couldn't guarantee a punctual action, a satisfying cleaning degree and/or the required surface integrity.

3.2.3 *Physical methods*

Further trials were carried out with three different types of lasers: Q-Switch, 8 ns pulse duration (Thunder Art, Quanta System, Milan, Italy), Long Q-Switched, 100 ns (EOS 1000 LQS El.En, Calenzano, Italy) and Short Free Running, 30 – 110 μs (Smart Clean II, El.En Group).

Tests were carried out both dry and wetting with isopropyl alcohol (water was avoided because of the carbonate-chalky preparation and the paint layer). Punctual wetting with a brush immediately before irradiation increased efficiency and uniformity, reducing any harmful thermal and/or mechanical effects (Brunetto, 2004).

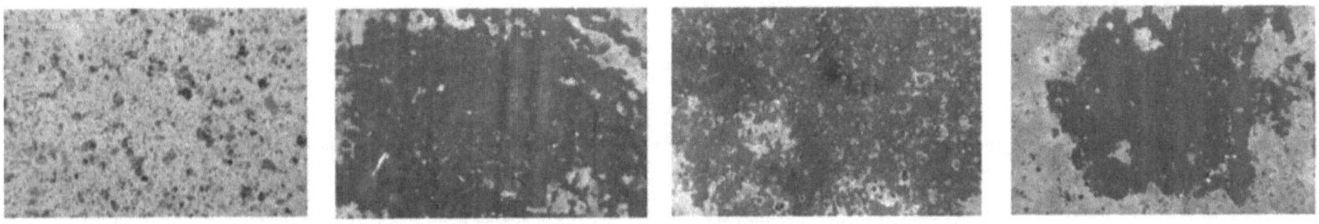

Figure 4. Samples surface (from left to right): before laser cleaning, after QS cleaning, after SFR cleaning and after LQS cleaning.

The parameters used for the tests were chosen in order to work in a gradual and selective way, employing low energies (in a range between 30 and 120 mJ) and varying the spot size, as shown in Table 2 (Bottacci et al. 2012; Gottardo, 2014; Brunetto et al., 2016; Mansi et al., 2016), thus remaining below the damage threshold (Di Stasio & Santamaria, 2014). The following regimes and parameters were tested on samples (Figure 4):

QS: the photomechanical ablative process of the Q-Switched laser was excessive for the type of deposit, causing localized micro-explosions of the deposit itself and leaving spot marks despite the low fluence used (0.27 J/cm^2, with an energy of 120 mJ). The cleaning performed wasn't uniform maybe due to the Gaussian beam; in addition, the spot size was too big and not enough selective.

SFR: Short Free Running laser didn't give an optimal result, as it didn't remove an appropriate thickness of coherent deposit and didn't achieve a uniform degree of cleaning. It was used a higher fluence than the other lasers tested to obtain successful removal of the deposit layer (1.98 J/cm^2). Moreover, the photothermal action, due to the long pulse duration, probably caused sporadic blackening of the red ochre.

LQS: Long Q-Swiched laser performed best, achieving a balanced cleaning degree while respecting the surface. The spallation effect was appropriate to remove the consistent layer of deposit, gradually breaking it up. A T25 filter was applied to reduce the outcoming energy by 75%, thus working at 32.5 mJ.

3.3 *Cleaning tests on artifacts*

The best results, after evaluating and comparing all different methods, were obtained thanks to laser technology, in particular LQS.

Tests on artifacts were carried out, which confirmed the excellent LQS results obtained on samples. Any residues of the disintegrated deposit remaining on the surface were removed by gently dabbing, without any mechanical action, with a Deffner & Johann sponge, which had given the best results among the dry-cleaning systems.

The photographs taken under optical microscope showed that the surface was always respected, especially in the case of painted surfaces with slight decohesion, resulting in a uniform outcome even where the thickness of deposit differed.

3.4 *Laser cleaning*

Based on the results obtained from preliminary laser cleaning tests, special attention was given to determine the operative fluence ranges and define a suitable irradiation protocol depending on the targeted problems.

The areas affected by similar conservation problems were treated with the same operating methods. The frequencies adopted allowed to obtain an even result without any risk of imprinting the spot mark (2 to 5 Hz for λ 1064 nm and 4 Hz for λ 532 nm). The treated areas were always wetted with isopropyl alcohol using a brush during laser irradiation.

By modulating the energy and the spot's diameter (Table 3), a punctual and gradual cleaning was achieved.

The operating conditions adopted for each type of degradation and the effect observed on the surface during cleaning are described below.

Table 3. Parameters used for each type of degradation.

Degradation	Laser (nm)	Energy (mJ)	Spot (mm)	Fluence (J/cm^2)
Coherent deposit <0,5 mm approx.	LQS 1064	32.5	4	0.26
Coherent deposit >0,5 mm approx.	LQS 1064	32,5	3	0.46
Coherent deposit above paint layer	LQS 1064	32.5	4,5	0.20
Carbonate concretio	LQS 1064	32.5	2,5	0.66
Stain	LQS 1064	32.5	3,5	0.34
	QS 532	88	2	0.48/0.55
Trace of biological attack	LQS 1064	62.5	2	1.99

3.4.1 *Thinner coherent deposit, <0,5 mm approx. (S. 3373, S. 3374, S. 3423, S. 3424)*

Generally, this type of deposit had less internal cohesion and adhesion to the substrate. In some cases, the deposit areas were minimal and with a jagged perimeter. Therefore, to be sufficiently selective, a fluence of 0.26 J/cm^2 with a spot of 4 mm was used.

The observed effect was the vaporization of the deposit, which was continuously removed from the surface with an extractor hood and a Deffner & Johann sponge.

3.4.2 *Thicker coherent deposit, >0,5 mm approx. (S.3373, S. 3423, S. 3424)*

The thickness of this layer was initially evened out by thinning mechanically with a scalpel. Generally, this type of deposit showed greater internal cohesion and adhesion to the substrate. Thus, it was necessary to use a slightly higher fluence (0.46 J/cm^2) and repeat the laser irradiation 2 or 3 times, to proceed carefully and step-by-step, carrying on a secondary spallation action (Figure 5).

3.4.3 *Coherent deposit above the paint film (S.3373, S.3412, S.3418)*

The cleaning of these areas was carried out gradually, repeating the laser irradiation as many times as needed with fluence of 0.22 J/cm^2 and constantly checking the surface to ensure that no damage was caused to the brittle paint layer (Figure 6).

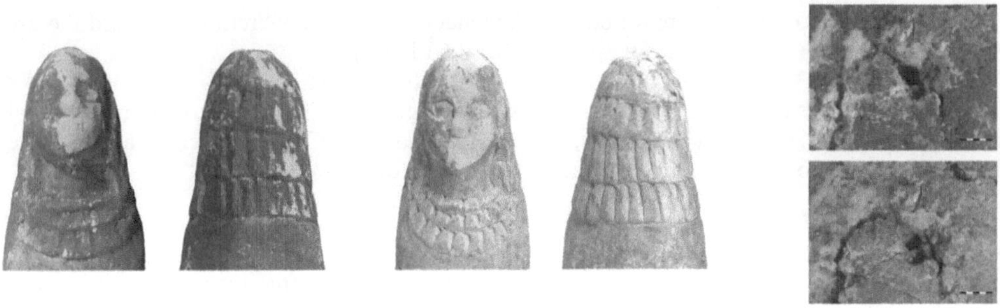

Figure 5. Left: S.3423 before and after cleaning. Right: microscope images (20) of a detail before and after cleaning.

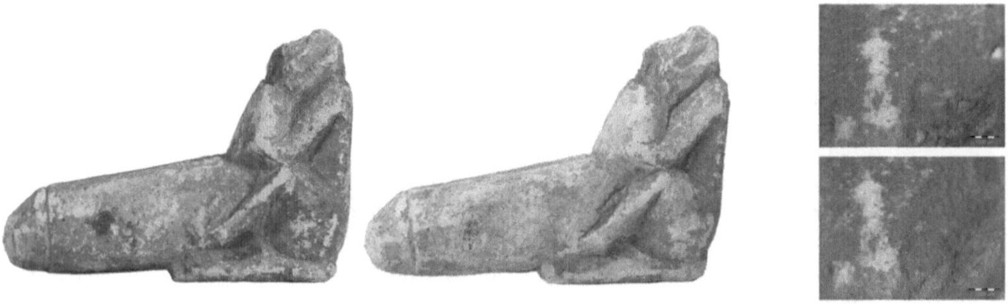

Figure 6. Left: S.3412 before and after cleaning. Right: microscope images (20) of a detail before and after cleaning.

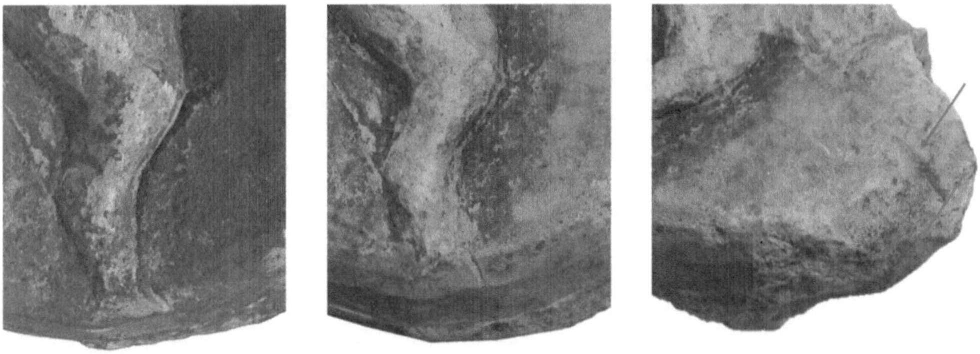

Figure 7. S.3418 (from left to right): the knight's saddle before and after cleaning and a detail of the horse's trappings and bridles defined with a black pigment.

In one case (S.3418), this process revealed new figurative details such as the horse's bridle and harness, previously completely hidden by the thick layer of deposit, allowing a new interpretation of the artifact's figurative elements (Figure 7).

3.4.4 *Carbonate concretion (S.3374, S.3418)*
The concretions were strongly adhered to the substrate, crystalline in appearance and carbonate in nature. They were never completely removed but gradually thinned due to the risk of tearing original material attached to the concretion itself (Figure 8).

The high degree of precision required and the tenacity of these concretions required the use of a small spot size (2.5 mm) and a higher fluence (0.66 J/cm^2).

In the case of S.3418, it was decided not to reduce the largest concretion on the left side, given its size and level of adhesion to the substrate (Figure 8). The risk of removing original material was too high and, additionally, its presence did not affect the interpretation of the shapes. The same considerations were made for the recto of S.3374.

3.4.5 *Stain (S.3424)*

Black stains of unknown nature were present on the surface of several artifacts. Gradual lightening was achieved by repeating the irradiation several times to soften the visual impact, particularly on the head of S.3424. After an initial treatment with the LQS, gradual clearing was carried out using the second harmonic (532 nm) of the Thunder Art QS without wetting with isopropyl alcohol and operating in a fluence range between 0.48 and 0.55 J/cm^2.

To further reduce the aesthetic impact of this black stain, the high-viscosity PVA/borax system (which was the best performing gel among the chemical methods) was locally applied, producing excellent results.

The extraction of the foreign material (now inside the limestone's porosity) was achieved by leaving the gel directly on the surface for 5 minutes and, once removed, gently rolling a swab soaked in a solution of isopropyl alcohol and demineralized water (70/30).

3.4.6 *Previous biological attack (S.3412, S.3424)*

The thinning of the biological traces was carried out last using the LQS at 250 mJ, applying a filter window of 25%, thus with an outgoing energy of 62.5 mJ (Figure 9).

An attempt was made with the Erbium laser (Er:YAG, 2090 nm, 150 mJ – 380 μs and 50/100/200 mJ – 250 μs, 5 Hz, ø 1 mm, fluences from 6.37 up to 25.48 J/cm^2) initially, buy it did not yield any significant results.

In some areas, an extension of biological traces below the surface was observed, possibly due to an endolithic organism (Bungartz *et al.*, 2004). Therefore, it was decided not to continue with laser cleaning as the risk of removing original material was considered too high.

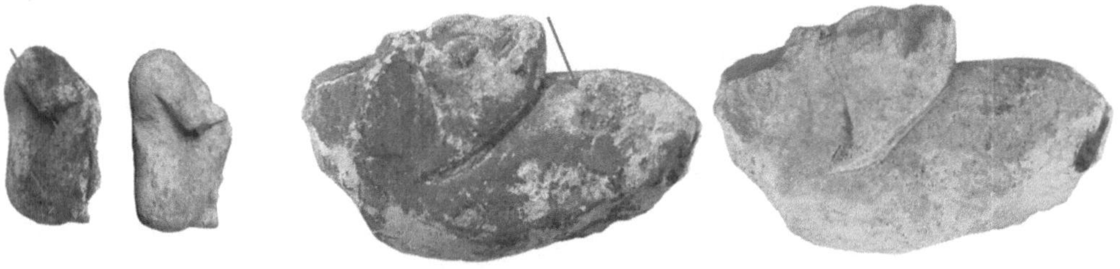

Figure 8. S.3343 (left) and S.2418 (right) before and after carbonate concretion laser thinning.

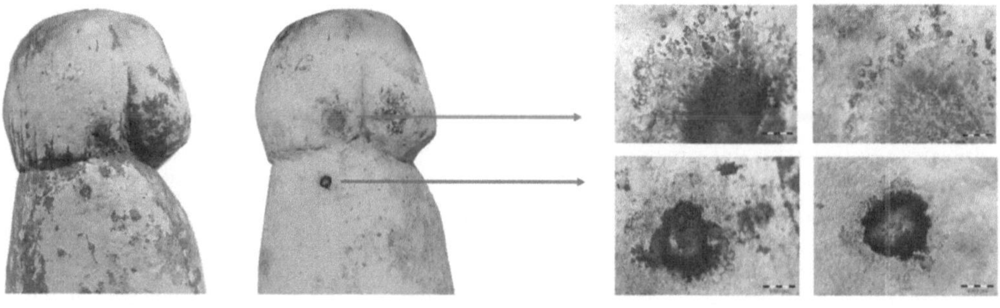

Figure 9. Left: S.3424 before and after cleaning. Right: microscope images (20) of biological attacks before and after cleaning.

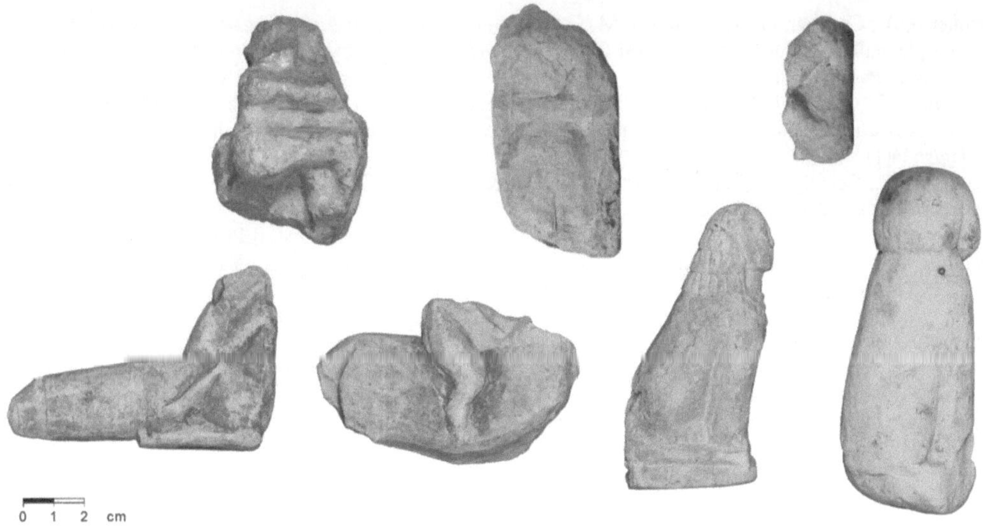

Figure 10. Heliopolitan statues after cleaning. Top (from left to right): S.3371, S.3373, S.3374. Bottom (from left to right): S.3412, S.3418, S.3423, S.3424.

4 CONCLUSIONS

In conclusion, the cleaning of the Heliopolitan statues presented significant challenges due to the delicate nature of the water-sensitive surfaces and the heavily degraded substrate. However, thanks to the use of laser technology, combined with traditional cleaning methods, a gradual, selective, and balanced approach was achieved, resulting in critical aesthetic restoration on a case-by-case basis.

The LQS laser method proved to be the best solution due to its versatility and ability to uncover both painted and unpainted surfaces with minimal risk of damage compared to mechanical or chemical approaches. The appropriate variation of parameters enabled the successful removal of dirt and grime without harming the underlying surfaces. The use of this innovative technique has paved the way for more efficient and effective restoration of delicate artifacts and artworks, providing valuable insights for future conservation efforts.

ACKNOWLEDGEMENTS

The Museo Egizio di Torino and the Soprintendenza Archeologia, Belle Arti e Paesaggio per la Città Metropolitana di Torino are gratefully acknowledged.

REFERENCES

Aston, B. G., Harrell, J. A. & Shaw, I. 2000. Stones (Chapter 2). In Nicholson P. T. & Shaw I. (eds.) *Ancient Egyptian Materials and Technology* (a cura di.). Cambridge University Press, Cambridge.

Baglioni, P., Berti, D., Bonini, M., Carretti, E., Dei, L., Fratini, E. & Giorgi, R. 2014. Micelle, microemulsions, and gels for the conservation of cultural heritage. *Advances in Colloid and Interface Science* 205: 361–371.

Bottacci, E., Felici, V., Lini, V., Mori, G., Sanches Del Pozo, R. & Spada, S. 2012. Applicazioni del laser nella pulitura di opere lapidee: esperienze recenti ai Musei Vaticani. In A. Brunettto (a cura di) *APLAR 5, Atti del convegno. Applicazioni laser nel restauro*. Padova: Il Prato.

Brunetto, A. 2004. *L'utilizzo della strumentazione laser per la pulitura delle superfici nei manufatti artistici*. Padova: Il Prato.

Brunetto, A., Giovagnoli, A., Laurenti, M. C., Mano, M-J. & Zanini A. 2016. Laser cleaning on the Mut Temple wall paintings, Gebel Barkal (Sudan). In *APLAR 6, Atti del convegno. Applicazioni laser nel restauro*. Padova: Il Prato.

Bungartz, F., Garvie, L.A.J. & Nash Iii, T.H. 2004. Anatomy of the endolithic Sonoran Desert lichen Verrucaria rubrocincta Breuss: implications for biodeterioration and biomineralization. *The Lichenologist* 36(1): 55–73.

Campani, E., Casoli, A., Cremonesi, P., Saccani, I. & Signorini, E. 2007. *L'uso di agarosio e agar per la preparazione di "gel rigidi"*. Saonara: Il Prato.

Cremonesi, P. 2012. *L'ambiente acquoso per il trattamento di opere policrome*. Il Prato, Padova.

Daudin, M., Van Keulen, H., Groot Wassink, M., Joosten, I., De Groot, S., Van Bommel, M. & Van Den Berg, K. J. 2012. Progetto "Dry cleaning". Metodi di pulitura a secco. Parte I/a - Panoramica sui materiali testati. In *Applicazione dei Metodi di Pulitura a Secco su Dipinti Moderni Non Verniciati*.

Di Stasio, F. & Santamaria, U. 2014. Determinazione della soglia di fluenza di danno nella pulitura delle superfici policrome. In A. Brunetto (a cura di) *APLAR 6, Atti del convegno. Applicazioni laser nel restauro*. Padova: Il Prato.

Domingues, J., Bonelli, N., Giorgi, R., Frattini, E., Gorel, F. & Baglioni, P. 2013. Innovative hydrogels based on semi-interpenetrating p(HEMA)/PVP networks for the cleaning of water-sensitive cultural heritage artifacts. *Langmuir* 29 (8): 2746–2755.

Gottardo, M. 2014. Coperchio di sarcofago egizio in arenaria dipinta: considerazioni sugli esiti e sulla messa a confronto di diverse modalità applicative dell'ablazione laser utilizzata per la pulitura. In *APLAR 5, Atti del convegno. Applicazioni laser nel restauro*. Padova: Il Prato,.

Harrell, J. A. 2012. Utilitarian stones. In W. Wendrich (ed) *UCLA, Encyclopedia of Egyptology*. https://escholarship.org/uc/item/77t294df.

Klemm, D. D. & Klemm, R. 2009. *Stone & Quarries in Ancient Egypt*. London: The British Museum Press,.

Mansi, S., Zenucchini, F., Croveri, P., Spagnoli, F. 2016. Pulitura laser della cassetta porta-ushabti (Cat. 2441) del Museo Egizio di Torino. In *APLAR 6, Atti del convegno. Applicazioni laser nel restauro*. Padova: Il Prato.

Sbriglio, A.M. & Ugliano, F. 2014. Re-excaving Heliopolis: unpublished archeological data from the archives of Ernesto Schiaparelli and M.A.I. In Pinarello P.S., Yoo J., Lundock J. & Walsh C. (eds.) *Current Research in Egyptology, Proceedings of the Fifteenth Annual Symposium*. Oxbow Books London, Oxford.

Thomas, R. 2015. Egyptian Late Period Figures in Terracotta and Limestone. In Villing A., Bergeron M., Bourogiannis G., Johnston A., Leclère F., Masson A. & Thomas R. (eds.) *Naukratis: Greeks in Egypt*. London: The British Museum.

Lasers in the Conservation of Artworks XIII – Siano & Ciofini (eds)
© 2024 The Author(s), ISBN 978-1-032-47995-8
Open Access: www.taylorfrancis.com, CC BY-NC-ND 4.0 license

Gradual cleaning of a seventeenth-century polychrome wood sculpture by Er:YAG laser

A. Andreotti & M.P. Colombini
Department of Chemistry and Industrial Chemistry, University of Pisa, Pisa, Italy

E. Cantisani & D. Magrini
Institute of Heritage Science, Italian National Research Council, Florence, Italy

A. De Cruz
Conservator Scientist, New York, USA

K. Nakahara
Conservator-Restorer of Cultural Heritage, Florence, Italy

ABSTRACT: This paper describes the application of the Er:YAG laser for the cleaning of a 17th-century polychrome wood sculpture which was originally opulently gilded, but had been totally overpainted around the nineteenth century. A series of diagnostic analyses and laser tests have been performed for a comprehensive understanding of the artistic-technique and the state of preservation, and for controlling the laser cleaning. Cross-sections of minute fragments and samples collected by the Er:YAG laser allowed us to understand the complex stratigraphy. PY-GC-MS, SEM-EDX, and XRF analyses highlighted the challenge of this cleaning work. It has been considered appropriate to remove gradually the overpaints on the clothing of the Virgin in order to uncover the original gilded portion, although it might be extremely degraded. The Er:YAG laser ablation allowed the selective removal of different layers of multi-colored overpaint and of the thick and resistant past restoration infilling which covered the thin degraded original layers.

1 INTRODUCTION

In the last 20 years, various laser technologies have been developed and applied for the conservation-restoration of cultural heritage. The cleaning of artworks by Er:YAG laser (2,94 μm) has been widely studied since late 1990's (De Cruz 1999), and this system has been proved efficient for the cleaning of paintings, particularly for -OH bond rich materials, i.e., for the removal of degraded varnishes as well as of overpaintings (Bracco et al. 2002; Bracco et al. 2003, Andreotti et al. 2007). Nonetheless, the application of the Er:YAG laser in cleaning polychrome surfaces is still quite limitedly reported (Pereira-Pardo et al.2018). Based on the previous positive experiences in Er:YAG laser for removing overpaintings on paintings on canvas/wood (Andreotti et al. 2016), it was considered very promising to continue cleaning test with Er:YAG laser on this case. The aim of this paper is to contribute to broaden comprehension and use of Er:YAG laser, which is suitable to delicate operations such as cleaning of painted surfaces thanks to its peculiar and intrinsic characteristic of high absorption of -OH bond. An irradiation of 1 mm spot and minimum pulse energy as low as 0,003 J/pulse (i.e., fluence 0,4 J/cm^2) permit to execute precise local cleaning, as well as reduced mechanical impact to the deteriorated polychrome or gilded surface during the cleaning, not to mention of its practical advantages such as better control and friendly-handling for restorers-conservators with small pen-guided hollow glass fiber, etc., which could be advantageous in resolving difficult conservation problems.

DOI: 10.1201/9781003386872-13

2 MATERIALS AND METHODOLOGIES

2.1 *Virgin and child*

A seventeenth-century (attr.) Italian polychrome wood sculpture (height ca. 66 cm) from a private collection, representing "the Virgin and Child" was examined and then treated using Er:YAG laser (Figure 1). Although the original provenance of this artwork is unknown, it is presumed that it could have been a devotional object in the area of Vicenza, northern-oriental part of Italy. The dating had been done according to a stylistic point of view, to the seventeenth century, which, however, was disguised by heavy and comprehensive overpaintings executed in the late nineteenth century onwards, as shown below, which covered completely the surface of the sculpture. During the cleaning, in fact, it was revealed that the sculpture had been originally gilded opulently, which might lead to revise the present dating hypothesis to an earlier period. Richly gilded polychrome sculptures, *Polittico*, a complex group of wood polychrome and gilded sculptures of the sixteenth century in San Cassiano di Quinto church near Vicenza, for example, might have a certain link to our sculpture. The sculpture is composed of various wooden parts assembled, carved, gilded, and painted. Under a closer observation, the wood species seems to belong to conifer. The right arm of the Virgin is attached to the torso by means of two dowels and adhesive. Various wooden components, such as the arms and the hands of Child and some fingers of the Virgin, have been lost. As previously described, the original modelling of the statue has been heavily disguised by thick *stucco* and thick and roughly executed overpaints all over the surface. So many detaching overpaint-layers that reveal the white *stucco* underneath, as well as old cleaning tests, can be observed in various parts (ex. blue and red Virgin's robe and faces).

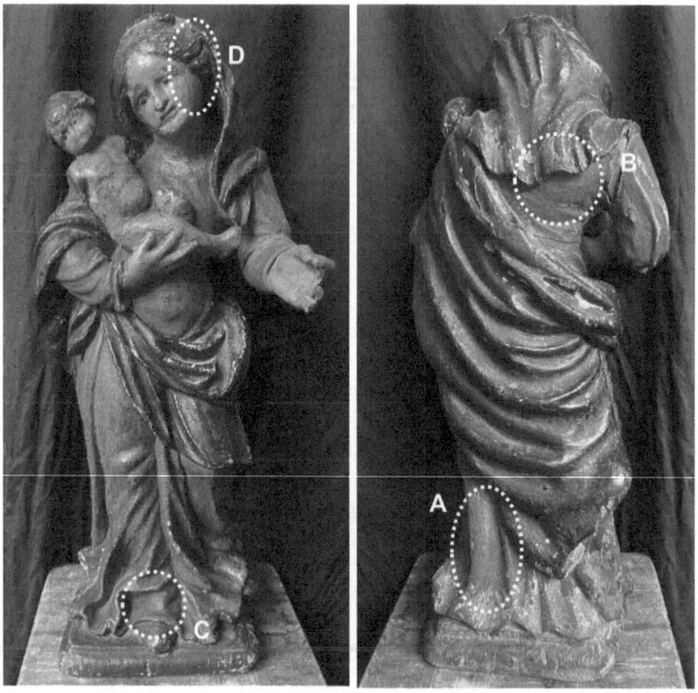

Figure 1. Virgin and Child, *recto* (left) and *verso* (right). A, B, C, and D indicate the areas treated by Er:YAG laser.

2.2 *Diagnostic analyses*

Heavy conservation interventions often occurred for the wooden polychromed sculptures, due to the deterioration that made the restoration necessary, or for the change of stylistic taste. In our case, the surface of the sculpture was totally covered by dust and dark resinous material, which should have been applied during old restoration. The reason for the outrageous overpaint is presumed to hide the degradation without respect for the original. The decision to remove or not the non-original part is a very critical issue and requires thorough diagnostics and discussion.

The results of preliminary close observation and diagnostic analysis permit to presume that there still exists the original gilded portion in various area, although it might be extremely degraded or fragmented. It has been considered, thus, appropriate to remove gradually the overpaints on the clothing of the Virgin in order to uncover the original gilded portion. As for the flesh portion, it was difficut to estimate the presence of the original layer under the thick and heavy overpaintings with non-destructive analysis executed, therefore it has been presumed appropriate to remove gradually those on the face of Virgin, near the prexisting evident lacuna on the left cheek in order to evaluate the possibility to retrive the original flesh color or original wood modelling.

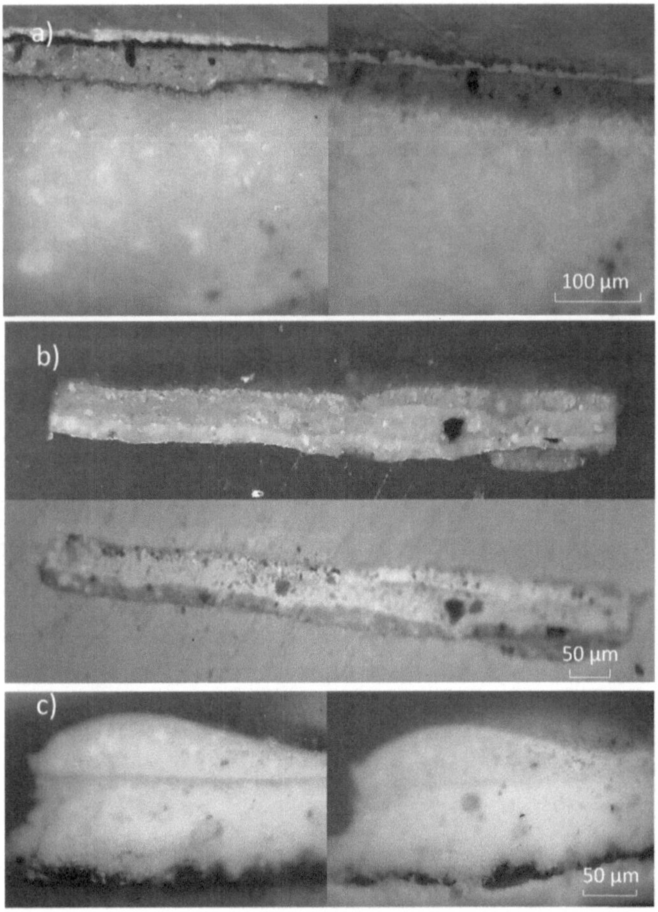

Figure 2. a) Cross-section analysis: thick blue and red overpaints over the 'stucco' on the Virgin's robe in VIS (left) and UV light (right); b) the red overpaints above the original gold layer in VIS (up) and UV light (down); c) the thick 'stucco' and overpaints on the Virgin's face in VIS (left) and UV light (right).

A series of analytical techniques have been performed for a comprehensive understanding of the artistic technique and the state of preservation of the materials, as well as for controlling the laser cleaning. Cross-sections of samples taken with scalpels have allowed us to understand the complex stratigraphy (Figure 2). The original flesh color has then been overpainted with different hues of reddish flesh colors after the application of very thick and hard white stucco. All the garment of the Virgin which is presently visible is not original.

X-Ray Fluorescence (XRF) is the technique used for acquiring information about the elemental composition of paint materials, penetrating the wooden inner structure of the statue (Bezur et al. 2020, Shugar et al.2013). XRF spectra were collected by means of a handheld Tracer III SD Bruker spectrometer, equipped with rhodium anode and solid-state silicon detector energy dispersion system. The used set-up was: 40 KeV and 12μA for 120 sec. The measuring area was an elliptical spot of 4 mm x 7 mm. For data elaboration, ARTAX software was used.

The XRF spectra of all areas belonging to the red-type3 group present the signals of mercury (Hg) and sulphur (S), thus confirming the identification of cinnabar (Mercury sulphide). In the blue overpaint visible on the Virgin's robe, some key elements of copper or cobalt-based pigment have been registered, together with lead likely used in shades with lead white. Elements characteristic of fillers used for tubed paints produced from 19th century such as Barium, Titanium or Zinc have been also revealed (Figure 3).

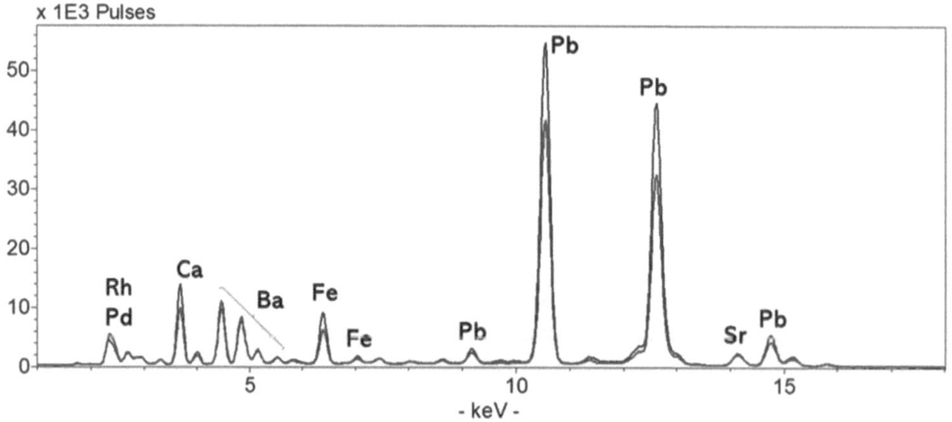

Figure 3. XRF spectra of a blue area on the Virgin's veil, and on the Virgin's robe.

Elemental analyses performed before the laser cleaning confirmed the presence of gold X-ray characteristic Lα and Lβ lines of Au in XRF spectra, respectively at 9.71 keV and 11.44 keV (Figures 4 and 6). After the laser removal of the thick and persistent blue overpaints and 'stucco' layers, in fact, fragmented original gilding appeared.

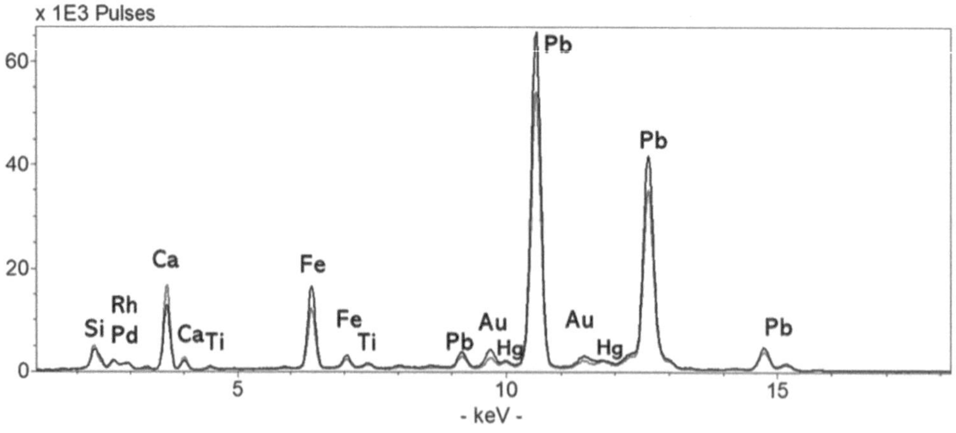

Figure 4. XRF spectra of two residuals golden traces.

The Pyrolysis-gas chromatography-mass spectrometry (PY-GC-MS) was performed for the identification of the organic material. A Virgin's robe sample showed the typical aromatic and heterocyclic compounds, characteristic of the pyrolysis profile of blue-green phthalocyanine (Figure 5). The Phthalocyanin blue, used for the blue overpaint, is a modern synthetic blue pigment which has been developed around the 1930s and reported as solvent/water resistant. Moreover, the marker pyrrole of animal glue was identified both in the ground and the stucco sample collected (Bonaduce et al. 2009).

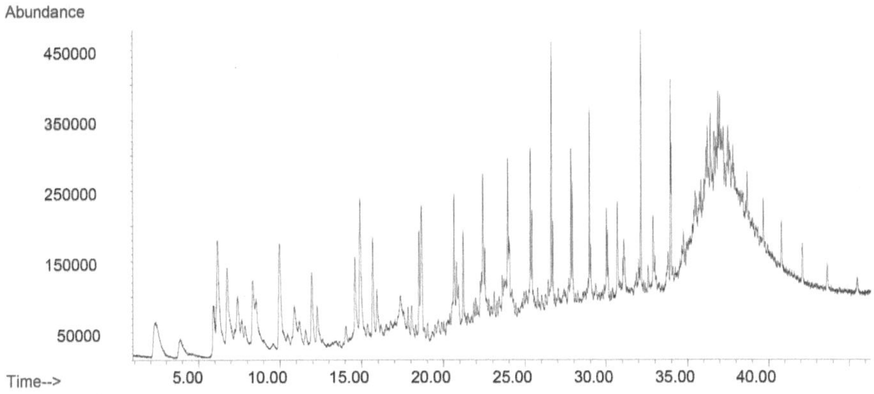

Figure 5. Pyrogramm of the sample from the blue overpaint of the Virgin's robe.

3 RESULTS

The tested areas selected for the laser gradual removal of various non-original layers with different stratigraphy were (Figure 1):

- Area A: Virgin's red robe overpaint on the blue underlayer which was executed on the thick white 'stucco' on the original gilding, located on the reverse side of the sculpture;
- Area B: Virgin's red robe overpaint which covers the original gilding on which there is a brown and stiff layer, located on the Virgin's back;
- Area C: dark brown overpainting on the Virgin's gold shoe.
- Area D: the flesh-colored overpaint which covered the very thin fragmental original flesh-colored paint (on the face of the Virgin, near the existing lacuna located to her left cheek).

A series of preliminary laser ablation tests has been executed to identify threshold and optimun operating cleaning parameters for each layer to be removed, increasing gradually from lower energy. Wetting agents - demineralized water and ethanol (1:1 v/v) or/and white spirit (added with etilenglicol), depending on the material condition - were always used in order to confine the surface temperature increase induced by laser within the safety limit (Andreotti et al. 2014), as well as to obtain better visibility during the cleaning. A microscope cover glass (18 x 18 mm) was always used for collecting ablated materials and preventing them from clogging up the glass hollow tube. The laser energy output was measured by Power Meter ('StarLite', Ophir Optronics Solutions Ltd.); the frequency of 15 Hz was considered appropriate; the spot diameter was fixed to 1 mm. The working laser parameters are reported in Table 1. Each ablation terminated quite fast, although it was necessary to repeat several irradiations over the same tested area (Figure 6–9).

For the gradual removing of very thick multi-colored overpainting and stucco on the Virgin's original gilded robe (Area A), the following laser parameters were successful: 6 consecutive laser pass (2,5J/cm^2 - 2,5J/cm^2 - 2,3J/cm^2 – 1,9J/cm^2 – 0,6J/cm^2 – 0,4J/cm^2), always prewetting the surface with wetting agent (demineralized water and ethanol (1:1 v/v) with a small cotton swab for the first 5 passages and white spirit for the last ablation in order not to damage water gilding (water sensitive material). After each irradiation, it was easy to clean up the remnants either by cotton swab or by surgical scalpel without provoking any damage to the underneath degraded and fragile original layer. The details of a few laser pass steps are shown in Figure 6.

Table 1. Measured Laser operation optimal parameters.

Area/Stratigraphy	Fluence	Prewetting agent	Clearing agent/method*
Virgin's robe - A /dust	2,5 J/cm^2 ×2	H_2O/EtOH (1:1)	idem in cotton swab
artificial patina/resin	2,3 J/cm^2	H_2O/EtOH (1:1)	idem in cotton swab
red overpaint	1,9 J/cm^2	H_2O/EtOH (1:1)	idem in cotton swab
blue overpaint	0,6 J/cm^2	H_2O/EtOH (1:1)	idem in cotton swab
white stucco	0,4 J/cm^2	white spirit	idem in cotton swab
gilding			
red bole			
ground			
Virgin's robe - B/dust	1.3 J/cm^2 ×2	H_2O/EtOH (1:1)	idem in cotton swab
artificial patina/resin	0.9 J/cm^2 ×2	H_2O/EtOH (1:1)	idem in cotton swab
red overpaint	0.8 J/cm^2 ×2	white spirit	idem in cotton swab
white stucco			
brown layer			
gilding			
ground			
Virgin's shoe - C/dust	1 J/cm^2×2	H_2O/EtOH (1:1)	idem in cotton swab
artificial patina/resin	×1	white spirit	idem in cotton swab
dark brown overpaint			
light brown overpaint			
white stucco			
brown layer			
gilding			
ground			
Virgin's cheek - D /dust	2.5 J/cm^2 ×2	H_2O/EtOH (1:1)	idem in cotton swab
artificial patina/resin	2.3 J/cm^2 ×2	H_2O/EtOH (1:1)	idem in cotton swab
flesh-colored overpaint	1.9 J/cm^2 ×2	H_2O/EtOH (1:1)	idem in cotton swab
white stucco	1.8 J/cm^2 ×2	H_2O/EtOH (1:1)	idem in cotton swab
brown layer	1.5 J/cm^2	H_2O/EtOH (1:1)	idem in cotton swab
flesh color (original)			
ground			

* For the removal of thick stucco, the surgery scalpel was occasionally used after ablation.

As for the gradual removal of red overpaints, stucco, and resistant brown layer on the Virgin's gold robe (Area B), the optimal laser parameters were: 6 irradiations (6.3 mJ/pulse – 10 mJ/pulse; i.e., 0.8 – 1.3 J/cm^2), prewetting the surface with wetting agent as above-mentioned (Figure 7).

The removal of brown overpaint and stucco on the Virgin's gold shoe (Area C) needed less irradiation: 3 passages at 1 J/cm^2 (for the area of ca. 2×2 cm^2), prewetting with the same wetting agent as descrived above (Figure 8).

To remove flesh colored overpaint on the cheek of the Virgin (Area D), the optimal laser parameters registered: 9 laser irradiations at max. 2.5 J/cm^2 – 2.3 J/cm^2 – 1.9 J/cm^2 – 1.8 J/cm^2 – 1.5 J/cm^2, prewetting the surface with wetting agent - demineralized water and ethanol (1:1 v/v) - with a small cotton swab (Figure 9).

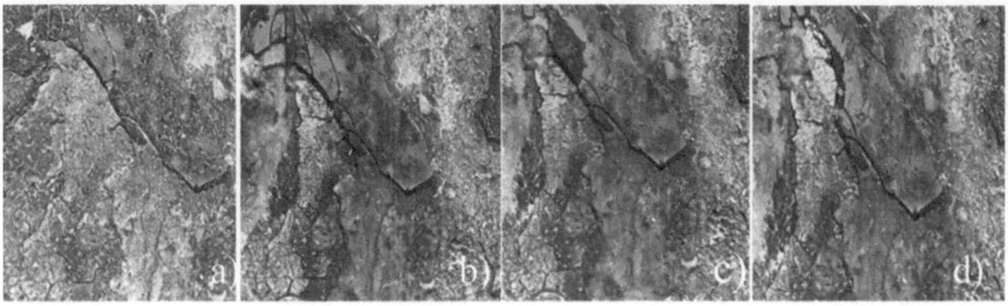

Figure 6. Procedure of Er:YAG cleaning (preliminary test), Area A: a) Before cleaning; b) 2.5 J/cm^2, H_2O/EtOH; c) 2.5 + 2.3 J/cm^2 H_2O/EtOH; d) 1.9 + 0.6 + 0.4 J/cm^2 H_2O/EtOH or WS.

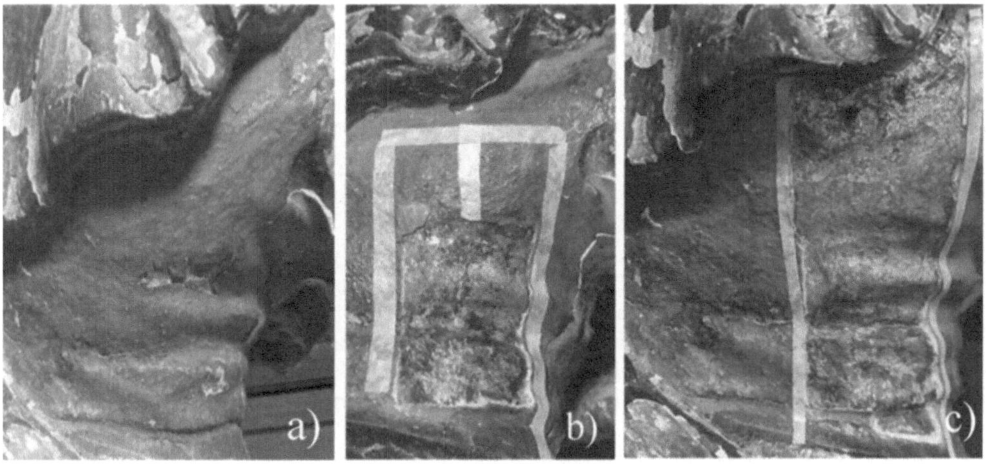

Figure 7. Procedure of Er:YAG laser ablation, Area B: a) Before cleaning; b-c) 0.8 – 1.3 J/cm^2 ×6 irradiations, H_2O/EtOH or WS.

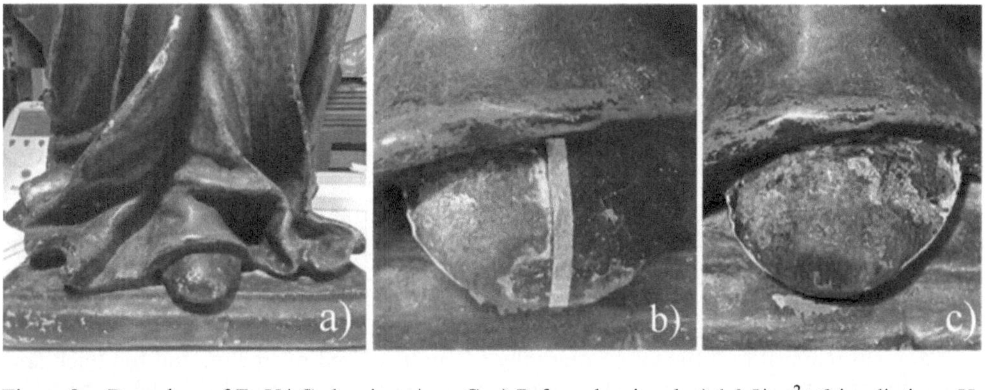

Figure 8. Procedure of Er:YAG cleaning, Area C: a) Before cleaning; b-c) 1.0 J/cm^2 ×3 irradiations, H_2O/EtOH or WS.

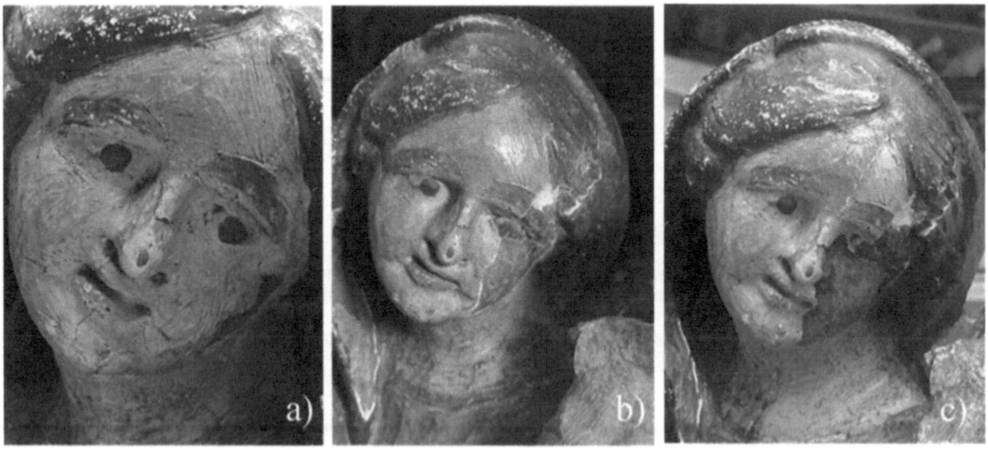

Figure 9. Procedure of Er:YAG cleaning, Area D: a) Before cleaning; b-c) During cleaning with Er: YAG laser at 2.5 - 2.3 – 1.9 – 1.8 – 1.5 J/cm^2, H_2O/EtOH.

4 CONCLUSIONS

The preliminary chemical tests with organic solvents have been performed before the Er:YAG laser cleaning one. None of the solubility tests based mainly on the polarity of organic pH-neutral solvent was successful. Acetone and Ethanol had a partial effect. Alkaline gel (pH 8-8,5) with a mixture of solvents (Ethanol and Benzyl alcohol) had a partial cleaning effect, although it was not selective and needed repeated application with a long posing time. The mechanical scalpel cleaning, preceded by the chemical one, had only a partial removal effect and was often dangerous for the stability of the original layer.

In this work, the preliminary laser ablation tests were performed in order to obtain the optimal laser conditions for each stratigraphy in various parts of the sculpture, after having had some difficulties with chemical and mechanical cleaning.

Fluences between 0.4 and 2.5 J/cm^2 (e.g., spot energy between 3 mJ/pulse and 20 mJ/pulse), at the frequency of 15 Hz with ethanol/demineralized water (1:1 v/v) and/or white spirit as wetting agent were successful. The thick layers of overpaintings and restoration infillings (stucco) which heavily disguised the original facial anatomy of the Virgin have been successfully removed.

The combination of laser cleaning and other methods (chemical, physical, and mechanical cleaning) in order to optimize the cleaning efficiency has been well studied and reported (Bracco 2001, 2002). In fact, in this case, various methodologies were adopted to execute an optimal cleaning. The choice of operating parameters/methodology depends on the nature of materials to be removed and the conservation condition of artwork. The pen shaped mirrored glass holder permits precise and controlled ablation even in the very interior part of the modelling, which is difficult to reach with a traditional scalpel or conventional laser which has a relatively big handling holder. It was observed that, after laser ablation, even a traditional surgical scalpel or a small cotton swab can perform positively to clean up the ablated residue on the surface without imposing excessive mechanical pressure to degraded fragile original material.

The mid-infrared (2.94 μm) Er:YAG laser ablation, thus, allowed the selective layer by layer gradual removal of different stratigraphy of multi-colored overpaint, and the thick and resistant past restoration infilling which covered the very thin degraded original layers. The Er:YAG laser, when used with appropriate operating parameters, becomes particularly efficient especially when very delicate operations are required for the sake of stability of fragile and degraded original substrates.

REFERENCES

Andreotti, A., Bracco, P., Colombini, M.P., de Cruz, A., Lanterna, G., Nakahara, K. & Penaglia, F. 2007. Novel applications of the Er:YAG laser cleaning of Old Paintings. *Laser in the Conservation of Artworks, Lacona VI Conference Proceedings 2005*: 239–247. Springer Proceedings in Physics 116. Berlin Heidelberg.

Andreotti, A., Brown, W. P., Camaiti M., Colombini, M. P. & DeCruz, A. 2016. Diagnosis of materials and effectiveness of Er:YAG Laser cleaning as complementary treatment in a panel painting attributed to Lluís Borrassà (15th Cent.). *Appl. Phys. A*, 122(6): 1–12. DOI: 10.1007/s00339-016-0100-1.

Andreotti, A., Ceccarini, A., De Cruz, A. & Colombini, M.P. 2014. Laser cleaning of works of art: evaluation of the thermal stress induced by Er:YAG laser, *Applied Physics B Lasers and Optics*, 117 (2): 533–541.

DeCruz, A., Hauger, S.A. & Wolbarsht, M.L. 1999. The role of lasers in fine arts conservation and restoration, *Optic and Photonics News*, 10: 36–40.

Bezur, A., Lee, L., Loubser, M. & Trentelman, K. 2020. *Handheld XRF* in *Cultural Heritage*, J. Paul Getty Trust and Yale University.

Bonaduce, I. & Andreotti A. 2009. Py-GC-MS of organic paint binders, *Organic Mass Spectometry*. In M.P. Colombini and F. Modugno (eds), *Art and Archeology*: 303-326. Research Signpost, Wiley Publication, Chippenham (Wiltshire).

Bracco, P., Lanterna, G., Matteini, M., Nakahara, K., Sartiani, O., deCruz, A., Wolbarsht, M.L., Adamkiewicz, E. & Colombini, M.P. 2001. Er:YAG laser: an innovative tool for controlled cleaning of old paintings: testing and evaluation, *Joutnal of Cultural Heritage* 4 (Suppl. 1): 202s–208s.

Bracco, P., Lanterna, G., Matteini, M., Nakahara, K. & Sartiani, O. 2002, L'esperienza dell'Opificio nella sperimentazione del laser ad Erbio per la pulitura dei dipinti. *OPD Restauro XIII*: 192–202.

Pereria-Pardo, L. & Korenberg, C. 2018. The use of erbium lasers for the conservation of cultural heritage. A review in *Jounal of Cultural Heritage* 31: 236–247.

Shugar, A.N., Mass, J.L. 2013. *Handheld XRF for Art and Archaeology*. Leuven University Press. Leuven.

Lasers in the Conservation of Artworks XIII – Siano & Ciofini (eds)
© 2024 The Author(s), ISBN 978-1-032-47995-8

Laser cleaning of the painted frames of the Issenheim altarpiece in Colmar

D. Martos-Levif & M. Lopez
C2RMF, Palais du Louvre, Porte des Lions, Paris, France

A. Brunetto
Restauri Brunetto, Vicenza, Italy

A. Pontabry
Private Conservators, Paris, France

A. Gérard-Bendelé
Private Conservators, Vesoul, France

V. Detalle
C2RMF, Palais du Louvre, Porte des Lions, Paris, France
CY Cergy Paris Université, UMR CNRS SATIE

ABSTRACT: The preservation and restoration of paintings are delicate tasks that involve removing dust, varnish, or overpainting without damaging the original material. Traditional chemical cleaning methods have limitations, especially when faced with frames containing sensitive paints. An alternative is laser cleaning. Laser cleaning involves the selective absorption of laser energy by contaminants, allowing for controlled removal. This study focuses on the restoration of the Issenheim altarpiece frames of Matthias Grünewald (1512-1516), painted in tempera and oil. bTwo types of lasers were tested for the cleaning process: Nd:YAG, LQS and SFR 1064 nm and Er:YAG 2940 nm. Optical coherence tomography (OCT) was used for the evaluation of the cleaning, providing valuable insights into the composition, structure, and conservation needs of the frames. The results of the laser cleaning tests showed that the Er:YAG laser in Very Short (150 µs) mode was effective in removing residues from the frames, but the original paint was sensitive to the laser treatment. The Nd:YAG laser in LQS (100 ns) mode demonstrated potential for selective cleaning. It al-lows for the restoration of the original appearance of the frames while minimizing the risk of irreversible damage. The study highlights the importance of selecting appropriate laser parameters, closely monitoring the cleaning process, and utilizing advanced imaging techniques like OCT for evaluation. Continuing research is being conducted to further understand the mechanisms of destructuring induced by lasers on paintings. In conclusion, this research contributes to the ongoing exploration of laser cleaning in heritage conservation and restoration, emphasizing the need for skill transfer and collaboration between professionals and heritage laboratories.

1 INTRODUCTION

The cleaning of paintings is a delicate and irreversible intervention of restoration that consists in a gradual reduction or removing dust, layers of varnish or overpainting that cover original material, without damaging them. Today, cleaning methods are diversified. The chemical methods that have been developed since the second half of the 20th century now allow a more specific action on the substances to be solubilized and the use of gels or emulsions limits their penetration into the pictorial material. If the materials to be removed have different physical and chemical properties from the paint, cleaning can be selective but chemical cleaning is impossible if they are sensitive to the same solvents. Restorers were confronted with this problem when they wanted to

DOI: 10.1201/9781003386872-14

remove the overpainting from certain frames of the Issenheim altarpiece conserved in the Unterlinden museum in Colmar (FR). A laser cleaning test campaign was then carried out.

Laser cleaning has been practiced in the field of heritage for the past 40 years, particularly in architecture and monumental sculpture. Nd: YAG Q-Switched lasers with a wavelength of 1064 nm and a fast pulse are used to remove dirt and black crusts. The high-energy infrared radiation sent to the surface is absorbed by the dark deposits and reflected by the white stone, which explains the self-limiting nature of laser cleaning. The interaction between light and matter is thermal and mechanical. Indeed, the localized rise in temperature leads to the formation of a plasma and propagation of shock waves in the matter have a photomechanical effect. On paints, made up of multilayer systems and more heterogeneous compounds than stone, the use of the laser remains marginal, especially in France. This is because the materials making up the paints, mineral and/or organic, are more or less photosensitive and of the same chemical nature as the layers of varnish, repainting or overpainting that must be removed. Nevertheless, research continues and technological innovations now make it possible to have different types of lasers that are better adapted to the problems of paint cleaning. The objectives and the methodological approach of paint cleaning, whether chemical or laser, are identical even if the mechanisms of destructuring of the material vary. In the first case, the aim is to obtain the dissolution or swelling of the materials to be removed by adjusting the solubility parameters of the solvents, whereas in the second case, the laser parameters, wavelength, energy, pulse mode, frequency and fluence are modified to modulate the photochemical, thermal and mechanical mechanisms.

Various optical evaluation methods, including optical coherence tomography, were used to verify the effectiveness and safety of the treatment, which have allowed to rediscovery the original polychromy of the frames.

2 MATERIALS AND METHOD

2.1 *Issenheim altarpiece: Stratigraphy study and matérials analysis*

The Issenheim altarpiece was probably executed in Strasbourg between 1512 and 1516 by the painter Matthias Grünewald and the sculptor Nicolas de Haguenau. This monumental polyptych which is composed of 11 wooden panels painted in tempera and oil, frames and polychrome sculptures was transported to Colmar in seventeen ninety-three (Figure 1). The altarpiece has been restored several times in the past. The last restoration campaign was carried out between twenty seventeen and twenty twenty-two, by a team of 21 conservators. It was supervised by a scientific committee of specialists in art history, materials and techniques. The C2RMF was present for the assistance and ensure the control of the state.

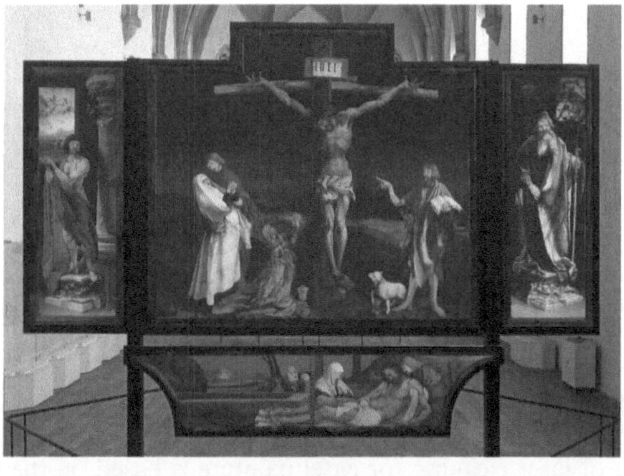

Figure 1. Alterpiece panels of Saint Sebastian, the Crucifixion and Saint Anthony. Painting panels Inv. 88. RP. 139; sculptures, Inv. SB. 69. ©Unterlinden museum, Colmar.

The frames had been repainted in the past. The investigations and the study carried out before the restoration showed that the original polychromy contributed to the aesthetic aspect of the work and that it was in a good state of conservation. So the scientific committee therefore decided to remove the overpaintings. The majority of the frames were cleaned with chemical and mechanical methods. But the polychromy of the frames of the Crucifixion, Saint Anthony, Saint Sebastian and the predella was too sensitive to solvents.

Investigations and analyses were carried out by Fourier transform infrared spectroscopy to characterize the different layers of the paint to be preserved and eliminated. The frames of Saint Anthony and Saint Sebastian were covered with a ground composed of a mixture of calcium carbonate and animal glue before being painted in tempera. The overpaint, undated, is composed with a ground of calcium carbonate in oil and a layer of oil paint (Figure 2).

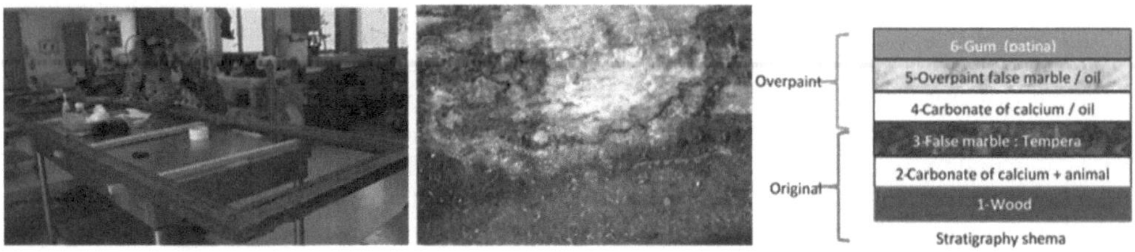

Figure 2. Frames of the side panels: Saint Anthony, Saint Sebastian. General view, binocular image and stratigraphy scheme. ©C2RMF Dominique Martos-Levif.

The crosspieces of the predella have been changed in the past and the polychromy is not original. The ground is composed of calcium carbonate and sulfate mixed with animal glue. The décor has been painted with oil and is covered with two layers of oil paint that would date from 1968 and 1986.The polychromy of these frames was quite fragile, and particularly sensitive to polar solvents (Figure 3).

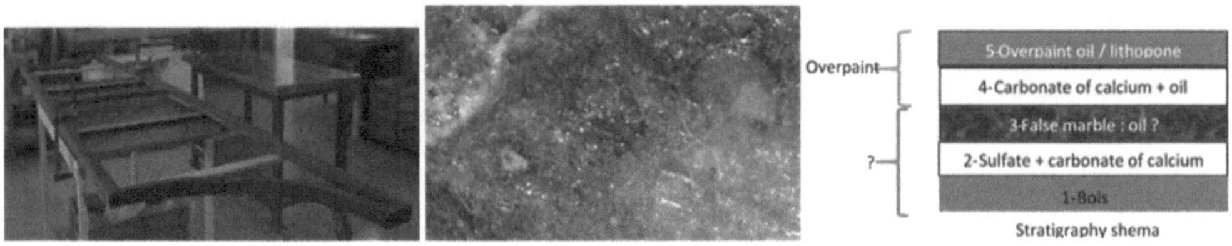

Figure 3. Frames of the side panels: Saint Anthony, Saint Sebastian. General view, binocular image and stratigraphy scheme. ©C2RMF Dominique Martos-Levif.

2.2 *Laser tests*

Two infrared lasers provided by ECP and El.En group were tested on the frames of the Issenheim altarpiece. The *Combo*, Nd: YAG, with a wavelength of 1064 nm, was tested in Short Free Running (SFR) mode, whose action is more photo-thermal due to pulse durations ranging from 30 to 110 microseconds and an energy of between 200 and 2000 mJ. The Long Q-Switch (LQS) mode with pulse durations of 100 nanoseconds and energy values of 150 mJ to 450 mJ for a more photo-mechanical effect. A laser *Light Brush 2*, Er:YAG, with a wavelength of 2940 nm, was also tested in Very Short and Short mode to break down hydroxyl bonds by thermal and/or chemically destructive action (Figure 4).

The main principle of laser cleaning for easel painting is based on the selective absorption of laser energy by contaminants on the painting surface. Different materials, such as dirt, varnish, or overpaint, have different absorption properties for specific laser wavelengths. By choosing the appropriate laser wavelength, conservators can target and remove the unwanted contaminants without damaging the underlying paint layers.

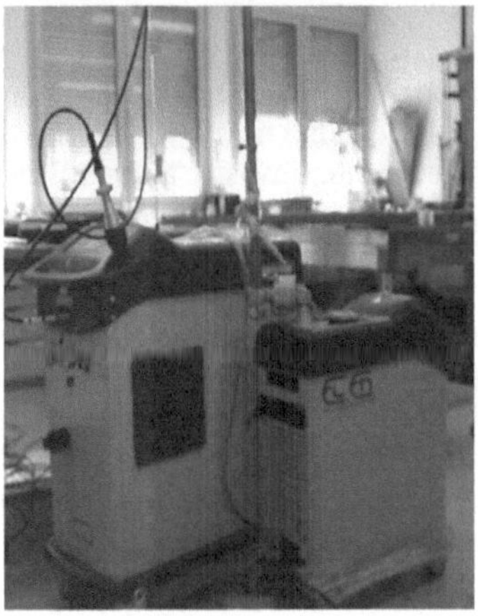
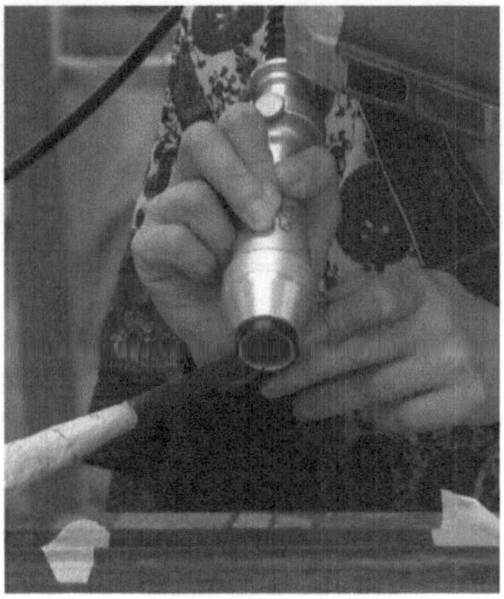

Figure 4. Laser Combo, Nd:YAG, λ 1064 nm; Light Brush 2, Er:YAG, λ 2940 nm, by El.En.Spa.

The laser cleaning process involves the following steps:

Selection of Laser Parameters: The laser system is configured with specific parameters such as wavelength, pulse duration, energy, and spot size. The selection of these parameters depends on the characteristics of the painting and the type of contaminants to be removed. The laser wavelength is chosen based on the absorption properties of the contaminants, aiming to maximize their absorption while minimizing the absorption by the paint layers.

Laser Irradiation: The laser beam is directed onto the painting surface in a controlled manner. The laser energy is absorbed by the contaminants, causing them to heat up rapidly. This localized heating results in the expansion or vaporization of the contaminants, effectively detaching them from the surface.

Contaminant Removal: As the contaminants absorb laser energy and undergo physical changes, they are either converted into gas or transformed into particulate matter. The gas or particles are then removed from the surface through a combination of mechanisms, such as thermal expansion, vapor pressure, and acoustic waves generated by the laser-induced ablation process.

Monitoring and Control: During the laser cleaning process, the conservator closely monitors the effects of laser irradiation on the painting surface. Techniques such as optical coherence tomography (OCT) or visual inspection are employed to assess the removal of contaminants and ensure that the underlying paint layers remain intact.

The success of laser cleaning relies on the careful selection and control of laser parameters to achieve optimal cleaning efficacy while minimizing the risk of damage. Factors such as the laser wavelength, pulse duration, energy density, and scanning speed must be tailored to the specific characteristics of the painting, including the type of paint, its sensitivity, and the nature of the contaminants.

By utilizing laser cleaning, conservators can selectively remove unwanted materials from easel paintings without the need for harsh chemicals or physical contact. It offers a precise and controlled method for restoration, allowing for the preservation of the artwork's original appearance while minimizing the risk of irreversible damage.

2.3 *Laser cleaning evaluation*

The assessment of the effectiveness and safety of the cleaning is necessary and essentially visual. It consists of observing the appearance and colour of the cleaned surface under different radiation or at varying levels of magnification and examining the residues present on the cleaning pads. Optical coherence tomography (OCT), a technique based on the physical principle of interference of light with matter, has also been used in recent years to assess the level of cleaning from a stratigraphic point of view. The use of Optical Coherence Tomography (OCT) in the examination of cultural heritage objects, specifically in the fields of art history, conservation, and restoration. OCT is a non-invasive imaging technique that allows for the visualization of the internal structure and stratigraphy of artworks. It can provide valuable insights into the authenticity, provenance, conservation conditions, and effectiveness of restoration treatments.

The study of stratigraphy in cultural heritage objects is crucial for understanding their composition and construction techniques. Traditionally, stratigraphic studies involved taking samples and observing them under a microscope. However, OCT offers a non-invasive alternative by providing cross-sectional images of the examined object's internal structure. It can reveal important information about the number and thickness of layers, the materials used, and the condition of each layer, aiding in the assessment of conservation needs and restoration interventions.

OCT operates based on the principle of light interference. It uses coherent light to measure the optical path length within a sample. There are two main OCT technologies: Time-Domain OCT (TD-OCT) and Fourier-Domain OCT (FD-OCT). TD-OCT involves mechanically scanning a reference mirror to modify the optical path length and obtain an interference pattern. FD-OCT, on the other hand, uses a fixed reference mirror and utilizes spectral or swept source methods to measure reflections simultaneously, allowing for faster imaging. OCT offers micrometer-scale resolution, making it suitable for studying multi-layered artworks like paintings, where layer thickness can range from a few microns to tens of microns. The axial and lateral resolutions are decoupled, with axial resolution defined by the light source's spectral width and lateral resolution determined by the focused Gaussian beam's minimal radius. The depth imaging capability of OCT is limited by the penetration depth of the light source, typically a few millimeters, and scattering within the sample. OCT has found various applications in cultural heritage, including the characterization of paintings, analysis of varnish and glaze layers, monitoring of varnish removal processes, identification of old and new varnish layers, examination of surface coatings, imaging underlying drawings, inspection of semi-transparent materials like jade and glazed ceramics, and examination of glass surfaces.

In conclusion, OCT has emerged as a valuable non-invasive technique for examining cultural heritage objects, providing insights into their composition, structure, and conservation needs. It offers detailed imaging capabilities and has been applied in various areas of art history, conservation, and restoration and especially for the control of laser cleaning restoration operations. It gives for the first time a new guaranty of non-invasive or in a control way to the curator, conservator and heritage scientist (Figure 5).

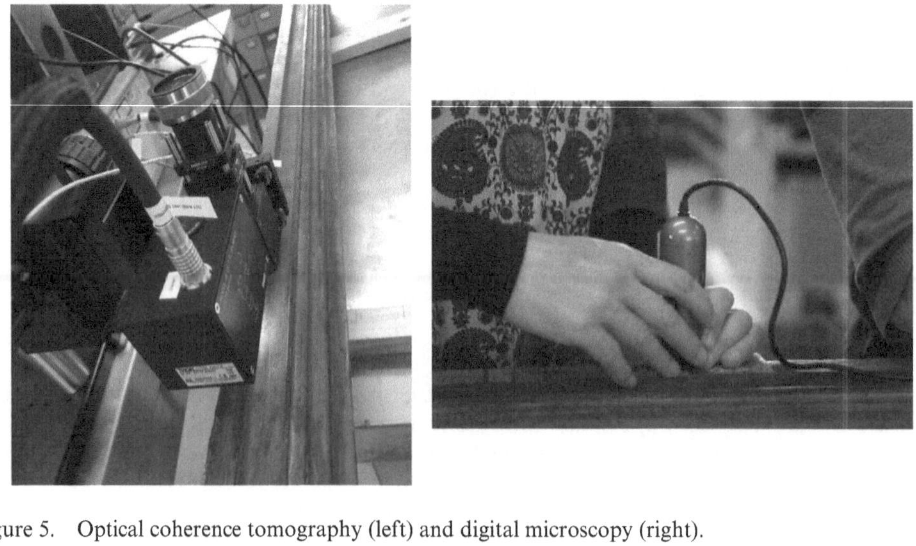

Figure 5. Optical coherence tomography (left) and digital microscopy (right).

3 RESULTS

Initial experimentation involved employing solvent-based and mechanical cleaning methods, but due to the intricate composition of the materials, the complexity of their mixture prompted us to explore an alternative approach utilizing laser technology. Consequently, laser cleaning tests were conducted, initially focusing on the Er:YAG laser. The rationale behind selecting this specific wavelength lay in its anticipated strong interaction with the -OH and -NH bonds, making it suitable for effectively removing the most recent repainting layers.

The test campaign with laser devices started with the Nd:YAG source type in the LQS (100 ns) and SFR (30-110 μs) pulse modes. Later, the laser was also tested on one of the frames to Er: YAG source type in the Very Short (150 μs) and Short (250 μs) pulse modes.

The tests are summarized as follows:

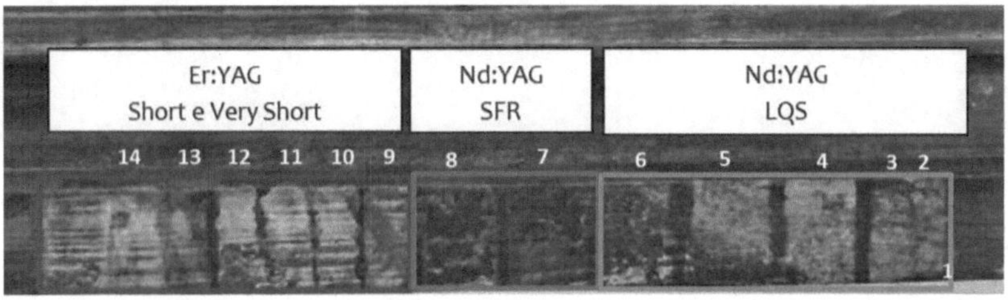

Figure 6. Results of laser tests in three different laser mode or wavelength.

Table 1. Parameters and results of the tests carried out in the painted area of Saint Sebastian.

Test N°	Laser	Wav.nm	Mode	E.gy mJ	Filter %	⊘ mm	Fluence J/cm²	Frequency Hz
1	Nd:YAG	1064	LQS	150	50	6,5	0.21	1
	Results: Insufficient action							
2	Nd:YAG	1064	LQS	150	50	5	0.39	1-2
	Results: Uneven action							
3	Nd:YAG	1064	LQS	150	50	6	0.26	2
	Results: Uneven action							
4	Nd:YAG	1064	LQS	150	50	3-4-5	1.07 – 0.39	2-3
	Results: Moistening of the surface with water before laser treatment - Effective							
5	Nd:YAG	1064	LQS	150	50	3-4-5	1.07 – 0.39	2-3
	Results: Effective – cleavage overpaint/ original layer - Use of scalpel punctually for residue removal							
6	Nd:YAG	1064	LQS	450	75	4-5	0.9 - 0,57	2-3
7	Nd:YAG	1064	SFR	400	50 -75	2,5 - 3	4 - 2 2.85 – 1.4	2
	Results: increasing the fluence improves the removal of the overpaint – The original paint is sensitive							
8	Nd:YAG	1064	SFR	350	50	3	2.5	2
	Results: moistening of the surface with water before the laser. The original paint is sensitive							
9	Er:YAG	2940	Very Short	150		5	0.76	5
	Results: residues removal with H2O /isopropanol (50/50) – 1 pass - The original paint is sensitive							
10	Er:YAG	2940	Very Short	200		5	1.02	5
	Results: residues removal with H2O /isopropanol (50/50) – 2 pass - The original paint is sensitive							
11	Er:YAG	2940	Very Short	250		4-5	2 – 1.27	5
	Results: residues removal with H_2O /isopropanol (1/1) – 3 pass -The original paint is sensitive							

The OCT images (Figures 7–8), which make it possible to visualize the overpainting and the original pictorial layer thanks to their difference in optical index, show that the cleared part presents an irregular surface aspect and an important scattering of IR radiation, due to the migration of solvent in the colored layer. The images taken under high magnification confirm that the cleaning is incomplete. In addition, the original paint is very sensitive to solvents.

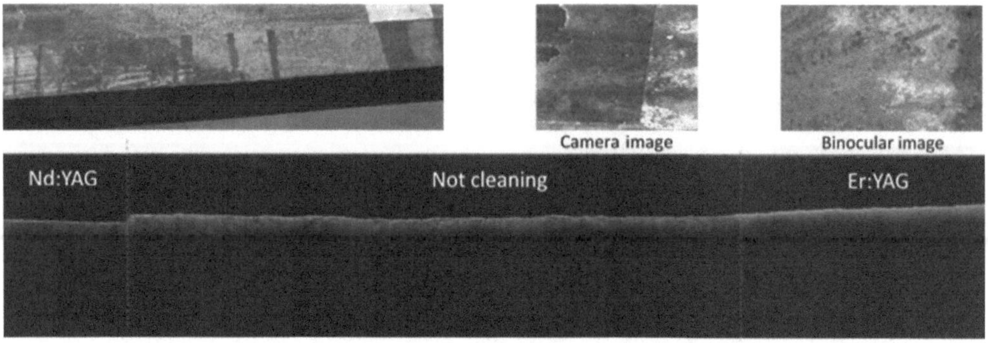

Figure 7. Side panel frame test, Saint Anthony. Laser Er:YAG, Very Short (150 µs), E=150 mJ, f=Hz 5, ø 5 mm. After removal with swab of H_2O + isopropanol (1/1). Progressive removal of the overpaint, sensitivity of the original paint layer to solvents. Laser Nd:YAG, LQS (120 ns), E=150 mJ, f=2-3 Hz, ø 5 mm, filter 50%. Evidence of spallation of overpainting and cleaning original paint.

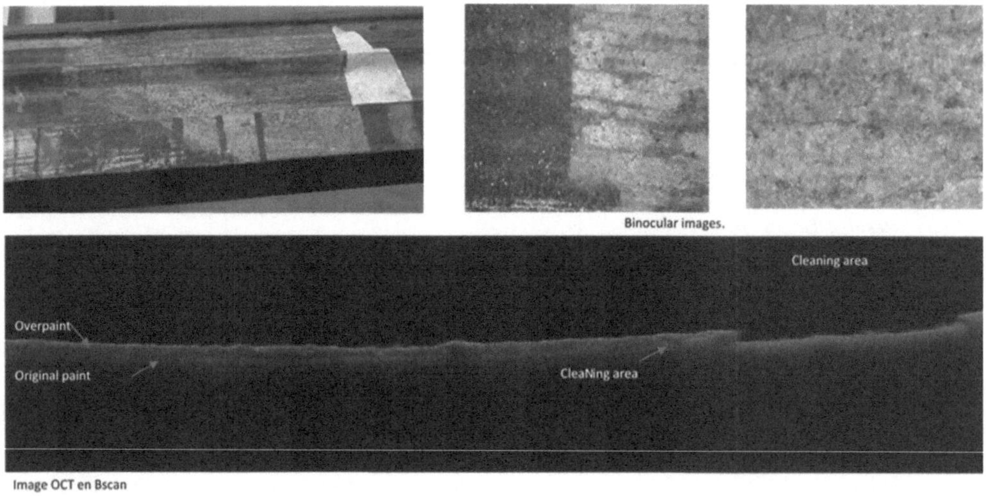

Figure 8. Side panel frame test, Saint Antoine. Laser Nd:YAG, LQS (100 ns), E= 150 mJ, f= 2-3 Hz, ø 5 mm, Filtre 50%. Demonstration of the spallation of overpainting.

4 CONCLUSIONS

In recent years, laser cleaning has emerged as a promising and innovative technique in the field of heritage conservation and restoration. The study of laser cleaning of the painted frames of the Issenheim altarpiece in Colmar presented in this paper represents a significant step forward in the application of this technology. One of the key factors that made laser cleaning a preferred option for this restoration project was its ability to selectively absorb laser energy by contaminants on the painting surface. This selectivity ensured that the original polychromy of the frames remained preserved while effectively removing overpaintings and residues. The researchers meticulously tested various laser parameters to optimize the cleaning process. The use of Nd:YAG and Er:YAG lasers in specific modes proved to be particularly effective in achieving the desired cleaning results. Notably, the Nd:YAG laser in Long Q-Switch mode demonstrated impressive success in selectively removing contaminants between the original layer and the repaint. To evaluate the safety and efficacy of the laser cleaning process, the researchers employed optical coherence tomography (OCT), a non-invasive imaging technique that provided valuable insights into the composition, structure, and conservation needs of the frames. The OCT imaging capabilities allowed for a detailed examination of the cleaned surface, aiding in the assessment of the preservation of the original paint layers. The study also emphasized the importance of closely monitoring the laser cleaning process. Researchers found that in some cases, moistening the surface with water before laser treatment enhanced the effectiveness of the process, highlighting the need for precision and control when applying laser technology. The successful outcomes of this research have sparked further investigations in the understanding of the mechanisms of destructuring induced by lasers on artworks, particularly paintings. The ongoing research at the C2RMF (Centre de Recherche et de Restauration des Musées de France) in this area underscores the continuous commitment to refining and expanding laser cleaning techniques. However, while laser cleaning holds great potential for heritage restoration, transferring these skills to the community of restorers is an essential challenge. The effective utilization of laser technology as a complementary cleaning tool requires specialized training, both in continuing education for experienced restorers and as part of initial training for aspiring professionals. Collaboration with heritage laboratories is crucial in providing scientific assistance to professionals seeking to incorporate laser cleaning into their restoration practices. Such partnerships foster a deeper understanding of the laser's capabilities and limitations and promote responsible and effective use of this advanced technology. In conclusion, the research presented here not only contributes to the preservation and restoration of the Issenheim altarpiece frames but also paves the way for future advancements in laser cleaning within the broader context of cultural heritage conservation. As technology continues to evolve, laser cleaning, coupled with non-invasive evaluation methods like OCT, offers immense promise for safeguarding the world's precious cultural treasures for generations to come.

ACKNOWLEDGEMENTS

Our thanks to the Musée Underlinden for the concrete collaboration.

REFERENCES

Bromblet, P. & Vieweger T. 2005. Le laser de nettoyage de la pierre et la restauration des sculptures, In: *Pierre actual*, n° 829: 86–94.

Maxime Lopez, Évaluation et développement d'une technique de nettoyage des peintures par procédé d'interaction laser-matière. Thèse de doctorat de CY Cergy Paris université, école doctorale sciences et ingénierie, 2020.

Menu, M., Ezrati, J.J., Laval, E., Pagès, S., Principaud, A., Rioux, J.P., Walter, P., Welcomme, E. & Nowik, W.2007. Analyse de la palette des couleurs du Retable d'Issenheim par Matthias Grünewald, C2RMF, Colmar: Musée d'Unterlinden.

Koch Dandolo, C., Lopez, M., Bai, X. & Detalle, V. 2019. Examen de la structure des objets de patrimoine culturel: tomographie à cohérence optique. *Photoniques* 95 (janvier-février): 24–28.

Lasers in the Conservation of Artworks XIII – Siano & Ciofini (eds)

A new approach for the restoration of gilded surfaces: Revealing original decors of the "Bargueño" (16th century) by Er: YAG laser processing controlled by Optical Coherence Tomography

S. Courtier, M. Lopez & X. Bai
Center of Research and Restoration of Museum in France (C2RMF), Paris, France

V. Detalle
CY Cergy Paris Université, SATIE CNRS UMR8029, Neuville sur Oise, France

ABSTRACT: During the meticulous restoration of "the Bargueño," an exquisite gilded travel cabinet dating back to the 16th century and hailing from the historic Château de Pau in France, an intricate conundrum surfaced. It was a puzzle involving the presence of two superimposed layers bonded by the same adhesive, a quandary insurmountable by conventional restoration techniques. The physical-chemical attributes of the materials posed a formidable challenge in the quest for an effective remediation strategy. Analysis revealed a duality in gilding techniques; the wooden portions were adorned with water gilding, while the ivory sections exhibited oil gilding. To address this exceptional restoration challenge, a groundbreaking method was embraced. A feasibility study was conducted employing a pulsed Er: YAG laser operating at 2940 nm in a µs regime, meticulously applied to gilded ivory prototypes. This laser methodology facilitated the precise ablation of layers, completely circumventing direct material contact. The degree of precision could be finely tuned by manipulating laser parameters. The laser ablation assessments on prototypes, varying in gilding layer thickness, were executed under the vigilant watch of optical coherent tomography (OCT). This innovative approach permitted the meticulous removal of layers during the restoration process of the ornate twisted and gilded columns, unequivocally affirming its efficacy in the preservation of gilded surfaces.

1 INTRODUCTION

During the transitional period spanning the fifteenth and sixteenth centuries, a distinctive style known as the plateresque emerged. This style drew inspiration from various artistic currents, namely the Italian Renaissance, the late Gothic, and Mudejar art. Initially conceived as ornate altarpieces, the plateresque design extended its exuberant embellishments to encompass facades of both furniture and edifices. Notable examples of this ornate aesthetic include the façade of the Santa Cruz Hospital in Toledo.

The plateresque style exhibited a lavish array of ornamentation, skillfully blending elements of architecture and goldsmithing, with the Spanish term "platero" denoting a goldsmith. During this era, the frontages of furniture pieces underwent a transformation, adorned with sculptures, columns, medallions, and other intricate details. Additionally, sumptuous crimson velvet found its way beneath the metallic embellishments. The Italian influence on this style continued to gain prominence.

By the sixteenth century, Spanish travel furniture, particularly the renowned "Bargueño," became an essential commodity. Consequently, it comes as no surprise that artisan guilds formed to meet the demand during colonial expeditions to the New World.

DOI: 10.1201/9781003386872-15

The Bargueño, housed within the National Museum and Castle of Pau, embodies the essence of Mudejar art, primarily produced in the province of Toledo. Exhibiting a plateresque demeanor, it derives inspiration from the Italian Renaissance, late Gothic, and Mudejar artistic traditions. This remarkable piece of furniture comprises two key components: a base housing a wooden box, sealed with a flap. The chest's interior is designed akin to an altarpiece, featuring registers of elements distributed across five distinct sizes and functions. Four doors, each equipped with key locks, serve to embellish the four corners, while 22 drawers, in various configurations, include seven secret compartments. The front of the Bargueño is resplendent with a diverse array of decorations, including sculptures, arcades, columns, medallions, miniature frames, inlays, painted motifs, shells, and religious symbols such as Christian crosses and celestial clouds—inspired by the portable chests of pilgrims en route to Santiago de Compostela.

However, an unintended application of oil gilding led to a shift in the gilded work's appearance and artistic expression. Thus, it becomes imperative to undertake its removal while ensuring the preservation of the original surface's integrity.

The mechanical technique of shrinkage poses a considerable challenge in this context, owing to the difficulty of distinguishing between layers in the stratigraphy visually. This challenge is exacerbated by the close colorimetry of the preparations. Furthermore, the extremely thin binders, measuring between 0.18 to 0.21 μm, introduce an additional layer of complexity to the shrinkage process. In addressing these issues, laser cleaning techniques have been applied in cultural heritage preservation since 1992 (Lopez, et al. 2020). By meticulously selecting laser parameters such as wavelength, pulse duration, and fluence, the laser-material interaction can selectively remove undesirable substances like dirt, corrosion layers, alterations, paint, and accretions from artworks (Wilkie-Chancellier & Detalle, 2020, Lopez 2020, Sawicki et al. 2009). The most commonly employed laser for conservation cleaning has been the Nd: YAG laser with wavelengths of 1064 nm, 532 nm, and 266 nm. However, when dealing with similar chemical compositions between the material to be removed and the material to be preserved, challenges arise.

This paper introduces a novel approach that employs the Er: YAG laser with a wavelength of 2940 nm for clearing gilded surfaces. Given the nature of the laser cleaning process, it allows for more controlled and less damaging cleaning operations. The objective of this study is to investigate the interaction of Er: YAG laser radiation with oil gilding layers and determine if clearance can be achieved at different levels within the same layer.

In antiquity, various techniques influenced by Carolingian and Roman methods were practiced to varying degrees across regions. Alchemists, clerks, and clerics actively exchanged and disseminated their knowledge, as evidenced by historical treatises like that of the monk Theophilus from the early 12th century, Jean le Begue's "Different receipts on the colors," and A. Félicien's "Dictionary of the terms proper to architecture, painting and other arts dependent on it." During the Middle Ages, two prevalent gilding techniques were employed on gilded wooden surfaces of decorative art furniture: gilding with mordants and gilding with distemper (de l'Escalopier 1863, Le Begue 1431, Félicien, 1676). These techniques led to a plethora of recipes and required the use of natural binders and mixtures, including substances like gum Arabic, gum tragacanth, animal and fish glues, casein, egg white, vegetable oils, egg yolk, garlic, and honey, either individually or in combination.

Beginning in the 14th century, political and commercial ties between Italy, Catalonia, the Kingdom of Aragon, and Valencia facilitated the exchange of knowledge and techniques. Cennino Cennini, among others, documented these ancient processes, outlining the use of mordants for gilding (Cennino Cennini, 2009 edition). The technique of gilding with mordant involves the application of one or more layers to create adhesion on the support while isolating it and providing a surface for attaching the gilded material. Mordants, whether aqueous or fatty, required a drying period to achieve material hardening upon exposure to oxygen. The longevity of the mordant varied, lasting from 8 days to 4 days or overnight before the gold leaf was applied.

This research presents a novel restoration approach to clean gilded surfaces on wood. By employing Optical Coherence Tomography (OCT), it becomes possible to differentiate between various layers within the gilding mixture stratum. This innovative technique offers a solution previously unattainable through traditional clearing methods. With the aid of this protocol, one can thoroughly investigate and control the effects of the laser beam on both the gold leaf and the organic layers comprising the gilded surface on wood. This protocol serves the purpose of optimizing cleaning processes without entirely removing the layer, thereby preserving any concealed evidence of repaints that may elude the naked eye.

2 MATERIALS AND METHODS

2.1 *Observations of the gilding surface*

In a previous study, the stratigraphy of this furniture was investigated and shown in Figure 1. The entire golden surface consists of a gilding mixtion. The film forming is organic, opaque, orange, and topped with a gold leaf. The study of the miniature frames, the carved and gilded wooden ornaments, and the turned and gilded ivory ornaments reveals the presence of important gaps on the gilded surface.

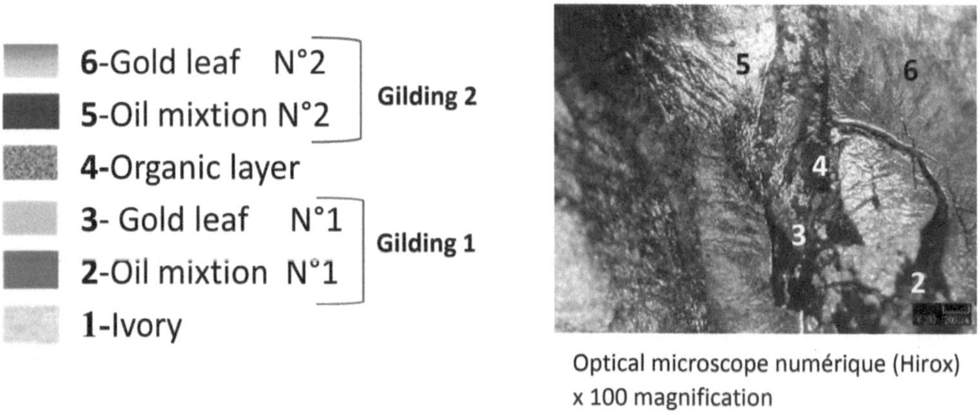

Figure 1. Stratigraphy and observations of the gilding surface.

Mechanical wear of the gold leaf, loss of cohesion of the mix on the gilding in tempera and oil, gaps in the preparatory layers and structural damage. These alterations are linked in this case to the superposition of different materials treated with different techniques. The study of the twisted columns in gilded ivory indicates the presence of a single gilding mix on the relief of the flutes and of two gilding mixes superimposed in the hollows of the twists. Thus, under the modern mixtion, which we will call "mixtion N°2," we can distinguish a mixtion that we call "mixtion N°1."

The mixtion gilding in the hollows of the twisted columns is made up of a gold leaf, worn in places but in a very good state of conservation. The adhesive is of the same nature, so chemical removal is impossible, and mechanical removal is dangerous, given the thinness of the layers, if the original gilding is to be preserved. It should be noted that the original decoration is made with mixed gilding techniques. It can be seen that the tempera gilded surfaces are covered with the same gilding in mixtion N°2, which indicates that this intervention is not original.

2.2 *Discussions and propositions for the conservation and the restoration*

The original gilded surfaces are treated with different techniques that give the decorations an abundance of play of material, relief, brilliance, and color that give a preciousness to the art object. It should be noted that during the dismantling and disassembly of the three columns that had been reassembled inverted during a later intervention, we were able to observe the reverse side of them and note that no intervention was visible. The gilder had redecorated the surface without dismantling the columns, so the gilding with the mixtion, which was only located in the concave zones of the columns, was the original decoration, shown in Figure 2. This testimony allowed us to observe that the gold leaf N°1 was worn but was in a satisfactory condition for a museum presentation. In order to reconstitute the hidden decoration of the twisted columns, prototypes were made. This consisted of making prints with the "Gros blanc marqué" (Large White, definition) covered with gold leaf and painted. This reconstitution accompanied and validated a proposal for intervention which consisted of removing the mixtion and gold leaf N°2 on the reliefs of the ivory column. Initially, comparisons of removal techniques were necessary.

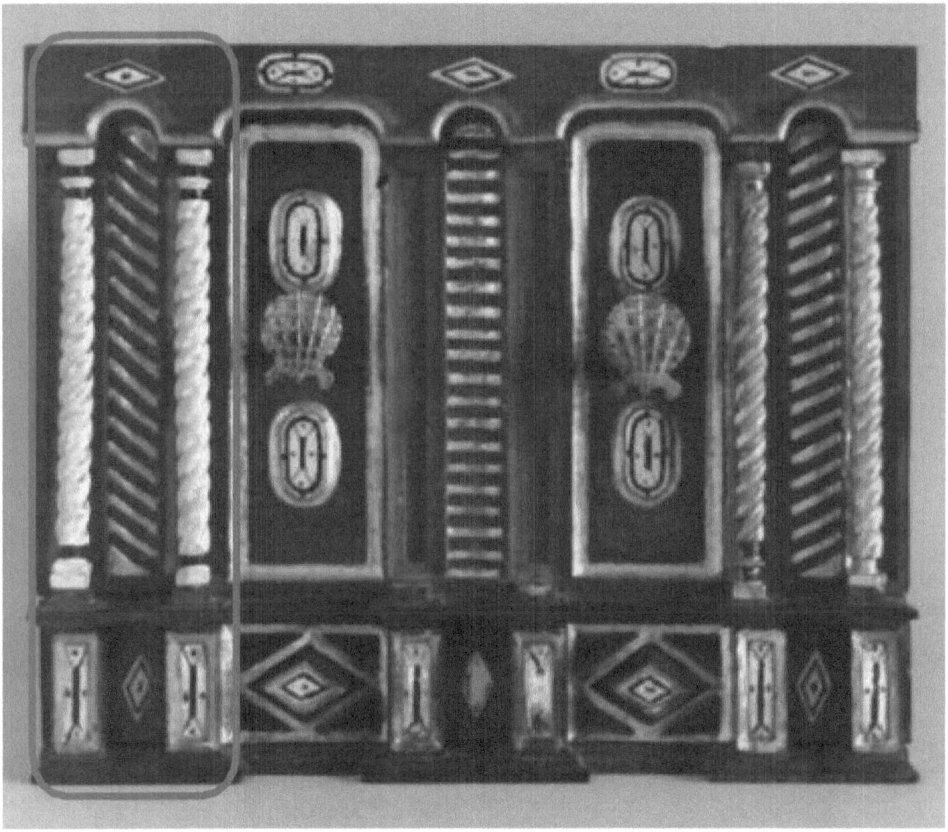

Figure 2. Drawer N°13, reconstruction of the original decoration on the ivory columns.

2.3 *Instrumentation and examination of cleaning protocol*

A pulsed microsecond Er: YAG laser (El.En.'s LightBrush2 500 mJ) of a wavelength at 2.94 μm was used. This laser can be operated with three pulse durations in the range of a hundred of microseconds. The output laser pulse energy can be tuned between 50 and 500 mJ. It is equipped a handpiece allowing the operator to change the distance between the output and the surface in order to control the ablated zones.

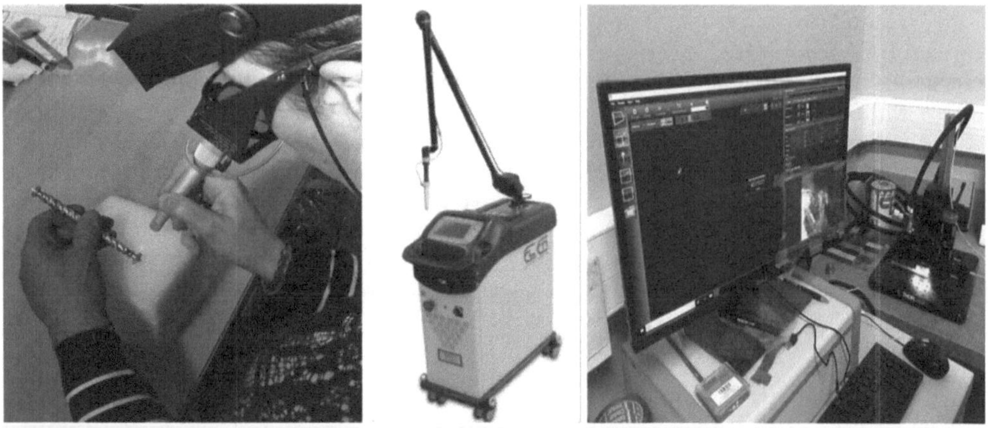

Figure 3. Conservator used the laser on the gilding surface, photographic images of the Laser ER:YAG and Optical Coherence Tomographic (OCT).

For getting the control, the optical coherence tomography (OCT) imaging was used to obtain a referential and examined the cleaning results. Presented in Figure 3. A commercial SD-OCT (Thorlabs GAN220-OCT Base Unit with a Thorlabs OCTP900M Scanner) equipped with a super luminescent diode (λc = 912 nm, $\Delta\lambda$ = 803.8 nm - 1021.5 nm). This system allows the measurement at high frequency (36 kHz) with an axial resolution of 3 µm in air and 2 µm in a transparent material such as varnish (n = 1.5) in a non-invasive way. By acquiring a line (side-by-side measurement points or A-scan), one can visualize a slice of the sample (called B-scan) and measure the thickness of the different layers of the stratigraphy. By acquiring side-by-side B-scans, one can obtain a volume representation in which the operator can choose any 2D plane in any direction and also study surface topography with a lateral resolution of 6.5 µm in Y and 3.3 µm in X.

2.4 *Samples*

Firstly, the specimens were made with the same stratigraphy and representative of the problem encountered on the Bargueño golden ivories, presented in figure N°4. The ageing of these surfaces was carried out in two stages, the first ageing was carried out on mix N°1 and the second, made up of the same parameters, on the two superimposed mixes. These samples deposited in the ageing chamber made it possible to obtain ductibility and reticulation of the N°1 mix. In order that the Er: YAG laser tests could be carried out on surfaces close to the gold surfaces of the object. In a second step, with the observation under optical coherence tomography (OCT), we were able to measure the thickness of each layer in order to establish a reference. The gold leaf N°1, the mixtion N°1, the interface, the gold leaf N°2 and the mixtion N°2 were characterized, as well as the two gildings and their stratigraphy. This is shown in Figure 4. In this case, it is important to note that these observations were possible thanks to the openings in the thickness of the material, present in the stratum allowing the penetration. These openings in the material correspond to gaps, cracks, crazing, oxidation and other types of alterations. The Stratigraphy of the mockup N°2 presented in Figure 5.

2.5 *Comparison of cleaning techniques*

Upon conducting a secondary evaluation, it became imperative to juxtapose the outcomes of material clearance tests employing both a scalpel and a laser. The results, as depicted in Figure 6 and scrutinized via Optical Coherence Tomography (OCT), unearthed distinctive disparities.

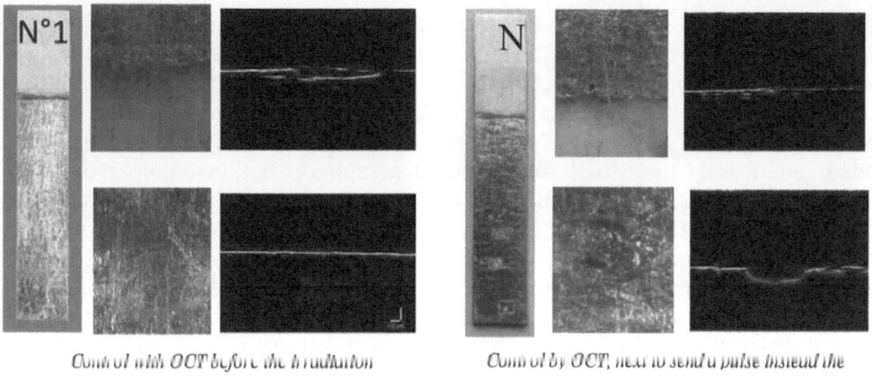

Figure 4. Control with the OCT, for compare the removal layers with ER:YAG.

Figure 5. Stratigraphy of two gildings mixtion for the ivory of the mockup.

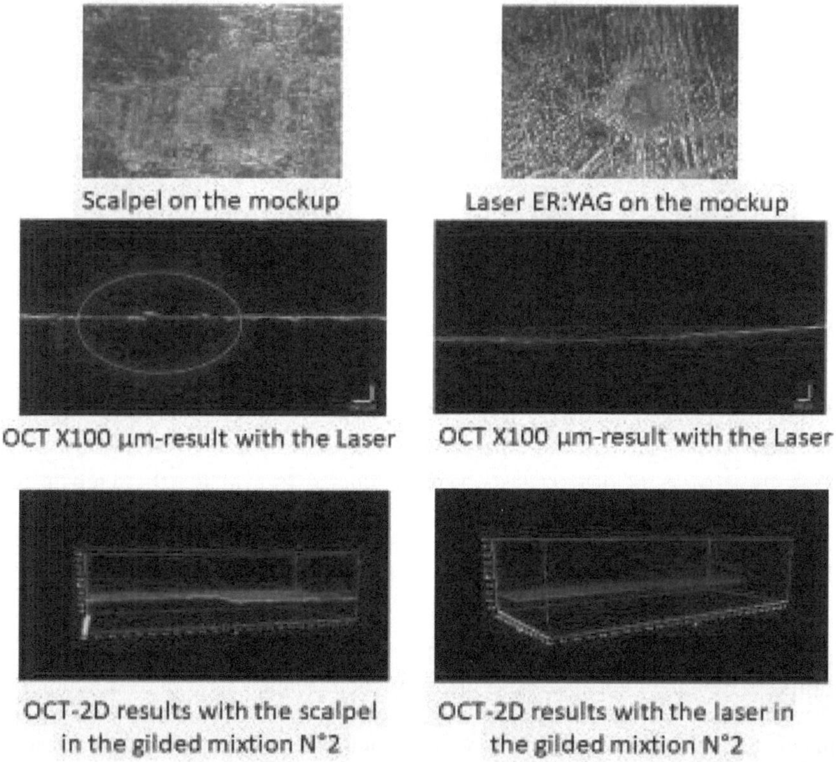

Figure 6. Observation of removal with the Laser Er: YAG and the scalpel by OCT.

In the case of scalpel utilization, we observed material fragmentation, irregularity in the removal process, and discernible tool marks. Conversely, laser-based clearance trials revealed circular releases measuring 2 millimeters in diameter, corresponding to the beam's shape, at a depth of 0.013 μm. The laser's pulsatile action ensured residue-free outcomes, with uniformity throughout the layer's thickness. Noteworthy is the temporal factor; the scalpel procedure demanded significantly more time and imposed greater restrictions compared to Er:YAG laser treatment. To illustrate, one centimeter of clearance with a scalpel, conducted under binocular scrutiny, necessitated 1 minute and 40 seconds, while the laser achieved the same in a mere 7 seconds, with the added advantage of not requiring return visits. Consequently, the laser technique was deemed compelling and selected for the release of Mixture No. 2.

The selection of parameters for laser testing was meticulous, involving three trials on a mockup utilizing the Er:YAG laser. These trials, detailed in Figure 6, were integral to our assessment of removal methods.

2.6 *Laser cleaning applied to the Mockup*

The Er:YAG laser, featuring a fixed beam diameter of approximately 50 mm and a pulsating mechanism at each focal point, each lasting 2 seconds, was subjected to close scrutiny. Monitoring at each pulse was accomplished through Optical Coherence Tomography (OCT) to ensure precision in beam impact. Xenon Power supply XPS-100 Xenon Lamp and a Nikon DS-RI1 camera, supplemented by NISS-Element Software, facilitated this meticulous oversight. Further examination was carried out using a Nikon Eclipse LV 100 ND optical microscope.

In the first trial, we employed "Short mode," characterized by a pulse duration of 150 μs, a 5Hz frequency, and 100 mJ energy, with fluence set at 3 to 5 J/cm², all maintained at a constant distance of about 5 centimeters. Results are eloquently depicted in Figure 7.

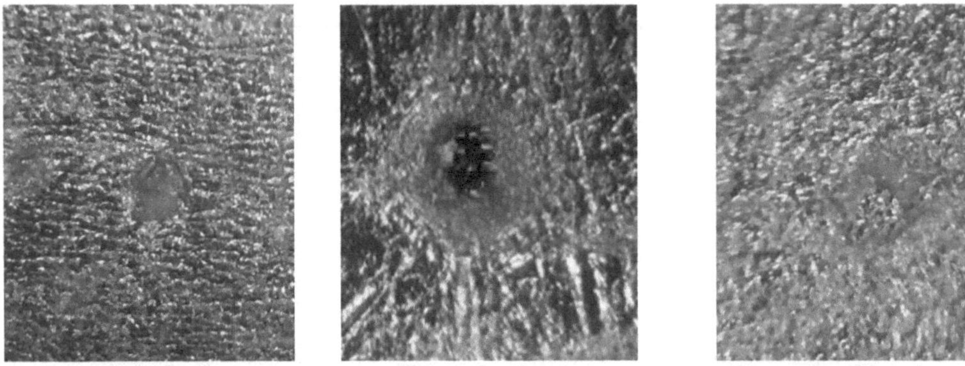

Figure 7. Er:YAG results with different modes.

Subsequently, a series of distinct tests were conducted to evaluate the removal of gilding No. 2. Test 2 employed "Long mode," mirroring the parameters of Test 1, while Test 3, denoted as "Very short" mode, maintained the same pulse duration, frequency, energy, fluence, and distance conditions. These experiments illuminated variations in the removal process, particularly in the case of dark surfaces where calcination of binders became apparent due to heightened absorption.

The most compelling results emanated from Test 3, wherein "Very short" mode engendered an abrasion process blending mechanical and thermal action. Employing a pulse duration of 150 μs, a frequency of 5 Hz, 100 mJ energy, and a fluence of 3 to 5 J/cm² at a consistent distance of approximately 5 centimeters, this approach yielded homogeneous ablation in a single pass, achieving a 90% removal of Mixture No. 2 while preserving a delicate 10% film. This 10% threshold was strategically retained to serve as a visible demarcation, ensuring the safeguarding of gold leaf No. 1.

It's important to note that the ablation thickness ranged between 0.35 and 0.73 µm, which underscored the meticulous control necessitated by the Optical Coherence Tomography. This protocol, integrating laser and OCT with an optical microscope, enabled precise measurement of ablation thickness.

3 RESULTS

Laser removal on the columns for the restoration procedure creates physical and chemical reactions visible to the naked eye. These changes are observed by the modification of the color, brightness, opacity and of pulsed particles result in visible deposits on clothing and surfaces. The impact of the laser beam is visible to the naked eye, however, thickness control is not possible with the naked eye and optical coherence tomography is the appropriate measuring device, it allows a measurement of the thickness of the ablated layers. The protocol set up to compare and measure the ablation thickness of the surface consists in coupling the laser with the OCT and the optical microscope. Presented in the Figure 8. The release tests of gilding N°2 have shown that different levels of ablation can be achieved.

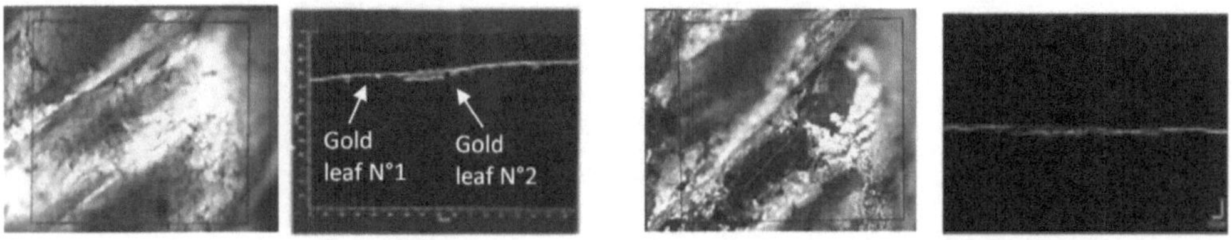

Figure 8. After pulsing with the laser Er: YAG, we observed under the OCT two gold leafs.

The observation under OCT, also allowed us to note that it is possible to ablate the layer on different levels, without changing the parameters of the Er: YAG, the factors: distance and time of pulsation are elements to be taken into account. Figure 10. During the first ablation, level only the gold foil is pulsed.

The second ablation level allows the gold foil to be pulsed and 20% of the mix n°2.

The third level allows the ablation of the gold foil and 90% of the mixtion n°2. The choice turns to a level 3 release, the veil of 10% of mixtion allows to obtain a threshold visible to the naked eye and protects the gold leaf n°1 (Castillejo et al. 2002, Ciofini et al. 2016, Castillejo, et al. 2003).

It is possible to obtain three different results A: we pulsing the gold foil only, or we pulsing the gold leaf. B: The gold leaf and 40% of the mixtion No. 2. C: we pulsing the gold leaf and 90% of the mixtion and the gold leaf No. 2. and without altering the underlying layers, the different layer thicknesses vary between 0.1 and 0.7 µm, visible in the Figure 9.

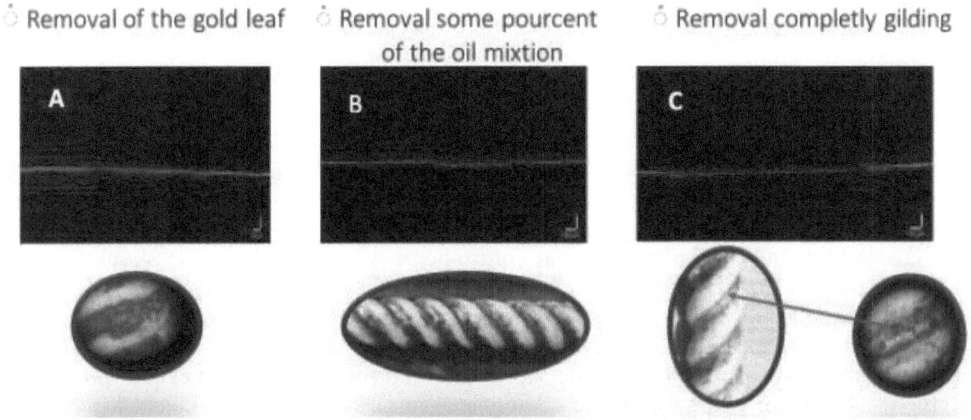

Figure 9. By the OCT, results of the different levels in the mixtion N°2.

The application of laser removal in the restoration process of columns brought about visible physical and chemical transformations, discernible to the naked eye. These transformations manifested through alterations in color, luminosity, and material structure.

He intervention yielded a nuanced revelation of the original decoration, unveiling stylistic references and showcasing the gilder's artistry on the gilded surfaces of the Bargueño. The condition of the gilded and painted surfaces post-removal of paint was deemed satisfactory, with selective retouching performed using gum Arabic, mica, or watercolors.

The juxtaposition of results before and after clearance is illustrated in Figure 10.

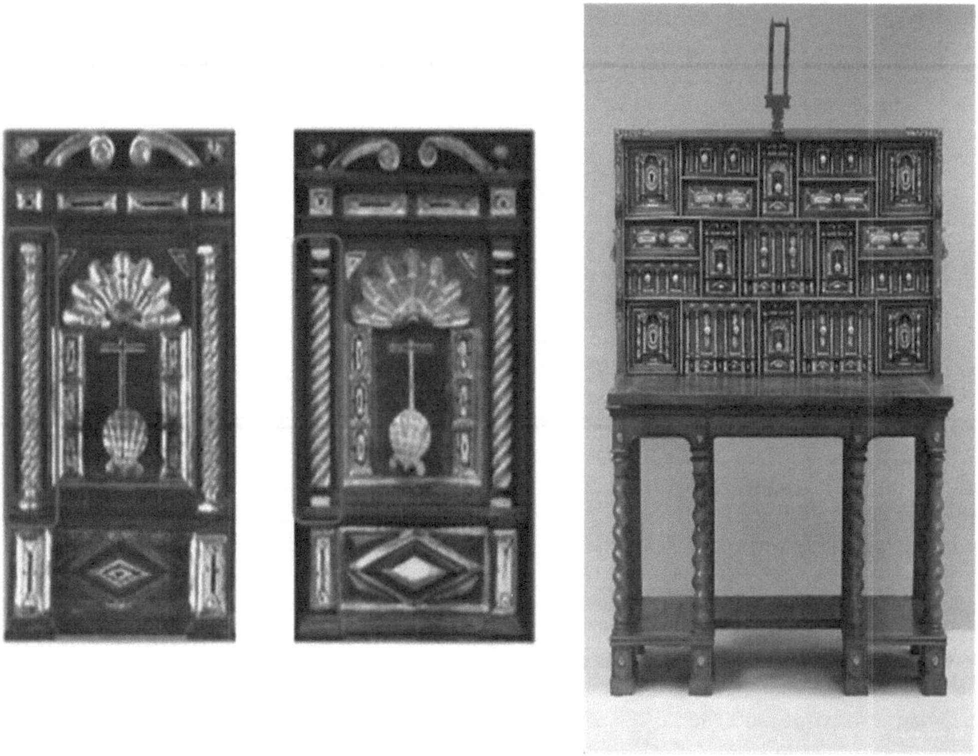

Figure 10. Results of the restoration of the gilding surfaces.

Er:YAG laser ablation yielded satisfactory outcomes, albeit with conspicuous alterations discernible through visual inspection. These changes were evident in terms of color modification and surface brightness.

The clearance procedure, while promising, revealed specific challenges yet to be addressed within traditional clearance methodologies. Future investigations should encompass the application of Er:YAG laser ablation to coatings, oily repaints, metallic repaints, and oxidized materials. Additionally, research aimed at controlling material destructuring should be pursued vigorously. The inclusion of solvents to enhance material interaction stands as a promising avenue for further refinement of clearance protocols for gilded surfaces.

The potential for collaboration between laser technology and Optical Coherence Tomography, as showcased in this study, underscores the need for systematic thickness control for precision. This alliance empowers the comprehensive measurement of ablated layer thicknesses, thereby advancing the field of art restoration.

4 DISCUSSION

The research presented in this document delves extensively into the application of Er:YAG laser technology for the intricate and challenging task of restoring cultural artifacts, with a particular focus on gilded surfaces. The comprehensive study has unearthed numerous noteworthy findings that shed light on the potential of this laser technology in the realm of art restoration.

To begin with, the laser clearance tests conducted as part of this research have vividly demonstrated the remarkable precision and efficacy of the Er:YAG laser when it comes to removing gilding and paint layers from delicate surfaces. What stands out prominently is the laser's focused 2-millimeter diameter beam, which exhibited minimal impact on the underlying substrates, showcasing its suitability for preserving the integrity of the original artifact.

The study embarked on an in-depth exploration of three distinct laser operation modes, each characterized by variations in pulse duration and energy levels. Among these modes, the "Very Short" mode emerged as the most effective, with its specific settings of a 150 μs pulse duration, 5Hz frequency, and 100 mJ energy. This particular mode proved to be adept at achieving homogeneous removal of gilding while minimizing any harm to the substrate. Notably, Optical Coherence Tomography (OCT) played a pivotal role in monitoring the laser's influence on the material, providing precise measurements of ablation thickness and enabling the differentiation of various ablation levels.

Furthermore, the "Very Short" mode exhibited a unique ability to preserve the gold leaf, retaining a remarkable 10% of the original gilding. This preservation was achieved while creating a visible threshold discernible to the naked eye, showcasing an unprecedented level of precision that traditional restoration methods struggle to match. Importantly, the utilization of laser-based removal led to material destructuration, characterized by physical and mechanical alterations, including the development of micro-fractures.

These micro-fractures, as it turns out, played a pivotal role in facilitating solvent penetration and the subsequent removal of residues, underscoring the multifaceted advantages of laser technology in art restoration. Further analysis through Scanning Electron Microscope (SEM) uncovered the presence of different fillers in the gold mixtures, highlighting the importance of considering the ductility and texture of layers when implementing laser technology in restoration efforts.

In summary, this research not only showcases the immense potential of Er:YAG laser technology in the restoration of cultural artifacts, especially those with gilded surfaces but also provides valuable insights into the nuanced techniques and considerations required for achieving optimal results in this delicate and important field.

5 CONCLUSIONS

This work emphasizes the Er:YAG laser's indispensable role in augmenting the array of tools at the disposal of conservators and restorers. Its unrivaled precision, impeccable cleanliness, and minimal utilization of chemical agents firmly establish it as a complementary asset to conventional restoration techniques. Furthermore, this research ardently promotes the comprehensive comprehension of an artifact's material history through non-invasive methods before embarking on the restoration journey. This knowledge should not only steer material research but also, when deemed necessary, guide invasive analyses. Although laser clearance techniques have undeniably excelled on gilded surfaces, this study hints at untapped research avenues, especially in the realm of other materials such as coatings, oily repaints, metallic repaints, and oxidized materials. Furthermore, it encourages the exploration of approaches aimed at managing material destructuration and optimizing interactions through solvent-based release protocols. In conclusion, this study vividly illustrates the enormous potential of modern laser technology in the field of art restoration while underscoring the significance of melding these innovative methodologies with established traditions to achieve the utmost precision and preservation in the restoration of our invaluable cultural heritage.

ACKNOWLEDGMENTS

The authors extend their sincere thanks and gratitude to the curator Madame Pebay Clote of the Domaine and château de Pau, France for the trust it has shown and her participation in the reflections. The restauration and the completely scientifique protocol could be consultated in the Intervention report on the restoration of the Bargueno by Stephanie Courtier (Courtier, 2019). Thanks to El.En. for lending us the Er: YAG laser to perform this operation.

REFERENCES

Castillejo, M., Martín, M., Oujja, M., Silva, D., Torres, R., Manousaki, A., Zafiropulos, V., Van den Brink, O.F., Heeren, R.M.A., Teule, R., Silva, A., & Gouveia, H. Analytical study of the chemical and physical changes induced by KrF laser cleaning of tempera paints, *Analytical Chemistry* 74(18): 4662–4671.

Castillejo, M., Martín, M., Oujja, M., Santamaría, J., Silva, D., Torres, R., Manousaki, A., Zafiropulos, V., Van den Brink, O.F., Heeren, R.M.A., Teule, R., & Silva, A. 2002. Evaluation of the chemical and physical changes induced by KrF laser. *Journal of Cultural Heritage* 4(Supplement 1): 257–263.

Cennino Cennini, 2009 (ed). *Il libro dell'arte, traité des arts, fin XIVe, l'œil d'or*, (part six, CLII): 190.

Ciofini, D., Oujja, M., Vega Cañamares, M., Siano, S., Castellejo, M. 2016. Spectroscopic assessment of the UV laser removal varnishes from painted surfaces. *Microchemical Journal* 124: 792–803.

Courtier,S., Broker, C. 2019. *Laser clearance report, C2RMF/Eros, C2RMF76786/ N°49223, 2019 and N°52115.*

de l'Escalopier, C. 1863. *Théophile, Moine, Traduit du latin. Le traité des divers Arts.* N/O de l'Allemagne.

Félicien, A. 1676. *Des principes de l'Architecture et dictionnaire des termes de l'architecture, à la peinture et aux autres arts qui en dépendent*. Chez Jean Baptiste Coignard, rue S Jacques, à la bible d'or (Chapitre XXII): 278.

le Begue, J. 1431. *Experimenta 118, decoloribus: praemittitur tabula ordine alphabetico digesta de vocabulis synonymis et aequivocis*, Source gallica.bnf.fr/Bibliothèque nationale de France. Département des manuscrits. Latin 6741, Greffier de la Monnaie

Lopez, M. 2020. *Evaluation and development of a laser cleaning tool for easel paintings, intended high precision*, Sciences and Heritage Foundation (FSP).

Lopez, M., Bai X., Martos-Levif, D., Zanini, A., Semerok, A., Serfaty, S., Menu, M., Wilkie-Chancellier, N., Detalle, V. 2020. Towards a new approach to cleaning paintings: using lasers to remove varnish, *Technè* 50: 80–91.

Sawicki, M., Bramwell–Davis, V., Dabrowa, B. 2011. Laser cleaning from a practical perspective: Cleaning tests of varied gilded-wood surfaces using Nd: YAG Compact Phoenix laser system, *AICCM Bulletin, Volume 32*, 2011 - Issue 1: Proceedings for the AICCM National Conference Fremantle, Western Australia, 21–15 September 2009.

Wilkie-Chancellier, N. & Detalle, V. 2020. Towards a new approach to cleaning paintings: using lasers to remove varnish. *Technè* 50:13.

Fibrous and membranous material artefacts

Lasers in the Conservation of Artworks XIII – Siano & Ciofini (eds)

Study and experimentation for a controlled laser cleaning of feathers

C. Mammoliti
Università degli Studi di Torino – SUSCOR, Venaria Reale, Turin, Italy

R. Genta, P. Croveri & F. Zenucchini
Centro per la Conservazione ed il Restauro dei Beni Culturali "La Venaria Reale", Venaria Reale, Turin, Italy

M. Castellino
Department of Applied Science and Technology, Politecnico di Torino, Turin, Italy

ABSTRACT: This work focuses on the scientific experimentation and practical application of a laser cleaning methodology for the macaw (Ara ararauna) feathers of an ethnographic bow with arrows from the Museum of Anthropology and Ethnography of the University of Turin. The fragility of the object and its unique degradation characteristics suggested the use of a very selective physical method, such as laser cleaning. Prior to the intervention, a comprehensive systematic experimentation phase was conducted including commonly used diagnostic techniques along with a rarely employed analysis for cultural heritage characterization: X-ray Photoelectron Spectroscopy (XPS). The aim of the experimentation was to gain a deeper understanding of the physical-chemical effects of laser irradiation. Through a dedicated experimental setup, optimal operating parameters for laser treatment of macaw feathers were determined. XPS analysis enabled the examination of the treated material and the assessment of potential surface alterations, thus allowing to precisely define the damage threshold. Following the comparison of different Nd:YAG(1064 nm) lasers, the Long Q-Switched mode provided better results and it was selected for the overall intervention. This laser ensures effective removal of deposits, selectivity and respect of the fragile substrate.

1 INTRODUCTION

This contribution summarizes the work carried out within a Master thesis in Conservation and Restoration of Cultural Heritage at the Università degli Studi di Torino, in agreement with the Centro per la Conservazione ed il Restauro dei Beni Culturali "La Venaria Reale" (Mammoliti et al. 2020 unpubl.). The thesis work concerned the restoration of a Bororo bow with arrows owned from the Museum of Anthropology and Ethnography of the University of Turin. Here, we present and discuss part of the entire work, which involved the definition of the experimental protocol for the laser cleaning of ararauna macaw feathers (Figure 1).

Case studies in the scientific literature report critical comparisons between traditional feathers cleaning methods and more innovative approaches, such as that based on laser cleaning (see Ciofini et al. 2022 and references therein). In the specific case of the artefacts that are the subject of this study, the latter was considered particularly interesting. In order to ensure a rigorous control of the effects of the laser irradiation of the feathers, both at the level of possible visible alterations and, more in-depth, of possible chemical-physical modifications, a protocol was developed for the experimentation on prepared samples, which was preliminary to the definition of the overall laser treatment. The examination of the state of the art (Pandozy, et al., 2014, Ciofini et al. 2022) along with the skills of the working group on the application of laser treatments allowed for a thorough systematic investigation and the final application of the laser technique within the conservation intervention on the present artefacts.

DOI: 10.1201/9781003386872-16

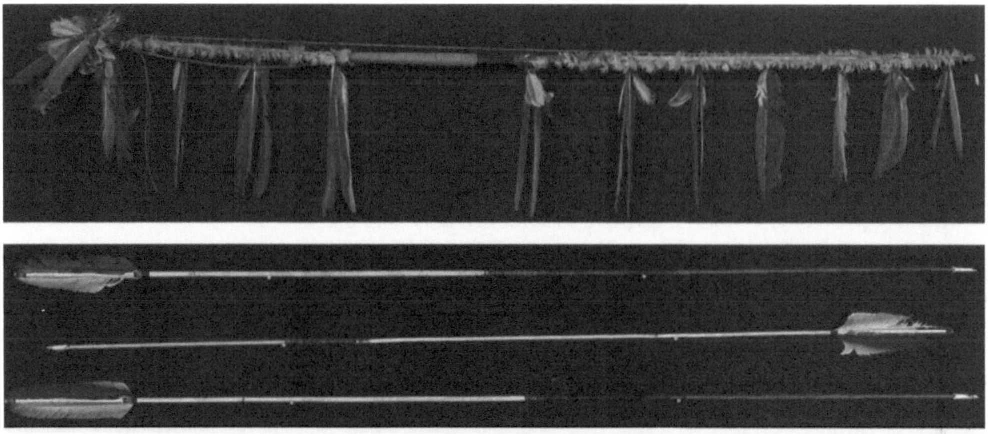

Figure 1. Bororo bow (top) and arrows (bottom) from the Museum of Anthropology and Ethnography at the University of Turin.

2 EXPERIMENTATION ON PREPARED SAMPLES

This work aimed at removing the deposits while safeguarding all the characteristics of the feathers, such as the following: 1) microscopic features producing optical effects, such as glossiness, iridescence, and other; 2) strong chemical and physical stability associated with by the keratin; 3) complex order of fragile structural elements such as barbs, barbules, hooks.

Preliminary tests were carried out on the same types of yellow and blue feather samples as those of the present ethnographic objects. In order to point out any possible chemical and physical changes induced by laser cleaning, a diagnostic protocol based on electron micro scope (SEM) examinations, colorimetric analyses, and Reflectance Transformation Imaging (RTI) techniques was implemented and applied to all types of feathers decorating bow and arrows (Figure 1). Furthermore, laser-induced physicochemical changes were also studied on prepared samples of Ara ararauna using X-ray photoelectron spectroscopy (XPS), an analysis rarely used in diagnostics of cultural heritage.

The laboratory samples of Ara ararauna feathers were divided into the series 1a (samples 12a and 12b) and 2a (samples 13a, 13b, 14a, 14b), characterized by means of colorimetric and RTI analyses, aged in the UV chamber, and eventually soiled.

RTI images showed that the surface of the feathers was uniform and smooth: the barbs were perfectly aligned and there were no surface irregularities (Figure 2, left). Following artificial aging and adhesion of the dust, all investigations were repeated. In particular, colorimetric investigations were also conducted on the 14a and 14b samples at an intermediate stage, immediately after UV aging (before dust application) to have more comparison data. RTI analyses of the aged and soiled samples were conducted for comparison purposes after the laser treatment (Figure 4).

After the investigations and related documentation of the results, following the two aging stages, laser tests were carried out. The samples of series 1a (12a and 12b) were treated using QS Nd:YAG(1064 nm), while those of the series 2a (13a, 13b, 14a, and 14b) were treated using fibre coupled LQS Nd:YAG(1064nm) laser (Salimbeni et al. 2003).

The second harmonic of Nd:YAG laser (532 nm) was excluded due to the poor results reported in the bibliography on the present specific feather colors, which were confirmed through preliminary tests conducted on a sacrificial samples. The surface exhibited significant inhomogeneity spot marks due to the laser ablation, leading to a sort of abrasion already at relatively low fluences.

The results of the tests performed on the series 1a were in agreement with the those reported in reference case study (Ciofini et al. 2022): the laser QS did not prove to be suitable for the treatment, inducing damage at fluences higher than 0.20 J/cm^2, which was still insufficient for the effective removal of the soiling (lack of discrimination).

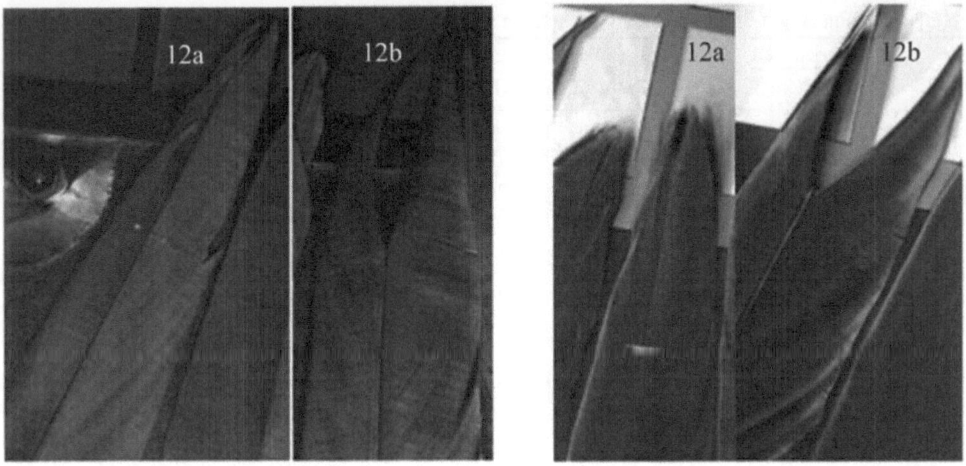

Figure 2. RTI image in Specular Enhancement (Diffuse color = 0; Specularity = 100; Highlight Size = 26): left) before aging (note the characteristic smooth and reflective surface of the intact feather), right) after artificial aging (surface matting and more irregular feather texture can be seen).

All tests with QS were carried out by keeping the pulse repetition frequency at 2 Hz to obtain more control and to be able to promptly verify the effect of the irradiation on the substrate. This allowed even a single irradiation to be carried out, and thus, to verify that surface damage occurs even with a single pulse.

Figure 3 show the samples in visible light (Figure 3a–b) and in RTI (Figure 3c) after laser treatment using the parameters listed in Table 1. Above the threshold fluence it was possible to observe a kind of surface abrasion produced by the inhomogeneity of the laser beam intensity distribution (This effect can be eliminated through the use of a homogenizer).

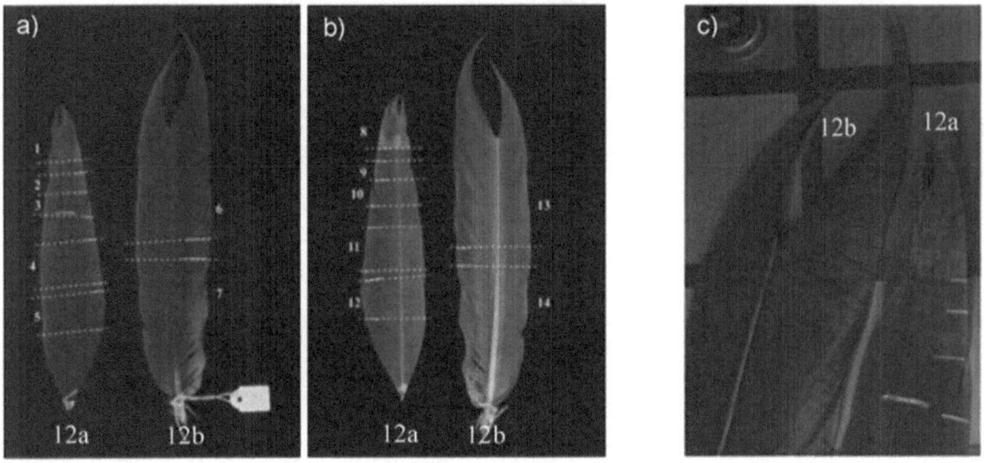

Figure 3. After laser treatment with QS: front (a) and back (b) of the samples 12a and 12b (treated zones marked by sequential number referred to in Table 2) and RTI image (c) in Specular Enhancement (Diffuse color = 59; Specularity = 70; Highlight Size = 75).

In the areas that macroscopically appeared to be intact after laser treatment, analyses were carried out in order to verify both the degree of damage and the effectiveness of cleaning reported in Table 1. In particular, the cleaning was monitored by colorimetric analysis, which made it possible to individuate the changes in colour from one stage to the next, thus starting from the analysis of the sample at level 0 (intact), going through level 1 (artificially aged specimen), and then analysing the laser-treated areas (level 2).

Table 1. Summary of tests performed with laser in QS mode on 12a and 12b specimens.

Sample	Area	Fluence (J/cm^2)		No. of pulses	Frequency (Hz)	Evaluation	
		Recto	Verso			Degree of damage (0-5)	Effectiveness (0-5)
12a	1	0.62	-	1	2	5	0
	2	0.48	-	2	2	5	0
	3	0.31	-	2	2	4	0
	4	0.44	-	2	2	5	0
	5	0.33	-	3	2	4	2
	8		0.40	1	2	4	0
	9		0.28	1	2	4	0
	10		0.27	1	2	4	0
	11		0.25	1	2	4	0
	12		0.24	1	2	3	1
12b	6	0.30		2	2	3	1
	7	0.25		3	2	2	1
	13		0.31	3	2	2	1
	14		0.20	3	2	2	1

Here, we report the most significant results of the colorimetric analysis, namely those including areas 4 and 5 of the blue side of the sample 12a and areas 12 and 14 of the yellow side of the same sample.

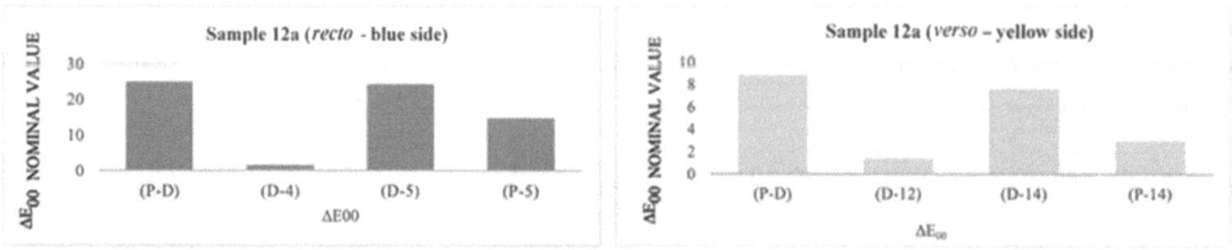

Figure 4. ΔE_{00} of the various treatment levels of the sample 12a (blue side and yellow side). P: before any treatment (level 0). D: after artificial ageing and soiling (level 1). 1; 2; 3 . . .: laser-treated areas (level 2, Table 3).

From the graph in Figure 4, relating to the blue side of the sample 12a, a strong colour change can be seen in the first ΔE_{00} (P-D): this is due to artificial aging and soiling. The ΔE_{00} (D-4), which allows comparison of laser-treated (level 2) and sample at level 1 (i.e. artificially aged), does not show a significant colour change, which means the cleaning was insufficient (the deposit was not removed). On the other hand, in ΔE_{00}(D-5), corresponding to the colour change between the sample at level 1 and area 5, a strong colour difference was measured, which can be attributed to two factors: surface damage and/or partial removal of dirt. The same hypothesis holds true for ΔE_{00}(P-5); in fact, in this analysis a rather large colour change can be seen, which can be attributed to both the dust still present over the surface and the laser-induced damage.

The same trend was observed in the analyses for the yellow side (Figure 4) of the sample 12a, although ΔE_{00} showed lower values than on the blue side. This could be attributable to the structure of the *verso* of the feather, which is less apt to retain particulate dirt and exhibits higher mechanical resistance. These two characteristics could make it less susceptible to colour change associated with deterioration and soiling.

The above described results suggest to rule out QS laser treatments because of their ineffectiveness and risk of damage at very low fluences, as also verified through SEM examinations (Figure 5).

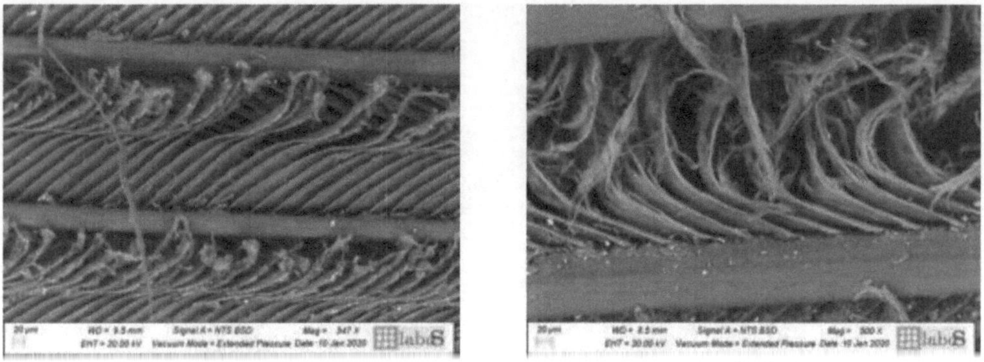

Figure 5. On the left: SEM image of a portion of Ara ararauna feather treated with laser QS at 0.25 J/cm^2 fluence. On the right: Example of unsuitable treatment in which a 0.48 J/cm^2 fluence QS laser was used. Note the severe damage created by the radiation on the barbules (©LaboS).

In particular, structural and textural damage were observed. The structure of the feather after the irradiation was devoid of its perfect periodicity, the barbs became defibrated and dehydrated, major deformations and breakings appeared in the originally smooth and intact areas (Figure 5).

Table 2. Summary of tests performed with laser in LQS mode on the samples 13a and 13b.

Sample	Area	Fluence (J/cm^2)		No. of pulses	Frequency (Hz)	Evaluation	
		recto	*verso*			Degree of damage (0-5)	Effectiveness (0-5)
13a	1	3.02		1	2	5	0
	2	1.03		1	2	5	0
	3	0.69		1	2	0	4
	4	0.10		3	4	0	0
	5	0.25		3	4	0	2
	8		0.69	1	4	0	3
	9		0.70	3	4	0	3
	10		0.75	3	4	0	5
13b	6	1.03		1	2	1	1
	7	0.69		3	6	0	5
	11		0.77	3	2	0	5
	12		0.77	3	6	0	5

This behaviour can be attributed to the short pulse duration of the QS laser (8 ns) and to the relatively high optical absorption of the blue feathers at 1064 nm (Ciofini et al. 2022).

It was interesting to note that in areas where visual or RTI observations showed little or no surface damage (e.g., area 5 of 12a), severe deterioration of the feather structure was produced at the micrometre scale, as it can be seen in Figure 5.

Given the inadequacy of laser QS, we proceeded with series 2a in which LQS laser was tested.

The samples 13a and 13b were irradiated (Figure 6) using the parameters listed Table 2 and characterized by means of colorimetric analysis, whose results are summarised in Figure 7 for the areas 7 (*recto* of 13a) and 10 (*verso* of 13b). These laser-treated areas were further investigated using XPS analysis in order to detect surface possible surface chemical effects.

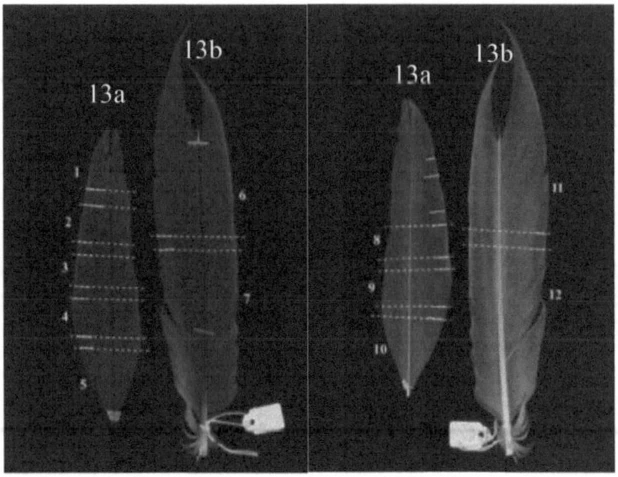

Figure 6. Recto and verso of the samples 13a and 13b after laser treatment with LQS laser. Treated zones marked with a sequential number referred to in Table 4.

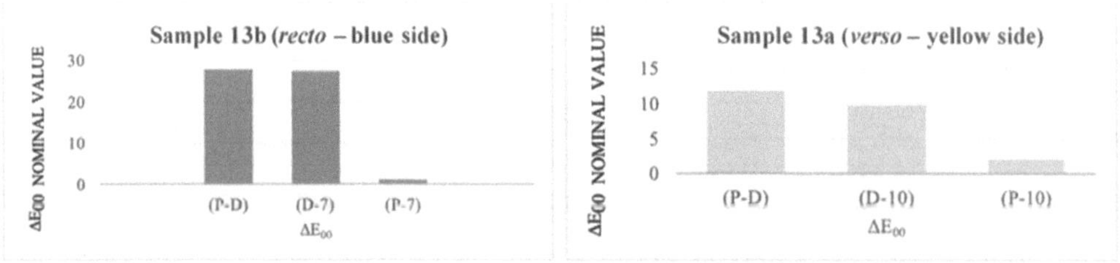

Figure 7. ΔE_{00} of the area 7 of sample 13b (blue side) after LQS laser irradiation at 0.69 J/cm^2 and of the area 10 of the sample 13a (yellow side) after LQS laser irradiation at 0.75 J/cm^2. P: before any treatment (level 0). D: after artificial ageing and soiling (level 1). 7, 10 = irradiated areas (Table 2).

Once the absence of colour change was established, based on the choices explained in the previous paragraph, feather specimens of the same sample taken before and after laser cleaning were analysed by SEM (Figure 8). The images showed that at the fluence of 0.69 J/cm^2, LQS (pulse repetition frequency 4-7 Hz) it was possible to effectively remove the deposits without any detectable injury to the feather barbules. Through SEM analysis of other samples, from areas treated at higher fluences, the damage threshold fluence was determined, which was 0.90-1.00 J/cm^2. The early damage that radiation induced at such energy density is displayed in Figure 9.

Following the microstructural observation of the samples through electron microscopy, we proceeded to assess possible undesired chemical-physical effects by analysing the areas representing the best cleaning results in terms of macroscopic and and microscopic observations (areas 3, 10, 11, 12).

2.1 *XPS analysis for chemical bonds diagnosis in Ara ararauna feathers*

The high sensitivity of XPS instrument made it necessary to subject the samples to be analysed prior to removal of dust, presenting a high risk of poor readability.

XPS measurements were carried out in two different points of each specimen listed in Table 3 in order to obtain information regarding the homogeneity of the chemical composition. Following the identification and quantification of the elements present in the analysed zone ("survey"), "high resolution" (HR-XPS) measurement was carried out only on the elements of interest for the study (S, N), which that allowed to trace the oxidation state of individual elements.

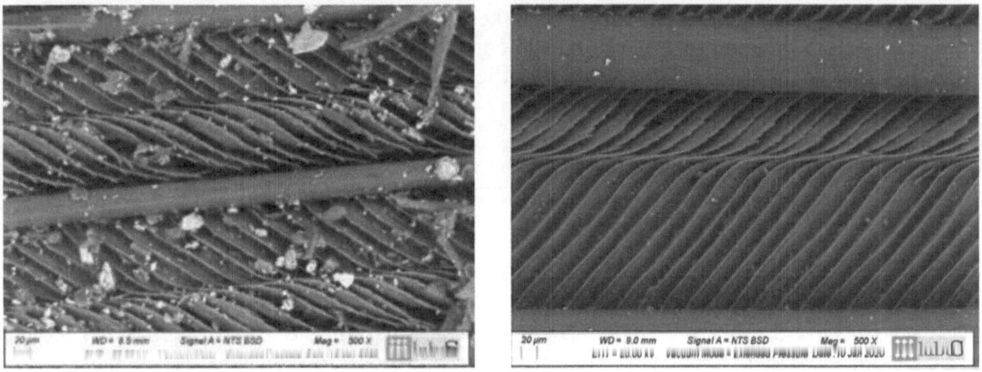

Figure 8. SEM details of 13b (blue side) before (left) and after (right) cleaning with LQS laser at 0.69 J/cm^2 (area 7) (Labos©).

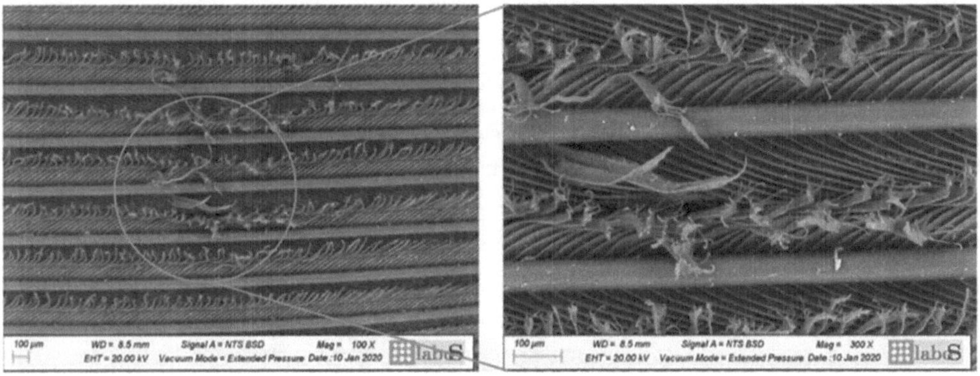

Figure 9. SEM images showing effect of a single laser pulse with a fluence of 1.03 J/cm^2. (Labos©).

Table 3. Specimens investigated using XPS analysis.

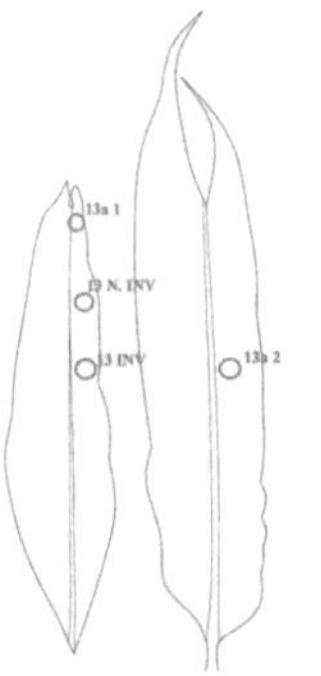

SAMPLE	DESCRIPTION
13 N.INV	Intact feather specimen, collected before any treatment
13 INV	Feather specimen artificially aged with UV Solar Box soiled with particle dirt from the MAET cabinets (aged on the blue side)
13a 1	LQS Nd:YAG(1064 nm) laser treated feather at 3.02 J/cm^2
13a 2	LQS Nd:YAG(1064 nm) laser-treated feather at 0.69 J/cm^2

The survey analysis showed the presence of carbon (C), oxygen (O), nitrogen (N), and sulphur (S), with contamination of silicon (Si), attributable to the surface particulate deposit. In particular, the latter is present in the specimens 13 N.INV and 13 INV, the former contaminated naturally while the latter was artificially soiled. In the specimen 13a 1 treated with laser above the damage threshold, no contaminants were found due to the deep ablation that totally removed exogenous material. As for sample 13a 2, on the other hand, silicon residues were still noticeable. However, the cleaning result in the area from which the specimen 13a 2 was taken should not be considered unsatisfactory. In fact, the high sensitivity of the instrument (0.01 at.%) succeeds in detecting minute amounts of substances that are invisible to the naked eye.

Table 4. XPS "survey": atomic concentration of the elements (at. %).

Specimen	Site of analysis	C1S	O1S	N1S	S2P	Other
13 N.INV	Zone 1	84.4	9.8	4.2	1.0	Si: 0.6
	Zone 2	80.5	12.3	4.7	1.0	Si: 1.5
	Average	**82.4**	**11.0**	**4.4**	**1.0**	**Si: 1.0**
13 INV	Zone 1	84.3	10.4	4.1	1.0	Si: 0.2
	Zone 2	82.4	11.9	4.6	0.7	Si: 0.4
	Average	**83.3**	**11.1**	**4.3**	**0.8**	**Si: 0.3**
13a 1	Zone 1	80.5	15.4	3.6	0.5	-
	Zone 2	74.8	17.7	5.9	1.6	-
	Average	**77.6**	**16.5**	**4.7**	**1.0**	-
13a 2	Zone 1	81.6	12.5	4.2	0.8	Si: 1.0
	Zone 2	83.3	12.4	3.5	0.7	-
	Average	**82.4**	**12.4**	**3.8**	**0.7**	**Si: 0.5**

In HR-XPS the S2p sulfur and N1s nitrogen signals provided information on possible laser-induced side effects. In particular, the S2p signal was rather uniform in not laser treated specimens (13 N. INV and 13 INV). In fact, the curves of the two distinct zones were similar each other (Figure 10), practically superimposable (the spectra were normalized for easier comparison).

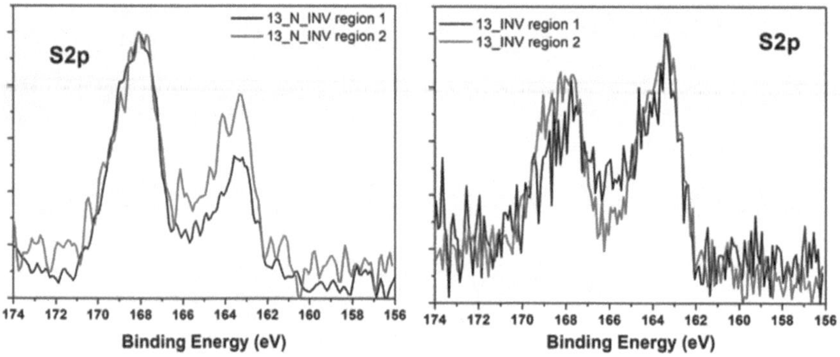

Figure 10. XPS spectrum of S2p for the specimen 13 N.INV (left) and 13 INV.

The XPS spectrum of S2p included two peaks corresponding to cysteine (163 eV) and oxidized cysteine (168 eV). The latter was more intense than the former in the naturally aged specimen (13 N. INV), while they got equivalent for the artificially aged specimen (13 INV), as well as for the laser treated specimen, 13a 2, at low fluence (LQS, 0.69 J/cm^2, 6 Hz). Conversely, both peaks got very noisy and their shape was barely recognisable at high fluence irradiation (specimen 13a 1, 3.02 J/cm^2) (Figure 11).

3 OPERATIVE PHASE

The preliminary experimentation to the select laser cleaning without carrying out risky tests directly on the artefact, having already determined the damage threshold (0.80 J/cm^2) and optimal fluence for the operation (LQS, 0,69 J/cm^2, 1-3 Hz).

A commercial LQS Nd:YAG laser was therefore used to clean the Ara feathers of the present bow and arrows by operating below the damage threshold. The cleaning was assisted by a soft-bristled brush and a chirurgical micro suction unit with the nozzle attached to the laser handpiece in order to suck up the removed material and prevent its redepositing.

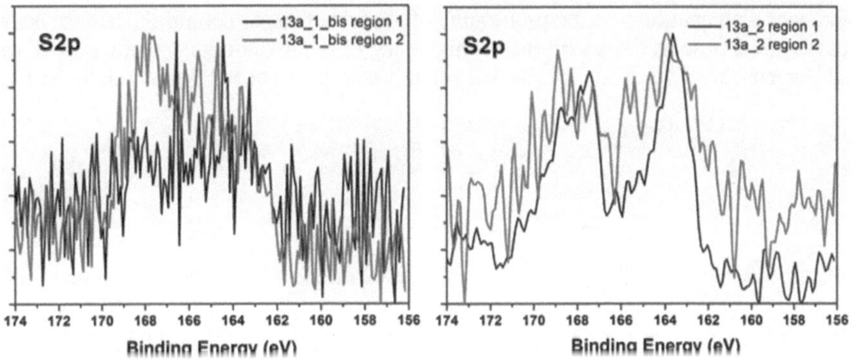

Figure 11. XPS spectrum for S2p in sample 13a 1 and in sample 13a 2.

After the validation test carried out on one of the ornithological elements most blackened by deposits, a small portion of the Ara ararauna feather on the bow was treated using LQS laser with a fluence between 0.65-0.69 J/cm^2. From the selected feather, two specimens were taken, one from laser treated and one from untreated area, respectively, in order to analyse them with XPS and verify the possible occurrence of chemical-physical alterations induced by laser irradiation.

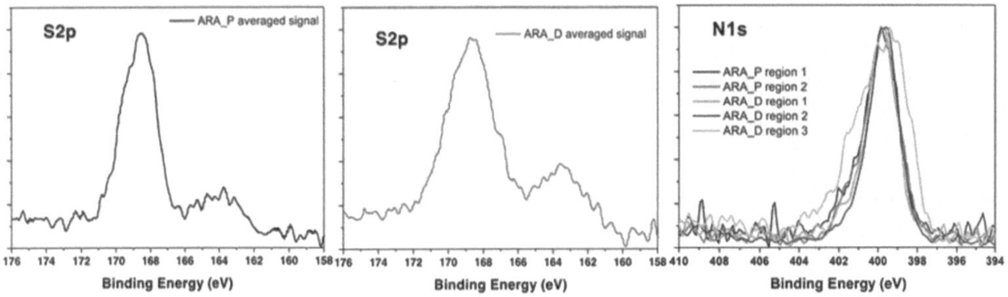

Figure 12. S2p signal of the non-irradiated specimen ARA_P (average from two measurement points), of the laser cleaned specimen, ARA_D, and N1s signal of all analysed specimens.

By comparing the S2p signal of the untreated sample, ARA_P, obtained as the average of the two measurement points, with that of the laser-treated sample, ARA_D, obtained as the average of the signals from the areas 1 and 3, no significant differences were pointed out (Figure 12). Thus, it demonstrates that the laser cleaning treatment did not induce any substantial changes in the oxidation state of S2p. By extending the comparison to the N1s signal (Figure 12), it was possible to further prove the absence of any changes of the cysteine, since this remains largely unchanged in all samples analysed. The laser treatment did not change the microstructure of the nitrogen component. Only a noisier and broader signal was noted in area 3 (Figure 12), due to the low intensity of the signal, but the position of the peak remained unchanged.

Following the assessment of the chemical-physical safety of the laser treatment, colorimetric analyses were carried out on all the cleaning tests, both on Ara ararauna feathers and on the feathers of the other bird species composing the artefact. This allowed to monitor in a non-invasive way the degree of cleaning, as compared to the results achieved in the experimental phase.

Cleaning was therefore conducted on all the coloured feathers, with the exception of the small red and yellow cover feathers attached to the shaft of the arch, which were particularly fragile and less dirty. These were treated with the sole purpose of dusting them with a soft bristle brush.

The laser cleaning operation yielded the most evident results on the yellow side of the Ara ararauna feathers, where the matting effect of soiling was well visible. Brush and micro-aspirator assists durin laser cleaning was very useful, as they facilitated the removal of the soiling concretion. (Figure 13)

A fundamental precaution was the positioning of absorbent paper behind the feather under laser treatment (Figure 13), so as to prevent dirt from passing onto the underlying feather, as all the ornithological elements are placed side by side and often bound together tightly, especially on the tassel.

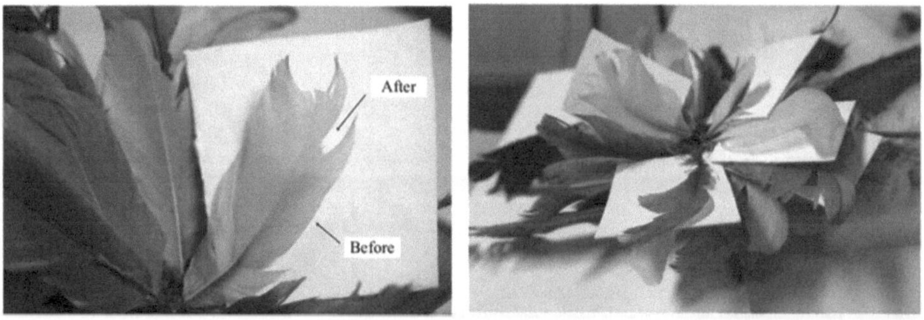

Figure 13. Left: a photographic detail of the ethnographic artefact during the laser cleaning showing the white paper used to protect the bow feathers durin. Right: a detail showing Ara ararauna feather during laser cleaning.

4 CONCLUSIONS

Based on the results of the experiments conducted on the feathers of different species and the outcomes obtained from the diagnostic protocol, it can be concluded that laser cleaning performed within a predetermined interval of fluence and frequency enables a satisfactory and controlled soiling removal that does not alter the microstructure and composition. In fact, it was observed in all the tests that the barbules of the feathers remained intact after treatment, and no changes in colour were detected.

More in detail, Ara ararauna feathers were found to be highly sensitive to QS Nd:YAG laser irradiation which can cause damage. At the same time, the same laser provided excellent results in cleaning feathers of other bird species.

LQS Nd:YAG laser (100 ns pulse duration) allows deposits to be effectively and safely removed from Ara ararauna feathers without any detectable alteration, when used at 0.50-0.70 J/cm^2 and maximum pulse repetition frequency of 6 Hz. Colorimetric values of the substrate did not change and XPS analysis allowed assessing the damage threshold and monitor the treatment at the molecular level. Here, XPS allowed to verify the absence of any modification to the S and N chemical bonds when operating in optimized irradiation conditions.

REFERENCES

Ciofini, D., Rossignoli, G., Tosini, I., Lanterna, G., & Siano, S. 2022. Laser ablation treatment of soiled featherworks: the first validation study. *Journal of Cultural Heritage* 56: 118–129.

Dignard, C., Lai, W. F., Binnie, N., Young, G., Abraham, M., & Scheerer, S. 2003. Cleaning of soiled white feathers using the Nd:YAG Laser and traditional methods. In K. Dickmann, C. Fotakis, & J. Asmus (eds), *LACONA V Proceedings, Lasers in the Conservation of Artworks*: 227–235. Springer Proceedings in Physics, vol 100. Berlin: Springer.

Huget, Y. 2011. Le nettoyage et la restauration des plumes: une sélection de méthodes. *CRBC. Conservation-Restauration des bien culturels* 29: 49–59.

Mammoliti C. 2020. Studio e restauro di un arco con frecce Bororo del MAET: un approccio etico, multidisciplinare e sperimentale alla pulitura e al consolidamento delle penne. Master thesis in Conservation and restoration of cultural heritage. University of Turin in agreement with Centro Conservazione e Restauro "La Venaria Reale". (Unpublished)

Pandozy, S., Rivière, C., Brunori, M., Nepote, F., Rivalta, A., Santamaria, U., Morresi, F., Fraticelli, F., Colantonio, C. & Brunetto, A. 2014. Sperimentazione sull'uso del laser per la pulitura delle piume presenti nella collezione etnologica dei Musei Vaticani. In A. Brunetto (A cura di) *Aplar 5 Applicazioni laser nel restauro. Atti del Convegno*: 423–441). Firenze: Nardini Editore.

Salimbeni, R., Pini, R. & Siano S. 2003. A variable pulse width Nd: YAG laser for conservation. *Journal of Cultural Heritage* 4(Suppl. 1): 72–76.

Lasers in the Conservation of Artworks XIII – Siano & Ciofini (eds)
© 2024 The Author(s), ISBN 978-1-032-47995-8

Development and application of a custom green laser system to remove decades old penciled graffiti on raw canvas from Morris Louis' masterwork, Beta Upsilon

A. Kerr
Smithsonian American Art Museum, Washington DC, USA

B. Dajnowski
G.C. Laser Systems Inc., Forest Park, IL, USA

ABSTRACT: In 1989, paintings conservators at the Smithsonian American Art Museum were confronted by the vandalism of Morris Louis' Beta Upsilon, a masterwork painting from his unfurled series (1960¬61). Comprised of colored bands of paint flowing inward from the edges on an unprimed canvas, the work was defaced with penciled graffiti lines on the raw canvas that mimicked the unfurling strokes of colors. Testing revealed the solution was not readily at hand at the time. Despite pressure to treat the artwork immediately, conservators persisted in finding the right solution. Laser cleaning was considered as a possible solution, but the correct parameters needed to be identified. Extensive testing was conducted on mockup samples using various pulse durations and wavelengths: 1064 nm, 532 nm, 355 nm, and 266 nm. After many years of preparation and testing, it was concluded that the 532 nm wavelength was appropriate for this particular cleaning problem. An entirely new laser cleaning system, the GC-¬532, was designed and built specifically to clean the Beta Upsilon. In order to allow great flexibility in laser parameter optimization the new bespoke laser system was built to have the 2nd harmonic 532nm, tunable pulse duration, pulse frequency, and pulse energy, an ultra¬high speed hot¬spot free circular scanner with tunable RPM, and various focal lens options. The research and testing leading up to the creation of a new purpose built green 532 nm pulsed laser system, the process of optimizing laser cleaning parameters, and the cleaning results achieved will be presented. This new laser technology will have future applications in cleaning canvases and organic materials.

1 INTRODUCTION

In 1989, paintings conservators at the Smithsonian American Art Museum were confronted by the vandalism of Morris Louis' Beta Upsilon (Figure 1), a masterwork painting from his unfurled series (1960¬61). Comprised of colored bands of paint flowing inward from the edges on an unprimed canvas, the work was defaced with penciled graffiti lines on the raw canvas that mimicked the unfurling strokes of colors. Testing revealed the solution was not readily at hand at the time. Despite pressure to treat the artwork immediately, conservators persisted in finding the right solution. Laser cleaning was considered as a possible solution, but the correct parameters needed to be identified. Extensive testing was conducted in various stages from 2013 to 2022 on mockup samples using various pulse durations in nanoseconds and picoseconds and wavelengths: 1064, 532, 355 nm, and 266 nm. After many years of preparation and testing, it was concluded that the 532nm wavelength was appropriate for this particular cleaning problem.

DOI: 10.1201/9781003386872-17

During prior years of testing it was observed that infrared 1064nm laser pulses, which are commonly used for cleaning inorganic materials such as metals or stone, are generally not safe for use to remove graphite pencil marks from raw canvas. Water absorbs 1064nm laser light and can instantly convert to steam. This absorption property of water is useful for cleaning inorganic materials such as masonry or metals as a micro steam cleaning effect takes place whenever a wet or moist surface is laser cleaned with 1064nm pulses. However, the absorption property of water in the 1064nm region can be very detrimental to organic materials such as canvas as the 1064nm laser pulses can cause immediate dehydration and potential browning of cellulosic fibers. 1064nm pulses were tested in both nanosecond and picosecond regions for removing pencil marks from mock-up pieces of canvas, and all of the tests resulted in discoloration of the canvas fibers while removing the pencil graphite. 355 nm and 266 nm UV laser pulses were also not successful at removing pencil marks from canvas and raising the fluence levels in an attempt to achieve cleaning caused damage and shearing of the fibers. The 532 nm second harmonic wavelength of laser light, however, can effectively remove carbon deposits and does not dehydrate the cellulosic fiber fibers of the canvas. Water does not absorb 532 nm light like it does 1064 nm light. The 532nm wavelength was identified as the optimal wavelength for removing pencil marks from canvas. It was observed that desirable results could be achieved using either picosecond or nanosecond pulses in the 532 nm region.

532 nm laser light is generally produced by introducing 1064 nm laser pulses to a second harmonic nonlinear crystal, which then converts approximately half of incoming 1064 nm light into 532 nm light, and also allows remnant 1064nm light to exit the crystal together with the 532nm light. This means 1064 nm light enters the crystal and both 1064 nm and 532 nm light comes out. 1064 nm light was clearly identified as undesirable for this application and so it was important to have an optical filtration method that cuts off 1064 nm light from being emitted so that only pure 532 nm light is allowed to exit the laser apparatus. It was concluded that a pulse duration tunable laser system emitting pure 532 nm pulses would be a good tool for this treatment.

Figure 1. Beta Upsilon by Morris Louis, 1960, acrylic on canvas, (260.4 x 618.5 cm), Smithsonian American Art Museum.

1.1 *Project sequence overview*

1) Data from prior years of testing was used to build a bespoke 532 nm prototype laser system, called the GC-532 P, that can provide a range of pulse duration and fluence options to dial in on optimal cleaning parameters for this particular project.
2) Tests were performed on aged canvas sample strips that had both new and aged pencil marks to identify optimal parameters for laser cleaning. These tests were successful.

3) The identified optimal parameters were then tested on individual loose strands of fibers from the Beta Upsilon painting. These loose fibers were collected from around canvas edges. The fibers were examined with microscopes before and after laser cleaning to verify that they were not being damaged. These tests were successful.
4) A laser cleaning test was performed using the verified optimal parameters on a small sample of the Beta Upsilon canvas, taken from the outer edge of trim. This test was successful and further verified the parameters were safe and effective. Raman and ATR IR spectral analysis also confirmed that there was no chemical alteration to the canvas.
5) Tests were carefully conducted on the outer edge of the actual painting canvas with the same parameters. These cleaning tests were successful as well.
6) Tests were conducted on sample pieces of canvas that were hung next to the painting to simulate the treatment approach with the laser scanner. These tests were successful
7) Laser cleaning of the pencil marks was performed and was successful. The pencil marks were removed and the canvas was noticeably cleaner after laser ablation cleaning.
8) Laser cleaning of the entire blank canvas was performed to even out the cleanliness of the canvas to a homogenous even appearance.

2 MATERIAL AND METHODS

2.1 *Treatment setup*

Laser system operators and observers wore properly rated laser goggles and followed all necessary laser safety protocols. The area where the laser cleaning took place was secured with 100% opaque laser shielding and proper laser signage posted and necessary laser PPE was in use at all times. Laser safety glasses were worn by all conservators and staff within the laser cleaning area during treatment. A portable laser cleaning fume extractor, which captured all the vaporized pencil marks and surface grime from the canvas was in use at all times during cleaning tests and treatment.

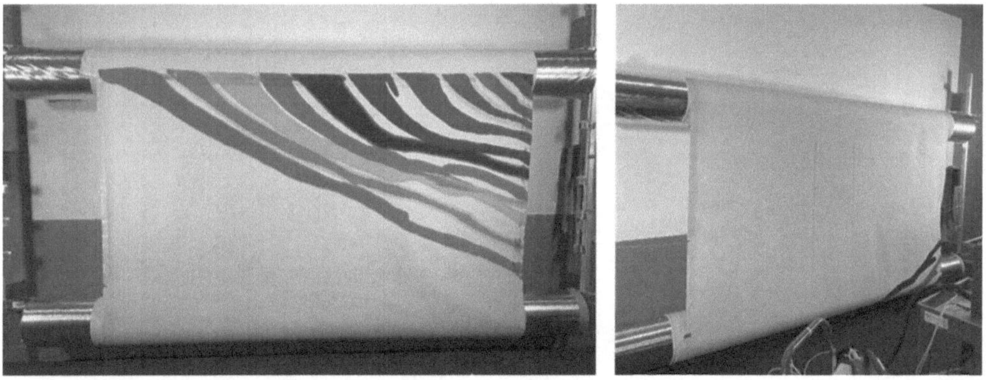

Figure 2. Custom roller setup that transfers canvas from one roll to another. Strings used to make a grid for keeping track of cleaning on the vast areas of blank canvas.

This painting measures 260.4 × 618.5 cm and has just over 16 square meters of surface area. Given the large size of the painting, a special apparatus was made and prepared by Smithsonian staff which allowed for the painting to be rolled and transferred vertically from one roll onto another, systematically revealing each section of the painting. The tacking edges of the painting had additional lengths of canvas safely secured to them so that the very ends of the painting could be accessed with this rolling system. This setup allowed for optimal ergonomic access to the entire canvas as any particular area could be selectively lowered or raised to a comfortable working height for both cleaning and inspection.

In order to ensure that every inch of the vast surface of blank canvas was laser cleaned evenly, and archaeological conservation approach was taken by dividing the canvas up into a systematic grid. This was done with strings that were strung in front of the painting in both horizontal and vertical orientation and divided the canvas up into rectangle a few square feet in size. The horizontal strings remained stationary and the vertical strings on the grid would be moved from left to right every time a section was completed. Then once an entire row was fully cleaned, the painting would be rolled down from one roller to the other until a new uncleaned area was positioned into the grid.

2.2 *The GC- 532-P*

A bespoke conservation laser system was designed and built specifically for this project by G.C. Laser Systems Inc.: The GC-532 P. This is a portable 50W 532 nm green pulsed laser system prototype that has tunable pulse duration from 1ns - 10ns, tunable pulse frequency from 1 kHz to over 1 MHz, tunable pulse energy, tunable circular scan speed up to 20,000 RPM, fiber optic delivery cable, and several interchangeable focal lenses. Additionally, the patented high speed circular scanning system which was invented for art conservation applications, allows for an even distribution of laser energy across the surface with no hot spots.

The laser system is on a cart and features the main laser system enclosure with a touchscreen interface that is connected to the handheld laser scanner via an umbilical cable. The system runs on standard 120V or 220V power, has an integrated green aiming beam, and a foot switch trigger. A portable fume extractor with prefilters, HEPA filters, and Carbon filters and a fume capture trunk was used in conjunction with the laser system to capture any vaporized contaminants during the cleaning process.

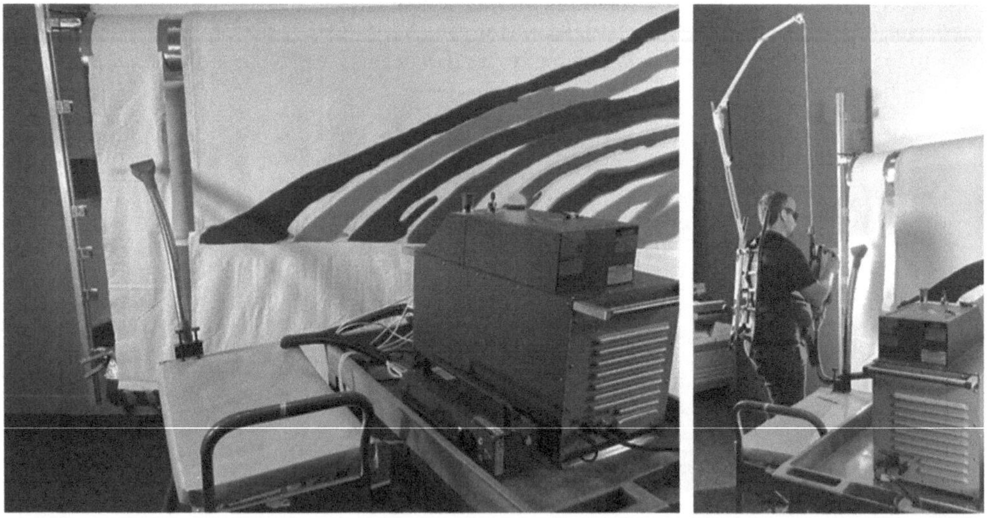

Figure 3. GC-532-P laser system with harness for holding scanner weightless in front of operator. On the right Bartosz Dajnowski is peforming cleaning tests on a mockup piece of canvas that was hung to the left of the painting.

After doing more tests on sample pieces of canvas, a long piece of sample canvas was hung to the left of the painting and was marked with pencil for additional testing. It was important that the cleaning tests took place on canvas that was suspended in the air and had space behind the canvas just like the painting. This allowed for a simulation of the actual treatment and the ergonomics of moving the laser around during cleaning. To ensure accurate and consistent cleaning a special laser cleaning harness was used during testing and during the treatment. The harness is worn like a backpack secured with several straps and features an articulating overhead arm with a pulley system that clips onto the laser scan head. This harness allows for the laser scanner to hang virtually weightless in front of the operator, significantly reducing arm fatigue from holding the scanner. A green aiming beam is used to confirm focal distance and where the laser is aiming before the higher power green laser cleaning beam is activated.

Different types of pencils with varying hardness levels were used to make marks on aged canvas samples which were examined visually and with digital microscopes during testing. The technique of moving the laser scan across the surface was practiced to achieve optimal cleaning as quickly as possible. The 20,000 RPM high speed circular scan of the GC-532-P effectively and quickly spreads the energy of the laser pulses over the surface with no hot spots, making the process safe for the canvas. The scan was always kept moving to prevent any excessive laser energy from remaining in any given area for more time than was necessary. 1ns, 5ns, and 10ns pulses were tested on mockup samples, with the best results being achieved by 10ns pulses. The 10ns pulses appeared to have an ideal balance of photomechanical, photo-thermal, and photochemical cleaning effects on the samples. The circumference of the circular scan path was set to 5.55cm. The scanner was set to 20,000 RPM and parameters were dialed in to provide a 25% pulse overlap pattern.

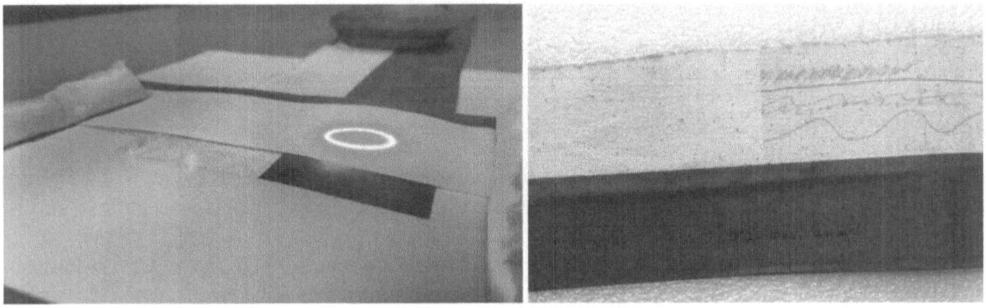

Figure 4. Laser cleaning fresh pencil marks on aged canvas samples. The left half of the sample was laser cleaned with 10ns 0.4mJ pulses. A piece of paper was used to mask off the area during cleaning.

Several weeks of testing resulted in the final parameters listed in Table 1, being chosen as safe for cleaning the pencil marks and the whole canvas overall.

Table 1. Optimized laser parameters.

Area cleaned	Focal lens (cm)	Pulse duration (ns)	Pulse energy (mJ)	Frequency (kHz)	Scan cm	Pulse overlap	Fluence (J/cm^2)
Pencil marks	16 cm	10	0.4	20	5.55	25%	1.27
Overall canvas	25.4 cm	10	0.4	20	10.5	25%	0.5

Tests on mockup samples indicated that 532 nm 10ns pulses with a fluence of 1.5 J/cm^2 were not causing damage to the canvas. A fluence of 1.06 J/cm^2 produced some cleaning and did not damage the fibers. Damage was observed at a fluence levels above 2.5 J/cm^2.

Figure 5. 50× magnified view in white light, UV, and IR imaging of a pencil mark with the left half laser cleaned. The mark is gone and the canvas is noticeably cleaner.

No damage was observed to the canvas fibers at a fluence level of 1.5 J/cm^2, suggesting that the damage threshold for this canvas material is somewhere above 1.5 J/cm^2. This is a critical parameter as too high of a fluence could damage the fibers, yet too low of a fluence would not remove the graphite. Given that the test data suggested that a fluence of 1.5 J/cm^2 is still fairly safe for this particular canvas, an extra buffer zone of precaution was incorporated into the testing protocols by using an even lower fluence value of 1.27 J/cm^2 for additional testing.

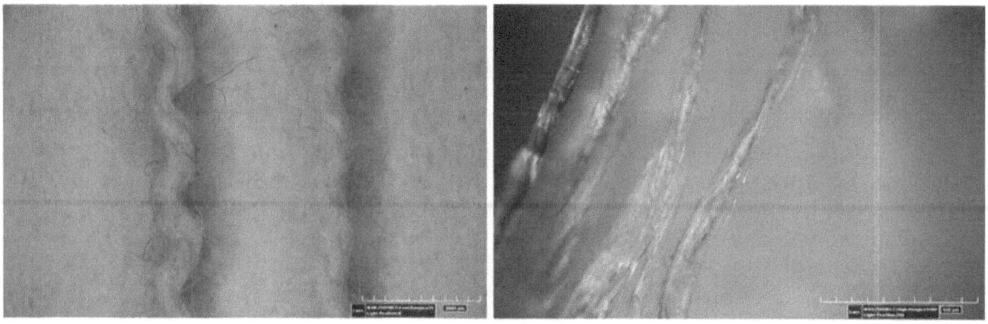

Figure 6. Detail of a laser cleaned fiber strand (right) next to an uncleand fiber strand. The image on the right shows a 1000×detail of the unharmed fiber strands after laser cleaning.

The strand on the left is uncleaned. The one on the right is laser cleaned with 10ns 532nm laser pulses with a fluence of 1.27 J/cm^2. Magnified inspection shows the laser treated strand is noticeably cleaner, brighter, and undamaged by the process.

The established optimal parameters using 10ns with a fluence of 1.27 J/cm^2 were tested on an actual sample of the painting canvas that came from and outer edge trimming. Tests using the optimal parameters were successful on the sample piece of the Beta Upsilon canvas. The canvas is noticeably cleaner and lighter in color in the test area.

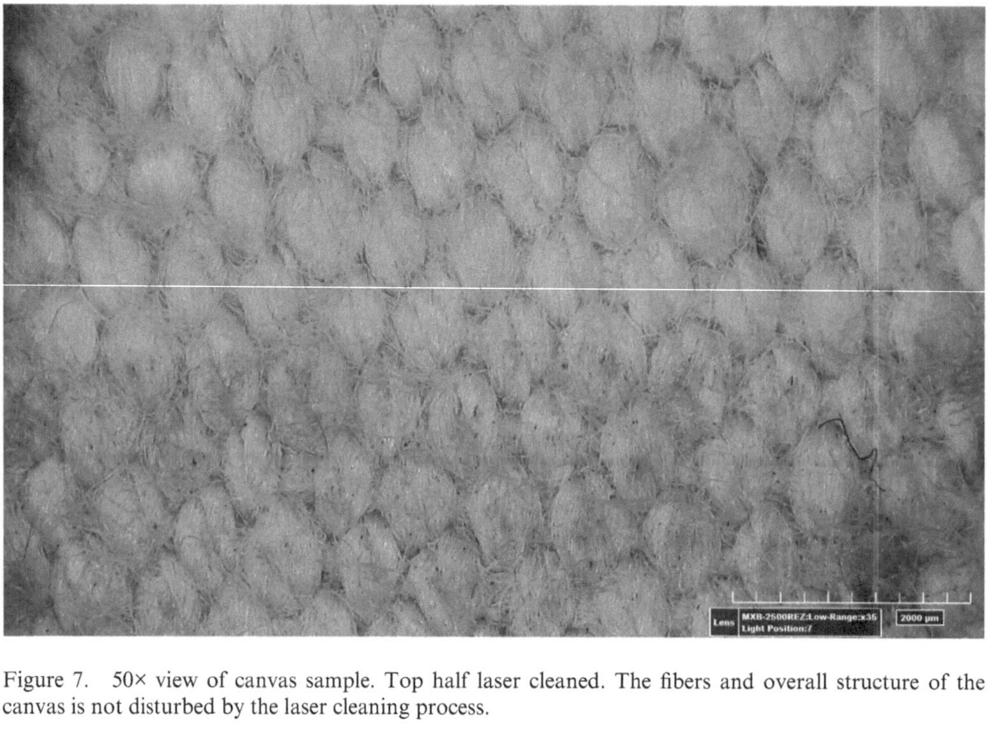

Figure 7. 50× view of canvas sample. Top half laser cleaned. The fibers and overall structure of the canvas is not disturbed by the laser cleaning process.

Additional data from ATR IR and Raman spectroscopy analysis was obtained on this cleaned sample of the painting. The analysis of the sample was conducted by Thomas Tague at Bruker Instruments. The analysis confirmed that there was no damage or chemical alteration to the canvas fibers after laser cleaning process.

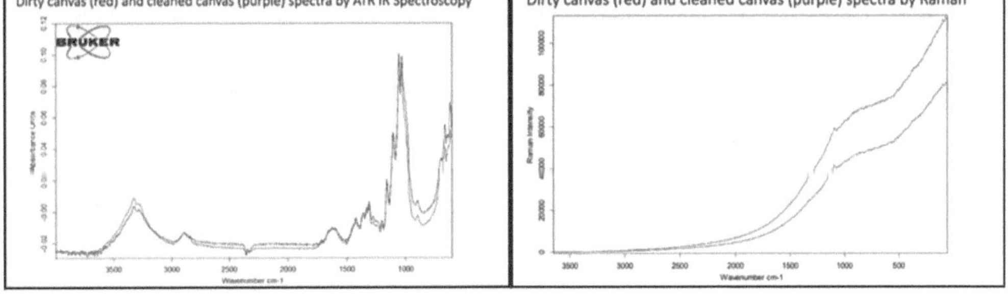

Figure 8. ATR IR and Raman spectroscopy support that there was no alternation or chemical change to the canvas caused by the laser cleaning process as clean and uncleaned canvas spectra are overlayed.

3 CLEANING THE PAINTING

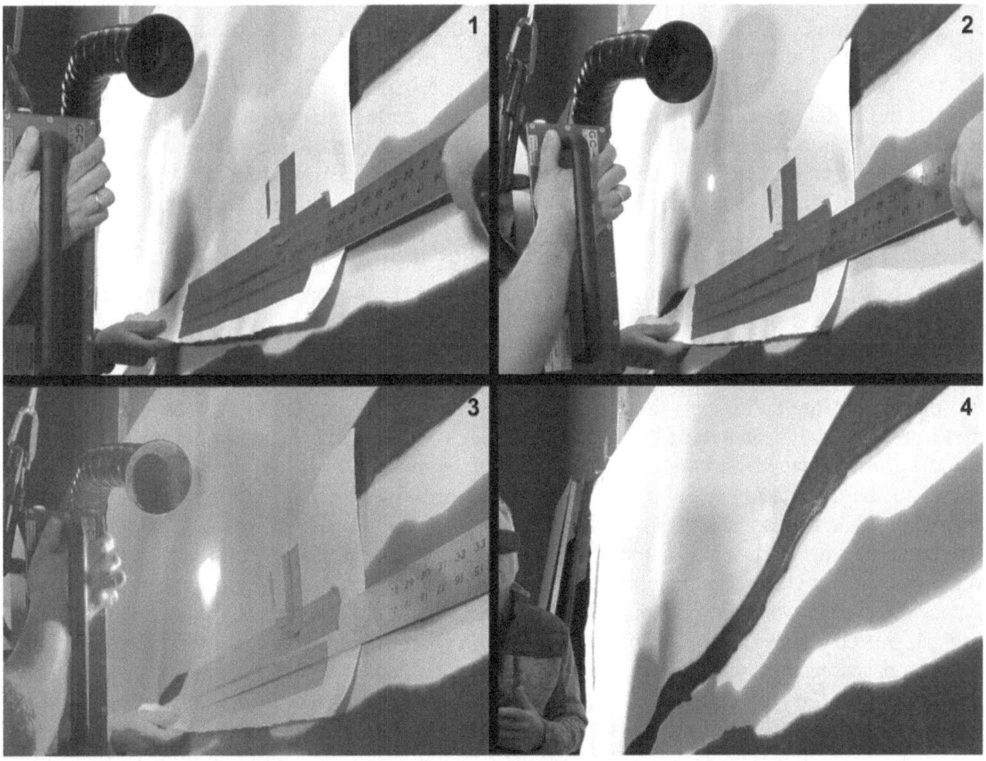

Figure 9. The laser cleaning sequence removing a pencil mark from the painting.

The above images show the sequence of removing a pencil mark from the canvas and show the following:

1) The area is prepared and secured. A fume extractor is held in position to capture the laser ablated contaminants. A custom cut and shaped piece of blotter paper, secured to a long ruler for rigidity, was used to mask off the painted areas of the canvas. There was no intention to have the laser beam go over the paint pigments, however in an abundance of precaution, the masking blotter paper was held in place as an extra layer of protection.
2) A low power but visible green aiming beam was activated and the movement of the cleaning laser scan over the pencil mark was rehearsed and the correct focal distance was confirmed.
3) The laser cleaning beam was activated and the surface was laser cleaned. The circular beam was moved in a circular motion across the surface, like a scrubbing motion, and was always kept moving. It took less than a minute to remove the pencil mark.

The laser beam is turned off to show the pencil mark removed. The canvas is also noticeably cleaner and brighter in tone where the laser removed the mark and also accumulated grime and dust from the canvas surface.

4 RESULTS

The laser cleaning process was a success. The pencil marks were removed and are no longer visible. Less than a minute was spent on each pencil mark on with the first laser cleaning pass. Each mark was laser cleaned and then examined with microscopes. A second laser cleaning pass, again with less than a minute of laser cleaning per area, was repeated to fully remove any potential visible traces of residue from the marks. A piece of white printer paper folded in half or a piece of thicker blotter paper can be used to mask off an area as the laser beam will not pass through it. This was important for protecting the pigments in the painted areas as it was not known if the parameters used to remove the pencil marks would also potentially remove or damage the pigments. Blotter paper adhered to a long metal ruler with blue painters tape was used mask off and provide added protection from any accidental laser exposure and to allow the beam to get close to the barrier between paint and raw canvas. Some quick exploratory tests were done on various colors in non-visible areas of the tacking edge of the painting and were not successful. More testing would be required to see if there were any laser parameters that would not compromise the paint colors. The laser was only used to clean the raw canvas.

This large painting is approximately 20 feet long and has an overall surface area of approximately 173 square feet. The majority of this paintings surface is blank canvas. The areas where the pencil marks were removed were noticeably brighter and cleaner as the surface dust and grime was also removed by the laser while it was removing the pencil marks. This was expected and it also meant that the entire canvas needed to be laser cleaned in order to achieve a homogenous even appearance. It was decided that the entire area between the painted ends of the painting would be laser cleaned and that the few strips of blank canvas in between the various colors would not be laser cleaned. Safe laser cleaning parameters were not explored or established to clean the painted areas, so it was decided the Smithsonian conservation staff would clean the painted areas with traditional methods such as cosmetic sponges. The amount of energy needed to remove surface grime and dust is lower than that needed to remove a pencil mark, so the laser fluence could be lowered for cleaning the overall canvas. This was done by using the same laser settings, but changing the focal lens to a longer one which would increase the coverage of the laser beam, increase the spot size of the pulses, and also decrease intensity and lower the fluence of the beam on the surface from $F=1.27\ J/cm^2$ down to $F= 0.5\ J/cm^2$. This fluence level is sufficient to remove dust and grime from the canvas and is significantly lower than the established safe fluence level that was needed to remove the pencil marks. The overall cleaning of the canvas took several days and was done systematically. In order to maintain consistency with the same technique and movements of the laser system, the entire canvas was cleaned solely by Bartosz Dajnowski. After each section was cleaned the canvas was inspected by Smithsonian conservation staff to verify nothing was missed. The end result was an even and homogenously cleaned canvas.

LAB colorimetry data was collected before and after laser cleaning the canvas and there was a color change LAB delta value of around 1.823. The human eye can typically perceive a change in color value of at around 1, so the 1.823 measured change was certainly perceptible. This was the result of the canvas getting cleaner and brighter, and was the reason why the entire canvas needed to be cleaned evenly.

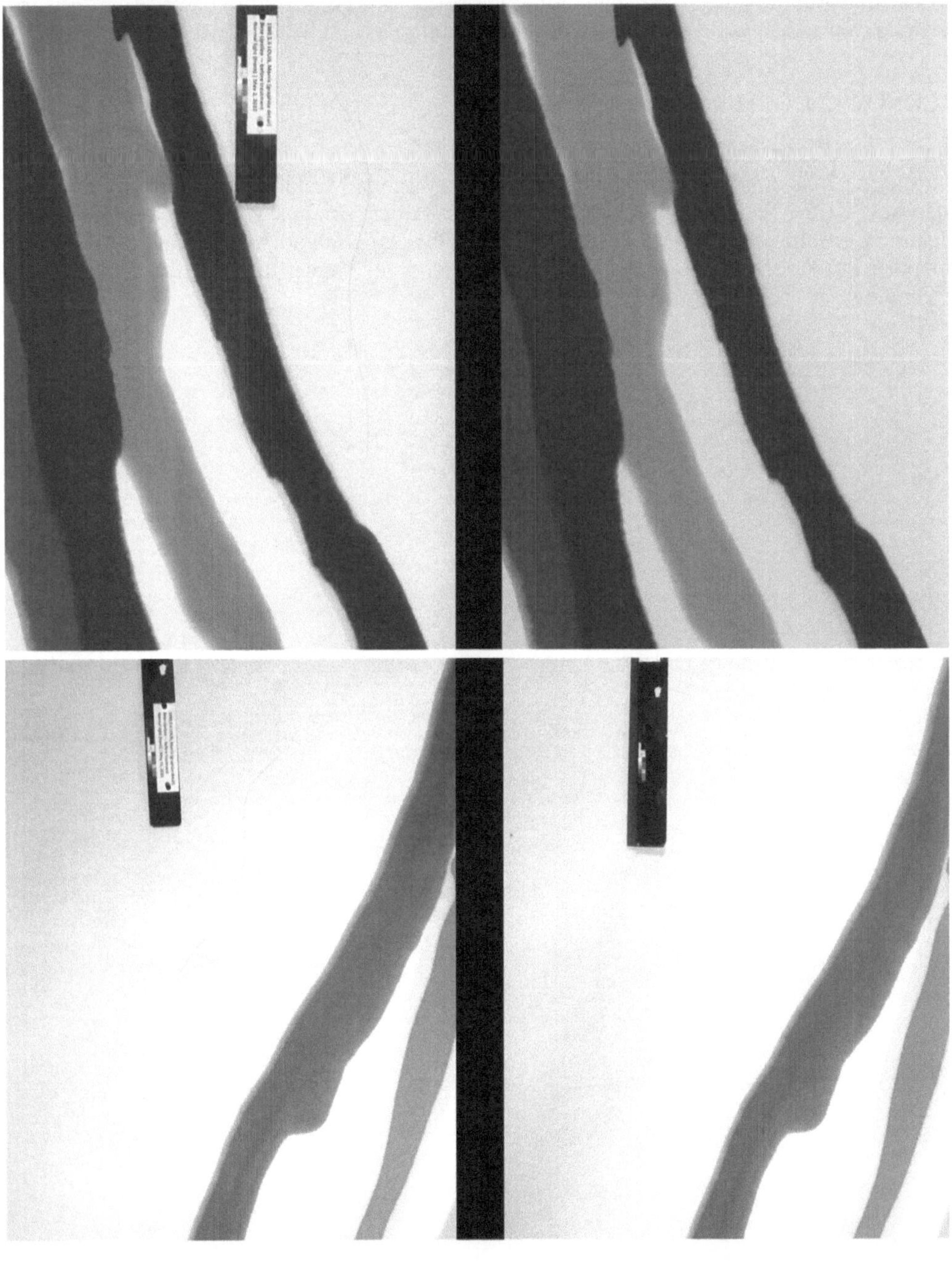

Figure 10. Detail area of pencil marks before and after laser cleaning. The images on the right side are after laser cleaning.

5 CONCLUSIONS

Several decades of patience, preparation, and due diligence ultimately resulted in a successful conservation laser cleaning treatment of Morris Louis' Beta Upsilon. The painting will finally be back on display at the Smithsonian American Art Museum after waiting for over 34 years for technology to catch up to the problem. The new highly tunable GC-532 P laser cleaning system that was developed and built for this project will be a useful conservation tool for cleaning canvases as well a wide range of organic materials and works of art.

ACKNOWLEDGEMENTS

Project Funded by Lannan Foundation, Michael Greenbaum, Michael Abrams, and SAAM Director's Circle. SAAM Staff: Stephanie Stebich, Virginia Mecklenburg, Melissa Ho, Gwen Manthey, Keara Teeter, Jenna Gustafson, Martin Kotler, Brianna Weakley, Josephine Ren. Historic Consultation with the conservators Stephano Scafetta and Anne Creager. Additional Support: Dr. Andrzej Dajnowski, Kazimierz Parys, Tom Tague, the Military University of Technology Institute of Optoelectronics.

Lasers in the Conservation of Artworks XIII – Siano & Ciofini (eds)
© 2024 The Author(s), ISBN 978-1-032-47995-8
Open Access: www.taylorfrancis.com, CC BY-NC-ND 4.0 license

Laser cleaning of an eighteenth-century waistcoat from the Civic Museums of Modena: Preserving silk and metallic threads

V. Scaglia
Università degli Studi di Torino (SUSCOR), Centro per la Conservazione ed il Restauro dei Beni Culturali "La Venaria Reale", Venaria Reale, Turin, Italy

F. Zenucchini, A. Piccirillo & C. Ricci
Centro per la Conservazione ed il Restauro dei Beni Culturali "La Venaria Reale", Venaria Reale, Turin, Italy

ABSTRACT: A laser cleaning treatment was specifically designed to remove distinct alterations from a waistcoat, conserved at the Civic Museums of Modena (Italy), which was manufactured between 1690 and 1720 but had been remodelled during the nineteenth century. A first set of tests was conducted to remove the oxidation patina of eight different types of metallic threads and sequins of the embroidery. The second set was performed to obtain a chromatic lowering of tidemarks and stains on the silk taffetas, both from the nineteenth and the eighteenth-century. Trials were performed using Nd:YAG lasers (QS 1064; 532 nm and LQS-1064 nm) and supported by an analytical campaign carried out by means of 3D optical microscope, SEM-EDS and colorimeter. The laser treatment of the nineteenth-century silk was unsatisfactory, while positive results were obtained on the eighteenth-century silk and on all metallic threads and sequins, which led us to clean the entire garment.

1 INTRODUCTION

The present contribution aims at presenting and discussing the cleaning treatments undertaken within the conservation intervention carried out on an eighteenth-century waistcoat, which has been the subject of a Master of Art thesis in Conservation and Restoration of Cultural Heritage at the Università degli Studi di Torino, in collaboration with the Centro per la Conservazione ed il Restauro dei Beni Culturali "La Venaria Reale" (Scaglia 2021).

1.1 *Historical background and description of the waistcoat*

The waistcoat (inventory number 10 CO. II) is currently preserved in the Civic Museums of Modena, which purchased it, along with a coat (9 Co. II), during a vast acquisition campaign promoted by Gabriella Guendalini, director of the Museums between 1963 and 1982.

The description by the costume historian Grazietta Butazzi, and also an exhibition catalogue (2013) entitled *C'è moda e moda... Dall'abito aristocratico all'abito "uniforme". Restauri di abiti dei secoli XVIII, XIX, XX delle collezioni museali*, reported the 23rd of August 1975 as the date of actual acquisition from "Elena Florio of San Cassiano".

This limited knowledge has not been sufficient to precisely identify the original manufacture. Nonetheless, the meticulous historical and material studies – conducted on each parts of the artefact and supported by a targeted diagnostic campaign using optical microscopy, digital radiography, Scanning Electron Microscopy with Energy-Dispersive X-ray spectroscopy (SEM-EDX) and X-ray Fluorescence (XRF) – have suggested its original tailoring had been manufactured between 1690 and 1720 and a further modification might have followed during the nineteenth century. The tailoring cut still preserves its early eighteenth-century shape

DOI: 10.1201/9781003386872-18

(Figures 1a, b): the front is defined by elongated skirts just above the knee, a central lacing with thirteen functional buttons, long sleeves ending with decorated cuffs and wide pocket flaps, while the back presents a central small slit and a tapered silhouette. The subsequent changes made the composition and stratification of the dress very complex and characterized by different elements (Figure 1c). The main fabric in silk ivory taffetas was manufactured in the early eighteenth century and is embroidered, along the hems and lacing, with a precious floral pattern. The motif was realized with seven different types of metallic threads and two kinds of sequins, along with inner paddings made of raw linen and cardboards made of cottons. The same embroidery also decorates the pocket flaps, all buttons and the two cuffs. This ornamental area is additionally supported, inside the tailoring, by an eighteenth-century white textile, a tabby weave, which was treated with a transparent substance, perhaps adhesive, to make it more rigid (difficulties have been found in taking a sample of this substance, which has therefore not been analysed). The front of the vest was reinforced with two industrial cotton-fibers cardboards, probably made in the nineteenth century. Furthermore, underneath the latter layer, there is a second lining made in the eighteenth century: a red cotton satin. The entire waistcoat tailoring is then completed with an outer lining made again with ivory silk taffetas, that also entirely forms the pockets, the two sleeves, and the back of the garment. This fabric is machine sewed by means of a chain point and, therefore, it is believed that it was applied in the dress tailoring during the nineteenth century, which represents the most important evidence of the rearrangement that the waistcoat has suffered.

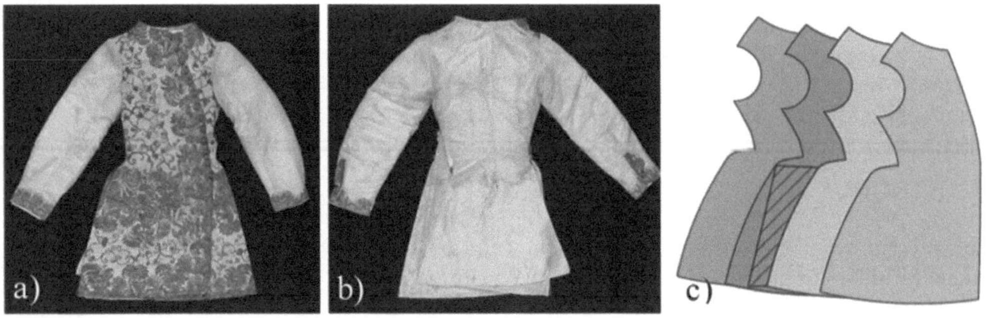

Figure 1. Waistcoat front (a) and back (b), after the treatments. Inner stratification (c): embroidered taffetas (blue), tabby wave (grey), cardboard (outlined), cotton lining (red), main lining (green).

1.2 *Analytical campaign and state of preservation*

The material heterogeneity influenced the deterioration phenomena: the most relevant losses and tears derived from the use of the dress and its incorrect manipulation. The garment also suffered from tidemarks and stains of different nature:

Type A) Grey tidemarks on the eighteenth-century taffetas likely produced by the solubilization of atmospheric particulates.

Type B) Brown tidemarks on the nineteenth-century taffetas of the sleeves and main lining due to the use of the dress and the consequent sweating.

Type C) Red tidemarks and stains located on the eighteenth-century and on the nineteenth century taffetas, produced by the solubilization of a polysaccharide substance (identified by means of FTIR analysis);.

The eighteenth-century metallic threads and sequins also had a widespread and dark oxidation, which characterized the entire embroidery. Morphology, elemental composition, and oxidation of these metallic parts have been investigated by optical microscopy, SEM-EDS and XRF. All threads, consisting mainly of a yellow silk core and a copper alloy lamina, gilded on both sides with a silver and gold alloy, have been found to be superficially altered by sulphides and chlorides.

1.3 *Research aims*

The alterations A, B, and C on textiles and the oxidation of the embroidery were so extensively spread over the artwork surface that it has been essential to seek a suitable cleaning methodology. The latter had to be aimed at preserving every historical evidence without unpick any of the tailoring and respect the conservation requirements of all garment's component. Furthermore, chemical compatibility and selective actions were needed, in order to prevent any alteration of the complex stratification of the garment. Additionally, since the complete removal of metal oxidation was unrealistic – due to the peculiar three-dimensional conformation of the laminas – a further aim of the intervention was to obtain an acceptable cleaning of the embroidery from an aesthetic point of view, by avoiding any mechanical damage to the threads and sequins.

To these goals, two systematic experimentations have been defined: the first has been focused on the removal of silver sulphide from the embroidery made of metallic threads, the second concerned the cleaning of tidemarks and stains on silk.

1.4 *Materials and methods*

Preliminary in-depth studies have been conducted to establish the scientific support for both the mentioned experimentations. Different methods have been considered, evaluating benefits and disadvantages from a scientific and conservation perspective: some of the most frequently used have been tested. In particular, chemical cleaning using rigid gels was tested, but it led to an ineffective removal of the alterations, either through macroscopical observation or colorimetric analysis. Following some positive result of the laser cleaning approach reported in the literature, preliminary tests have been conducted on the waistcoat in order to establish the best laser instrumentation and the most suitable treatments.

Three laser instrumentation, commonly found to be effective in similar case-studies, have been tested. 1) fibre-coupled Long-Q-Switched Nd:YAG(1064 nm, 130 mJ, 100 ns) laser (EOS 1000 LQS, El.En., Calenzano, IT), which allows to vary the spot diameter to the target between 1.5-6 mm. Pulse energy control by neutral filters (T50% filter, T25%). 2) articulated arm-coupled Q-Switched Nd:YAG(1064, 532, 355 nm) laser (Thunder Art, Quanta System), handpiece with fixed focal length, which allow the use of a homogenizer.

Tests have been conducted in dry conditions, by combining the laser ablation with a complementary removal of the residues using a micro-vacuum cleaner. Furthermore, those residues that still were present on the artefact surface after the cleaning, have been removed with polyurethane (PU) sponges.

2 RESULTS

2.1 *Cleaning test of the embroidery*

2.1.1 *Preliminary considerations*

According to the literature, the removal of oxidation from metallic threads can be addressed by means of different methods, depending on the characteristics of the threads themselves and of the artefact on which they are found. Mechanical approaches have been ruled out because of the associated risk to damage the gilding and the structure of the threads. Chemical methods are easily practicable when the object can undergo a rinse bath, which was not the case of the present waistcoat. Here, a preliminary experimentation was carried out based on a reducing and a chelating solution supported in different rigid gels, which were applied on a limited area of the garment.

The assessments relating to the types of gels, the choice of surfactants and their respective concentration have been published, along with the results obtained, in a previous article (Scaglia et al. 2022). Results have been characterized using a colorimeter that showed the oxidation was not satisfactorily removed, thus also this method was ruled out. For similar reasons, electrochemical methods were also excluded. Eventually, the studies reported along the last

Table 1. Operating parameters for laser physical cleaning, found in literature.

Reference	Laser	Wavelength (nm)	F(J/cm^2)	Frequency (Hz)
Patti & Siano 2017	Nd:YAG QS	532	0.10-0.90	2-5
Elnaggar et al. 2015	Nd:VAN	193-355	1.91	10
Degrigny et al. 2003	Nd:YAG	355	0.08	2-10
Sokhan et al. 2003	Nd:YAG	532	0.08	/
Jong-Myoung et al. 2003	Nd:YAG QS	266	2.60	10

twenty years in which laser in cleaning tarnished metallic threads with an inner silk core (Table 1) have been considered.

This literature encouraged the testing of the laser cleaning on the embroidery of the waistcoat using relatively low fluence, F.

Moreover, bibliographic references on laser cleaning of paper (Ciofini et al. 2013, Patti & Siano 2015, Ochocinska et al. 2003) reported that laser Nd:YAG at 532 nm, with fluence lower than 1 J/cm^2, does not cause photolysis, cellulose oxidation, photothermal alteration or paper yellowing. This suggested that no negative effect to the cardboard components of the embroidery and of the tailoring of the waistcoat would be expected.

On the basis of these considerations, the laser cleaning has been evaluated as potentially suitable for the treatment of the embroidery of the waistcoat, and for this reason a detailed experimentation plan has been designed and carried out, as described hereafter.

2.1.2 *Preliminary tests*

Preliminary cleaning tests (Table 2) have been conducted on a limited area of the embroidery to identify those operating parameter suitable for an efficient and safe cleaning. Laser systems have been tested by progressively increasing the fluence in order to identify optimal irradiation ranges.

Once these parameters have been set, all laser instruments have been tested again on an area decorated with leaves, each leaf characterized by the same kind of metallic threads and similar grade of alteration (Figure 2a). In this case, it has been decided to select the minimum fluence for each laser, which had previously provided an effective cleaning, so that it was possible to safely irradiate the yarns from multiple angles and to obtain a homogeneous removal, always associated with micro-aspiration of the residues. The upper part of each leaf was cleaned using PU sponges.

As a result (Figure 2b), the LQS 1064 nm (F=0.26 J/cm^2) removed the black oxidation but the metallic threads underneath lost their characteristic golden colour and showed the typical hue of the silver substrate. Also using the QS 1064 nm laser (F=0.16 J/cm^2) produced a similar invasive effect. On the contrary, the QS 532 nm laser (F=0.08 J/cm^2) has allowed to obtain an effective ablation of the oxidation layer, and uncover the original gilded surface without relevant discoloration/damage effects. PU sponges proved to be very useful to remove ablation and were hence integrated within the overall cleaning process (Figure 2c).

Lastly, to better evaluate the aesthetic difference between the effects of the QS 1064 nm and QS 532 nm treatments, two areas of the embroidery, characterized by similar oxidation grade, have been further compared by using the same parameters and operating conditions of the previous tests (Figure 3). This confirmed that 532 nm laser was more effective and decidedly safer in guarantying the integrity of the gilding layer. Contextually, some analytical measurements have been undertaken to obtain, on the one hand, a quantitative confirmation of the aesthetic improvement occurred and, on the other hand, a concrete information on the possibility of incurring into damage during irradiation. In particular, colorimetric analyses of the irradiated areas have been conducted, which showed clear increase in brightness (ΔL) and yellow component (Δb) after the treatment, as expected according to the visible recover of the gilding.

Table 2. Preliminary test to identify the optimal operating parameters range for each laser.

Test	Laser	F (J/cm^2)	Results
1	LQS 1064 nm	0.46	Significant ablation residues, silver color
2	LQS 1064 nm	0.26	Significant ablation residues, silver color
3	QS 1064 nm	0.16	Effective removal of oxidation, silver color
4	QS 1064 nm	0.21	Effective removal of oxidation, silver color
5	QS 532 nm	0.04	Partial removal of oxidation, golden color
6	QS 532 nm	0.08	Effective removal of oxidation, golden color
7	QS 532 nm	0.11	Effective removal of oxidation, golden color

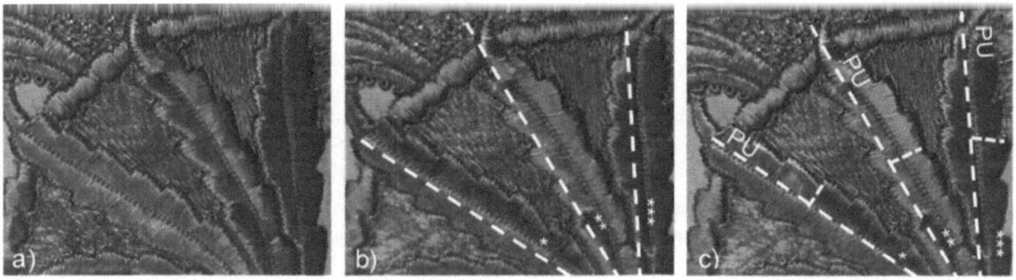

Figure 2. Embroidered leaves before the laser cleaning (a); after laser cleaning on the right part (b) and after the PU sponges cleaning on the upper part (c). *) LQS 1064 nm (1 Hz, F=0.26 J/cm^2). **) QS 1064 nm (1 Hz, F=0.16 J/cm^2). ***) QS 532 nm (1 Hz, F=0.08 J/cm^2). PU: sponges cleaning.

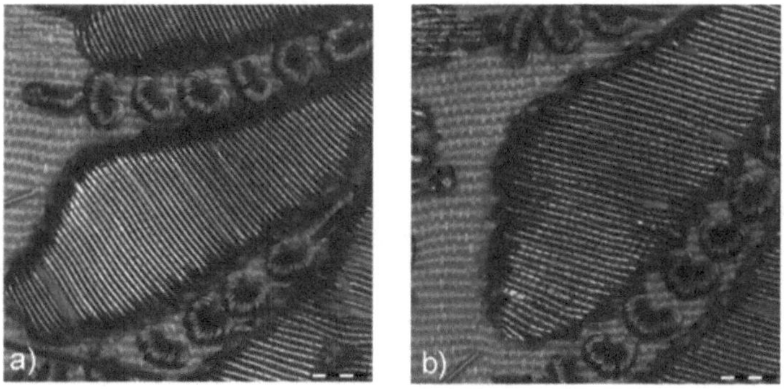

Figure 3. Laser cleaning of embroidery: a) QS 1064 nm, 1 Hz, F=0.16 J/cm^2; b) QS 532 nm, 1 Hz, F=0,08 J/cm^2. Pictures taken via 3D optical microscope.

Table 3. Colorimetric analysis values.

	Before treatment			After laser cleaning			After PU sponges cleaning		
	L*(D65)	a*(D65)	b*(D65)	L*(D65)	a*(D65)	b*(D65)	L*(D65)	a*(D65)	b*(D65)
QS 532 nm	49.79	3.83	13.54	66.76	7.11	30.16	72.7	4.08	21.11

QS 532 nm before treatment and after laser cleaning	ΔL	16.97
	Δa	3.28
	Δb	16.62
	ΔE00	17.48
QS 532 nm before treatment and after PU sponges cleaning	ΔE00	20.28

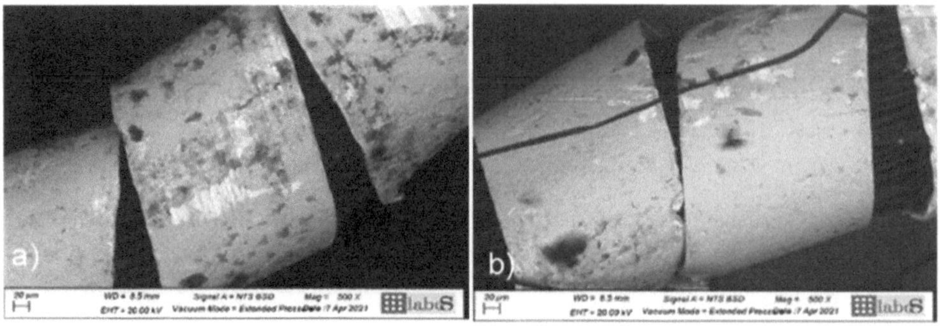

Figure 4. SEM images before (a) and after (b) cleaning with QS 532 nm laser, 1 Hz, F=0.08 J/cm^2.

Some erratic fragments of altered metallic threads have been also treated according to the selected parameters and operating conditions previously set (QS 532 nm, 1 Hz, F=0.08 J/cm^2), which have been observed through SEM (Figure 4). This allowed to prove the lack of any laser induced alteration of the metallic and textile components. In addition, it was noted that the removal of the oxidation products was not complete, thus confirming the correctness of the irradiation parameters used.

2.1.3 *Extending the test*

The positive preliminary results encouraged us to extend the cleaning to an area that presented all kinds of metallic threads. In this way it would have been possible to observe the overall result of the cleaning and thus evaluate the possibility of further extending this treatment to the entire embroidery. The QS 532 nm laser has therefore been operated using the same parameters and procedures of the preliminary test (1 Hz, F=0.08 J/cm^2), which confirmed the correctness of the approach (Figure 5).

2.2 *Cleaning test on silk taffetas*

2.2.1 *Preliminary considerations*

Undesired alterations located on the silk taffetas of the waistcoat have initially been subjected to three preliminary tests, aimed at identifying the most effective method for their removal.

Figure 5. Laser cleaning of embroidery: before (a) and after (b) irradiation (QS 532 nm, 1 Hz, F=0.08 J/cm^2).

The first test consisted in a cotton swab, slightly soaked in osmotized water. This provided a negative result in all the present cases, thus allowing to understand that none of the tidemarks and stains could have been removed only by means of water.

It has therefore been decided to test the use of surfactants, suitably supported by different rigid gels. None of the applied combinations of different concentrations of Saponin and Dehypon® LS54, supported respectively in Gellano, Agar or Nanorestore Gel® HWR, made it possible to obtain even a slight reduction of the alterations, from a macroscopic observation.

The ineffectiveness of all the previous tests led us to test QS 532 nm laser also on silk, a decision also supported by previous applications on similar conservation cases reported in the literature (Table 4). These references have shown practicable operational ranges below damage threshold of silk and have testified that it does not undergo mechanical or combustion damage.

A further information has been gained from an experimental setup developed at the CCR including a thermal camera used to estimate the maximum temperature and heat dissipation time during laser irradiation of nineteenth-century ivory-coloured taffetas fragments, characterized by an evident depolymerization of the fibres and a widespread deposit of greyish particulate. These tests were conducted with the QS 532 nm laser on an area of 1 cm^2, by gradually increasing the fluence and frequency, under the damage threshold, up to a maximum average temperature of about 45 °C (Table 5).

Table 4. Parameters and operating procedures for laser cleaning of textiles obtained from literature.

Reference	Sample	Alteration	Laser	Wav. (nm)	F (J/ cm^2)	Freq. (Hz)	Analysis
Patti & Siano 2017	Old silk waved in new taffetas	Artificial	QS Nd:YAG	532	0.9-1.6	4-5	Raman, SEM
	Brocade cotton (XVI)	Adesive deposits, stains	QS Nd:YAG + water.	532	0.3-0.6	2-10	Raman, SEM
Bagnoli et al. 2012	Silk taffeta (XII)	Oxidation stains	QS Nd:YAG	532	/	/	/

Table 5. Maximum parameters obtained during laser irradiation of nineteenth-century taffetas fragments.

Laser	F (J/cm^2)	Frequency (Hz)	Temperature (°C)	Dissipation time (s)
Nd:YAG QS 532 nm	1.50	2	37	10
Nd:YAG QS 532 nm	1.70	4	45	10

2.2.2 *Preliminary tests on silk*

Cleaning tests were conducted on the nineteenth-century silk that formed the sleeves, back and main lining of the dress, and was mainly characterized by type B and C alterations, described above. The purpose of these tests was to use QS 532 nm laser, in order to identify a possible cleaning method able to remove the alterations or at least decrease their chromatic intensity. QS 532 nm laser has been used with 1 Hz, 7 mm spot and F=0.39-0.59 J/cm^2. Both this early treatment, that caused yellowing of all tidemarks (Figures 6a, b), and the following one made using the handpiece with the beam homogenizer and F=1.03-1.88 J/cm^2 returned similar unsatisfactory results (Figures 6c, d). The laser removal of the alterations from the nineteenth-century silk has therefore been unfortunately discarded.

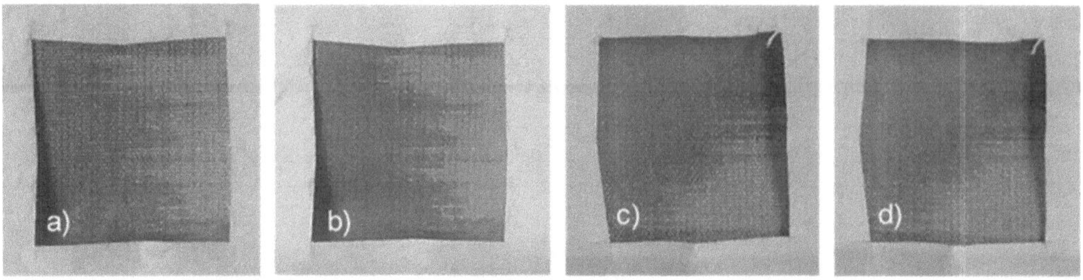

Figure 6. Laser treatment of altered nineteenth-century silk: before (a) and after (b) laser irradiation using F= 0.53 J/cm^2, before (c) and after (d) laser irradiation using F= 1.35 J/cm^2.

The second series of cleaning tests has been performed on the eighteenth-century silk taffetas in an area that was characterized by type A and B alterations (Figure 7a), and in a second area with type C alteration (Figure 7b); laser has been used at 1 Hz, 7 mm spot and the homogenizer handle. The range of fluence in which an effective result was possible for the removal of alterations A, B and C was F=0.59-0.76 J/cm^2. By effective removal, however, it is not meant a complete removal of the alterations, now strongly penetrated inside the fibres, but a significant lowering of their tone, which visually compromised the garment. The observation of the tests has been conducted through stereomicroscope and colorimetric analyses. The latter made it possible to notice a general increase in brightness (L*), for both areas treated. Moreover, the area characterized by type C alteration have led to a slight decrease of the red tone, passing from a*= 4.15 to a*= 3.09 (Table 6). The general lowering of the alterations was also confirmed by SEM analysis, for which yarns have been sampled from altered areas and have subsequently been irradiated at F=0.76 J/cm^2. With this analysis it has been clearly observed that no mechanical alterations were induced on the silk fibers (Figure 7c).

Table 6. Colorimetric analysis values.

	Before laser cleaning			After laser cleaning			ΔE00
	L*(D65)	a*(D65)	b*(D65)	L*(D65)	a*(D65)	b*(D65)	
Reference*	72.65	2.4	18.95	/	/	/	/
Type A-B	72.05	2.95	19.01	76.93	2.24	20.81	3.82
Type C	67.54	4.15	21.64	72.42	3.09	23.35	4.06
Reference and type A-B							3.3
Reference and type C							2.29

* Reference: Taffetas area characterized by a minimum alteration, chosen by unaided eye observation.

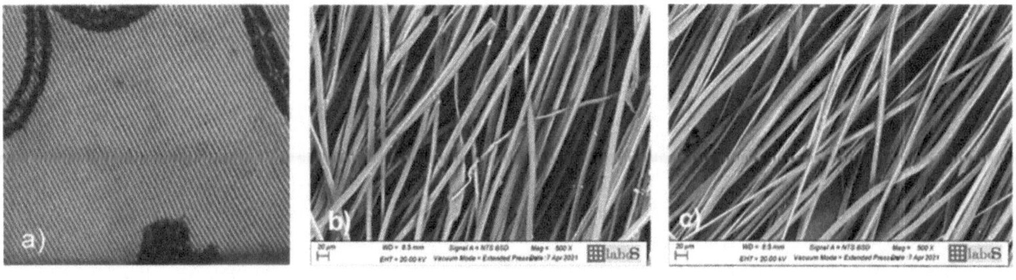

Figure 7. Laser cleaning test on type A and B alterations of silk taffetas (a) and SEM analysis images before (b) and after (c) laser cleaning of type C alteration.

2.2.3 *Extending the test*

Given the positive preliminary results achieved, the cleaning of a larger area of the eighteenth-century silk has been performed (Figure 8). The frequency value has been maintained at 1 Hz and the value of the fluence has been calibrated according to the different alteration phenomena: type C has been treated with F=0,76 J/cm^2, type B and A have been treated with F=0.69 J/cm^2 and, lastly, silk that presented only type A alteration has been treated with F=0.59 J/cm^2 (Figure 8b). In this way, it has been possible to mitigate the alteration caused by tidemarks and stains and restoring, as much as possible, the ivory hue that the silk exhibits in relatively well preserved areas.

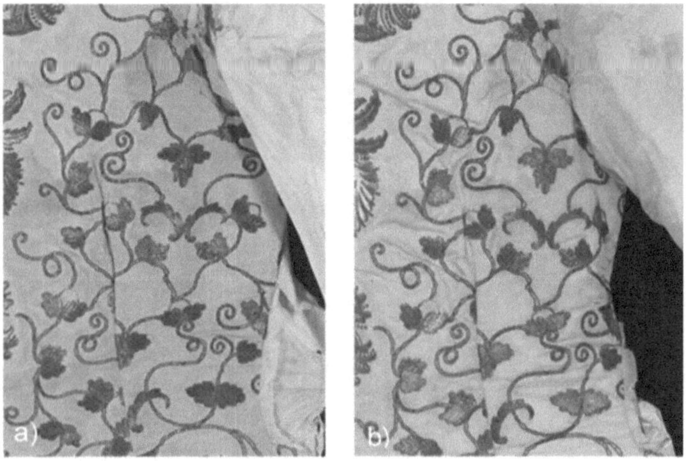

Figure 8. Before (a) and after (b) laser cleaning using optimized fluences.

3 FINAL RESULTS

Satisfactory results were obtained on embroidery and eighteenth-century silk taffetas using QS 532 nm laser for the cleaning of the artwork after preliminary consideration and trials (Figures 5 and 8).

The embroidery has been treated mostly maintaining the optimal parameters identified through the preliminary experimentation (QS 532 nm, 1 Hz, F=0.08 J/cm^2). However, according to the interaction effects observed, the fluence has been slightly modified during the treatment by varying the distance of the handle from the artefact, thus obtaining an estimated range of fluence between 0.05-0.11 J/cm^2. The frequency has also been gradually increased up to 4 Hz, to speed up the operation and make homogeneous the laser cleaning action.

Similarly, the eighteenth-century silk taffetas have been treated according to the parameters previously established (QS 532 nm, 1 Hz): F=0.76 J/cm^2 for alteration type C, F=0.59-0.69 J/cm^2 for alterations type B and A. In order to avoid accidental irradiation of silk or metallic threads with unsuitable parameters, paper masks have been specifically crafted in multiple shapes and used when necessary to suitably shield areas to be protected.

Finally, the embroidery and silk taffetas of the present eighteenth-century waistcoat have been successfully laser treated. The chromatic lowering of the alteration areas has made them less evident, and a more homogeneous and aesthetically acceptable appearance of the artefact was recovered (Figure 9a, b).

4 CONCLUSIONS

This study approached cleaning issues encountered during the conservation-restoration of an eighteenth-century waistcoat, which are very common for textile artefacts, which are very often untreatable due to the heterogeneous nature and fragility of their decorations and basic

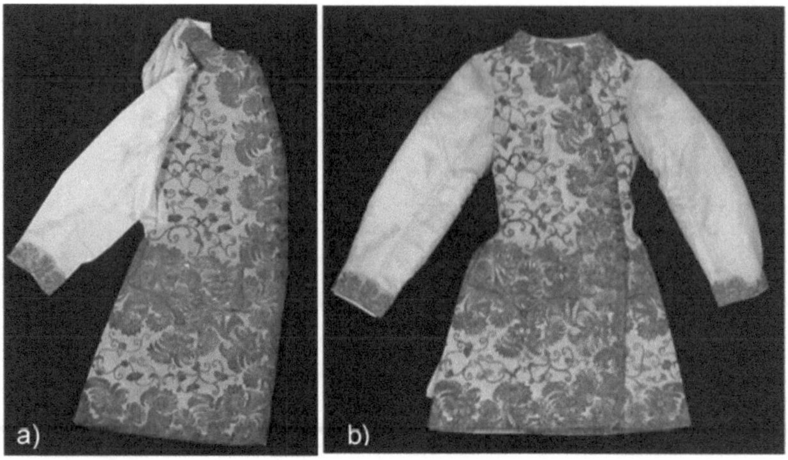

Figure 9. Result of the overall laser treatment of the waistcoat: before (a) and after (b).

materials. Different cleaning methods were tested on the embroidery in metallic threads, on the eighteenth-century taffetas, and also on the nineteenth-century textiles of the tailoring.

The physical cleaning methods based on second harmonic QS Nd:YAG(532 nm) laser irradiation in dry condition showed high effectiveness on the garment, as documented in the present work. This result demonstrates that laser cleaning may represent a viable innovative solution for the conservation treatments of complex textile artefacts.

REFERENCES

Bagnoli, D., Mascalchi, M., Nucci, L., & Siano, S. 2012. Trattamento laser di tessili: dall'abito funebre di Sigismondo Pandolfo Malatesta allo studio del paliotto della Beata Giulia della chiesa dei Santi Jacopo e Filippo di Certaldo. In A. Brunetto (ed) *Aplar 4 Applicazioni laser nel restauro*: 253–265.

Ciofini, D., Osticioli, I., Micheli, S., Montalbano, L. & Siano, S. 2013. Laser removal of mold and foxing stains from paper artifacts: preliminary investigation. *SPIE Conference Proceedings* 9065: 12–11.

Degrigny, C., Tanguy, E., Le Gall, R., Zafiropulos, V. & Marakis, G. 2003. Laser cleaning of tarnished silver and copper threads in museum textiles. *Journal of Cultural Heritage*: 152–156.

Elnaggar, A., Fitzsimons, P., Nevin, A., Osticioli, I., Ali, M. & Watkins, K. 2015. Investigation of ultra-fast picosecond laser system cleaning of metal decoration of 17th c. gloves of King Charles I. *e-Preservation Science* 12: 14–19.

Jong-Myoung, L., Jae-Eun, Y. & Yang-Sook, K. 2003. Experimental study on the effect of wavelength in the laser cleaning of silver threads. *Journal of Cultural Heritage* 4: 157–161.

Ochocinska, K., Martin, M., Bredal-Jorgensen, J., Kaminska, A. & Sliwinski, G. 2003. Laser cleaning of historical paper documents and diagnostics by means of LIF technique. *Proceedings of the SPIE* 5229: 296–300.

Patti, M. & Siano, S. 2015. *La carta, applicazioni laser*. Milano: Nardini Editore. ISBN: 9788840403007.

Patti, M. & Siano, S. 2017. *I tessili, applicazioni laser e altre indagini per i materiali fibrosi*. Noventa Padovana: Nardini Editore. ISBN: 9788840403038.

Scaglia, V. et al. 2021. *"Marsina" e sottomarsina settecentesche: analisi tecnica, storica e restauro di due abiti rimaneggiati provenienti dai Musei Civici di Modena*. Master Thesis, University of Turin, 2021 (Unpublished).

Scaglia, V., Palei, G., Gulmini, M., Failla, M.B., De Blasi, S., Piccirillo, A., Zenucchini, F. & Genta, R. 2022. Sperimentazione di pulitura chimica dei filati metallici di una sottomarsina settecentesca dei Musei Civici di Modena. *Lo Stato dell'Arte* 20: 81–87.

Silvestri, I. & Lorenzini, L. 2013. *C'è moda e moda... Dall'abito aristocratico all'abito "uniforme". Restauri di abiti dei secoli XVIII, XIX, XX secoli delle collezioni museali*. Modena: Museo Civico d'Arte.

Sokhan, M. et al. 2003. Initial results on laser cleaning at the Victoria & Albert Museum, Natural History Museum and Tate Gallery. *Journal of Cultural Heritage* 4: 230–236.

Lasers in the Conservation of Artworks XIII – Siano & Ciofini (eds)

PVC cleaning via different methods: Comparison of laser and CO_2 snow

M. Havlová, K. Kocová, J. Neoralová, D. Novotná & P. Vávrová
National Library of the Czech Republic, Collection Preservation Division, Prague, Czech Republic

L. Mašková
The Czech Academy of Science, Institute of Chemical Process Fundamentals, Prague, Czech Republic

ABSTRACT: In this work, we discuss two conservation methods, such as laser ablation and two-phase spray of CO_2 snow particles, evaluate their effectiveness in cleaning softened PVC samples and eventually we compare them to classical mechanical cleaning. First, we tested the mentioned methods on clean samples, which were later subjected to accelerated ageing. The results show that laser and CO_2 snow do not increase the aging of the PVC samples. The cleaning performances of laser and two-phase spray were assessed on samples with artificial encrustation made of dust and artificial sebum soil (palmitic acid). Afterwards, the mentioned methods were used for cleaning real PVC samples. The results presented are very promising and they show that laser ablation and CO_2 methods represent effective alternatives to the mechanical cleaning of library collections.

1 INTRODUCTION

Polyvinylchloride (PVC) is one of the most commonly used plastic material. It is also widely used in library collections in bookbindings and covers. For this reason, it is important to study cleaning and conservation procedures for this type of material (Shashoua 2008).

In the last pair of decades, new methods for the removal of contamination from modern library collections were studied and developed. One of them is laser ablation, which has become a very promising method for cleaning sculptures, paintings, or textiles (Shashoua 2012).

Another cleaning method, which was also examined in recent years, is based on two-phase spray of CO_2 snow particles in a carrier gas stream (Feng 2017). This method is also thought to be promising in various areas of conservation and preservation, for the stream of CO_2 particles and carrier gas (in this case nitrogen) is less abrasive than conventional mechanical cleaning methods and the CO_2 particles do not erode the surface of cleaning material (Silverman, 2008).

This work is one of the outcomes of the NAKI II project "Research and development of advanced techniques of cleaning of books and manuscripts" supported by the Czech Ministry of Culture. In this project, two cleaning methods were explored, namely the mentioned laser ablation of two-phase spray. Aim of this work was to compare the performance of these two methods in the case of softened PVC and compare them with the mechanical cleaning (via Swedish cloth).

2 MATERIALS AND METHODS

The studied artificial samples were made from softened pigmented PVC. There were two main types of artificial dirt investigated: commercial Ashrae dust and artificial sebum (palmitic acid). These two types of contaminants were chosen, because they imitate common types of dirt. In this case, palmitic acid was used to simulate the oily types of dirt, such as fingerprints.

DOI: 10.1201/9781003386872-19

For all prepared samples, contact angle, roughness and color change were measured before the application of artificial dirt (or before artificial aging), and after cleaning (or after artificial aging).

For the artificial aging experiment, three sets of samples were prepared (control group, samples for TEA CO_2 laser and for two-phase CO_2 spray). The two groups of samples were cleaned by respective methods and the artificially aged for 28 days under Xe lamp. For several samples, the FTIR analysis was also performed. Furthermore the glass transition temperature and the tensile strength were also measured.

Artificial sebum was prepared from palmitic acid and propanol (1:5) and for each sample 0.04 ml was used (Shashoua 2008).

For the TEA CO_2(10.6 μm) laser (Plovdiv University, Bulgaria), the energy density released to the target was 0.91 J/cm^2 and the number of pulses was 10.

The two-phase spray was generated using a SnoPen 2000 instrument (Cleanlogix) with TSA-CF-7538 nozzle with the N_2 as a carrying gas. The flow rate of N_2 and CO_2 were 0.6 l/s and 3.0 g/s, respectively.

3 RESULTS AND DISCUSSION

3.1 *Artificial aging*

The results show, that neither of the methods used have an influence on aging processes. The results from measurement of contact angle, color change, and roughness show no difference compared to the control group (which was not cleaned by any method before aging). For example, the differences in color change, contact angle or roughness were within the range of uncertainties of a measurement (less than 2° in case of contact angle, $\Delta E \leq 0.3$ in case of color change). The FTIR analysis and measurement of glass transition also show no significant difference between all three groups. Following these promising results, it was possible to continue with the cleaning of artificial dirt.

3.2 *Cleaning*

All three methods (laser, CO_2, and mechanical) were tested on prepared samples made using commercial Ashrae dust and artificial sebum (palmitic acid).

The character of the two types of the artificial soiling is very different and therefore we can derive from the present results much more than from the testing of only one type of dirt. Figure 1 shows the comparison of all types of cleaning methods and their performances for both types of artificial dirt. The top row shows the cleaning of artificial sebum. As it can be seen, all methods removed such a type of dirt quite well, however, there are some differences. These are even more visible in Figure 2, which shows all types of samples before and after cleaning. In this case, the most successful method was the mechanical cleaning and laser ablation; however, as the more detailed picture (Figure 2b) shows, the laser ablation causes a melting of palmitic acid crystals. This means, that the artificial sebum was not removed during the cleaning process, but it rather changed into a thin and more homogenous layer. This is the case only for the laser ablation, while the other two methods showed the removal of a palmitic acid and no change in the remaining dirt.

In the case of Ashrae dust (bottom row in Figures 1 and 2), we can observe similar behavior of cleaned dirt. In the case of laser ablation, the Ashrae dust parts were smashed by the laser into smaller particles, which produced a thin layer of dust on the sample surface. Therefore, for this type of dirt, laser ablation is also not suitable.

There is also a noticeable difference between the sample cleaned by Swedish cloth and by CO_2 snow. As we can see in both figures, the CO_2, snow cleaned samples more efficiently. In case of mechanical cleaning, some of the larger particles of Ashrae dust were broken and, similarly as in the case of laser ablation, a thin layer of dust was formed.

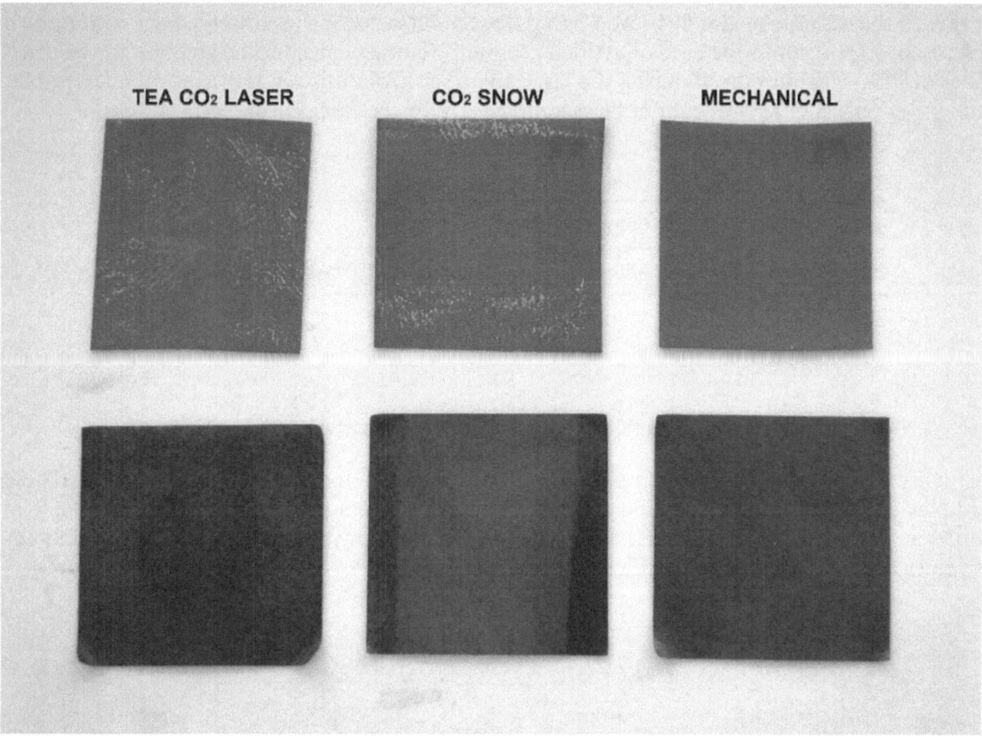

Figure 1. Comparison of samples after cleaning: artificial sebum (top) and Ashrae dust (bottom).

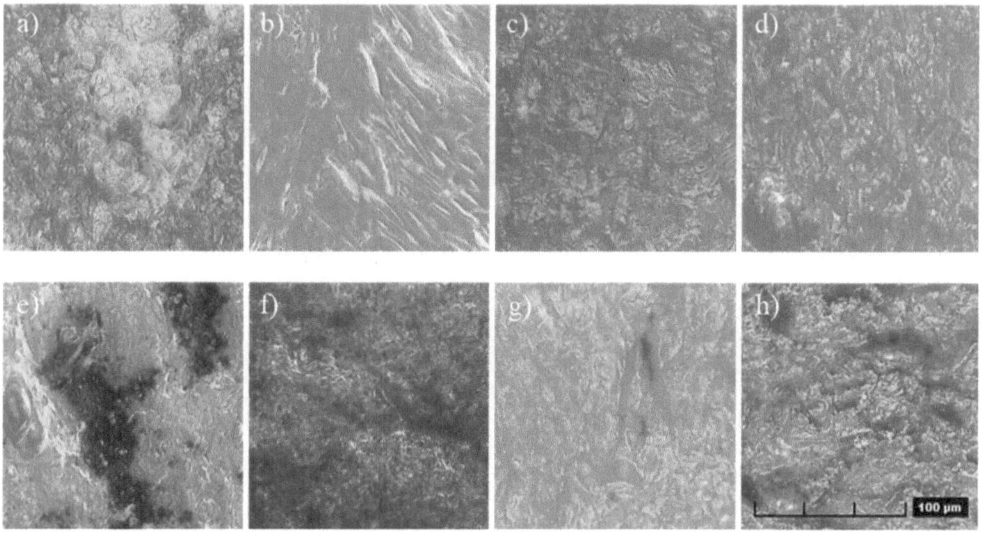

Figure 2. Comparison of a samples before and after cleaning (Hirox, 600x). Artificial sebum: a) before cleaning, b) laser, c) two-phase spray, d) mechanical cleaning; Ashrae dust: e) before cleaning, f) laser, g) two-phase spray, h) mechanical cleaning.

Table 1. shows the results from the measurement of color change. From this point of view, the most successful methods for cleaning the artificial sebum were laser ablation and mechanical cleaning. In the case of Ashrae dust, the most successful method was two-phase CO_2

spray. In the case of Ashare dust, we observed a similar behavior, as observed through optical microscopy (Hirox). In the case of artificial sebum, we must consider the changes in the character of dirt, which occurred during the laser ablation. From this point of view, the laser ablation using pulsed CO_2 laser cannot be recommended for the cleaning of artificial sebum.

Table 1. Color change ΔE in case of different methods.

	Before cleaning		After cleaning	
ΔE	Ashrae	sebum	Ashrae	sebum
TEA CO_2 laser	30.7	18.2	38.5	0.8
CO_2 spray	32.4	22.1	9.8	4.6
mechanical	26.6	23.9	23.7	0.7

Figure 3 shows the comparison of all methods described above on the real samples. These samples came from a bookbinding (circa 1970) made from softened PVC. This type of real dirt is more similar to Ashrae dust tested previously. As shown, the CO_2 spray and the Swedish cloth clean the sample more efficiently than laser ablation.

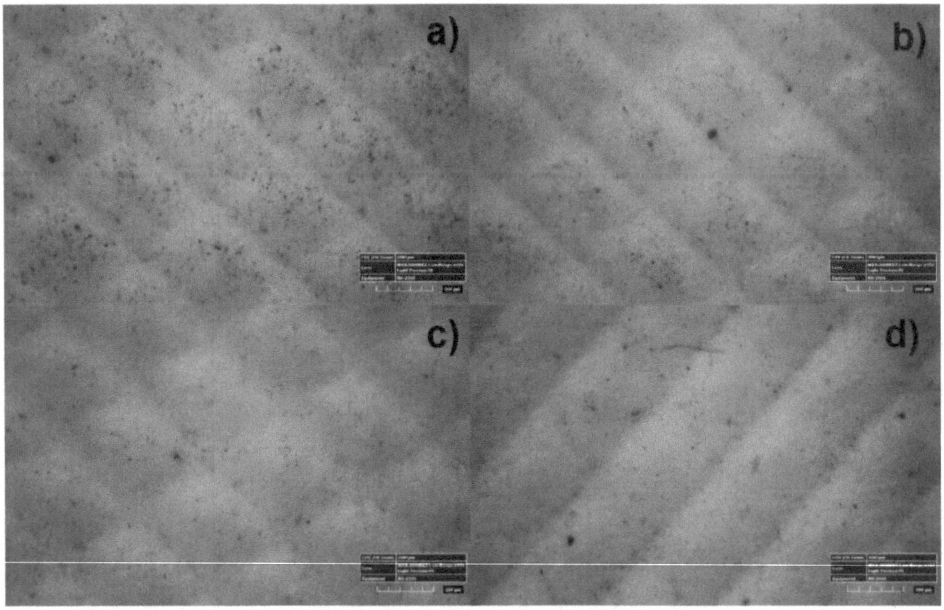

Figure 3. Comparison of the cleaning methods on the real PVC sample: a) before cleaning, b) laser ablation, c) CO_2 spray, d) Swedish cloth.

4 CONCLUSIONS

Results show, that neither TEA CO_2 laser nor two-phase spray affect aging of the samples. In the case of artificial dirt, all used methods showed better performance in case of artificial sebum (palmitic acid). In this case, the two best methods are two-phase spray and mechanical cleaning. In the case of Ashrae dust, the best results were also obtained by using mechanical cleaning or two-phase spray. In the case of real dirt, the results were similar. Thus, the present study demonstrates TEA CO2 laser ablation is not suited for cleaning treatments of PVC artefacts.

ACKNOWLEDGEMENTS

The financial support of this work has been provided by the Ministry of Culture of the Czech Republic under grant DG18P02OVV048.

REFERENCES

Feng, X. et al. 2017. Development of CO_2 snow cleaning for in situ cleaning of CMM stylus tips. *Measurement Science and Technology* 28: 015007.

Shashoua, Y. *Conservation of Plastics: Materials science, degradation and preservation*. Oxford: Elsevier, 2008.

Shashoua, Y. 2012. Studies in active conservation of plastic artefacts in museums. In, Bertand, L., Fournier, A., and Martin, G. (eds) Preservation of plastic artefacts in Museum Collections (pp. 219–269) Paris: Editions du Comite des travaux historiques et scientifiques.,

Silverman, R. Fire and Ice: A Soot Removal. Technique Using Dry Ice Blasting. Archival Products News. 2008, 15, 3, 1–8.

Lasers in the Conservation of Artworks XIII – Siano & Ciofini (eds)

Diagnostics and conservation of an archaeological 'coin purse' from the Vesuvian area

G. Rossignoli, A. Patera, G. Lanterna & I. Tosini
Opificio delle Pietre Dure, Florence, Italy

E. Gualandris
Freelance Textile conservator, Prato, Italy

J. Agresti, D. Ciofini & S. Siano
Institute of Applied Physics "Nello Carrara" (IFAC-CNR), Sesto Fiorentino, Florence, Italy

F. Miele
MANN, Museo Archeologico Nazionale di Napoli, Naples, Italy

ABSTRACT: The "coin purse" from the excavations of the Vesuvian area is an exceptionally valuable textile artefact, one of the few organic material artefacts that did not carbonize during the eruption of Vesuvius in 79 AD. The find was in a poor status of preservation due to a thick deposit of brownish-grey soil that entirely covered the surface and compacted the fibres. One of the openings showed remnants of bright green corrosion products coming from the metal objects inside. Preliminary investigations were carried out to understand the nature of the materials and the level of degradation by false-colour IR imaging, UV fluorescence, Scanning Electron Microscopy (SEM). Radiographic acquisitions were also conducted in order to study the characteristics of the purse and the metal coins inside. After the diagnostic phase, it was considered appropriate to proceed with the dry removal of soil deposits by micro vacuuming followed by laser ablation of the residual deposits firmly bound to the fabric. Finally, copper corrosion products were extracted via chemical methods.

1 INTRODUCTION

The "coin purse" from the Archaeological Museum of Naples (MANN) here characterized and subjected to conservation treatments was made using a fabric wrapped around itself several times, so called because of its typical cylindrical shape and the presence of coins inside (Melillo, 2020) (Figure 1). It was likely excavated in Pompeii and represent an exceptionally interesting piece both for its completeness and for the fact that it is one of the few organic material artefacts that did not carbonize during the eruption of Vesuvius in 79 AD (Fiorelli, 1873).

In general, it is important to remind that the organic material remains of the two main urban sites destroyed by the eruption exhibit different degradation and deterioration phenomena. In Pompeii they decomposed in the very permeable layer of ash, leaving an imprint in the ground, while in Herculaneum such remains were sealed within the pyroclastic mud that protected them from atmospheric agents and bacteria along the centuries, thus allowing the preservation of very fragile and perishable constitutive materials such as wood, fabrics and foodstuffs (Pappalardo 1990).

The present object was made using a beige vegetable fibre fabric with a plane weave, which appeared almost intact in some areas, although the overall status of preservation was rather poor. Earthy materials with grey-brown hue entirely covered its surface, thus compacting the

DOI: 10.1201/9781003386872-20

fibres and making them dry. In some areas, the deposit was in the form of small greyish pebbles and, especially at the openings, appeared compact and crystallised, where the weave was no longer visible. Earthy deposits consolidated the fibres of the fabric, which were very dehydrated, thus their removal appeared very difficult because of the risk of considerable losses, breakage or fraying of textile material, especially near the endings.

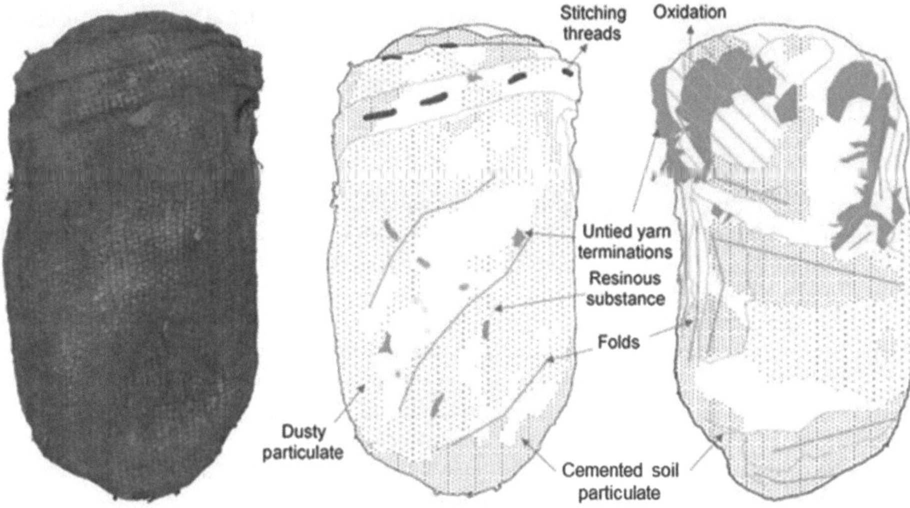

Figure 1. The 'coin purse' before the treatment, side C; conservation survey, side C and side A. (Gualandris 2019).

Technical data sheet		
Armour	Weave	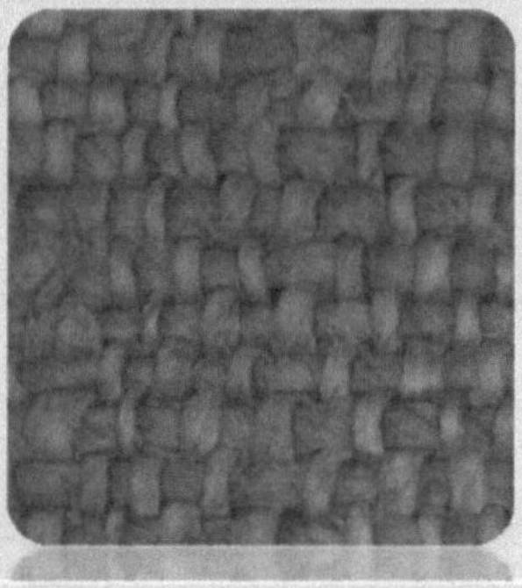
Warps	12-13/cm	
Weaves	11-12/cm	
Torsion	S	
Heads	unknow	
Material	Linen – *Linum usitatissimum*	
General features		
Weight	Before treatment 100.4 gr	After treatment 99 gr
Dimension	9.5 x 5.4 x 3.5 cm	
Site of discovery	Vesuvian area	

Figure 2. Technical data sheet (taken from Gualandris 2019).

The present purse has two openings: the shape is semi-circular on one side, in which nine layers of fabric are clearly visible, and the other one is barely discernible. Its sizes are 9.5×5.4×3.2 cm. Before the conservation treatments its weight was 100.40 g, which suggested the presence of some material content within the fabric (Figure 2).

The textile envelope showed a pronounced crushing at the middle part, while at one of the openings bright green mineral products where visible. The latter along with the mentioned significant weight provided clear clues of the possible presence of copper alloy objects inside.

2 METHODS

2.1 *Diagnostics*

UV-Induced Vis Fluorescence photography was carried out using Nikon D800 digital camera with a thermal filter removed and equipped with CMOS detector (36 mm × 24 mm FX format), nominal 36.3 MP CMOS sensor full range from 350 nm to 1000 nm. UV pass filter (365 nm), Philips L18/73 Wood tube lamp.

To study anatomical features of the fibres and the elemental composition of the soiling deposits a Scanning Electron Microscope (SEM, Zeiss EVO series) equipped with energy dispersive X-ray spectrometry (EDS) was used. For the analysis, samples were gold-coated and observed in high vacuum with low acceleration voltage (6 kV).

Radiographic examinations were performed using a commercial YXLON X-ray system (Cheetah EVO). It is equipped with a Multifocus X-ray tube operating at 20-160 kV and a flat plane detector with 1004×1004 squared pixels of 127 μm size.

2.2 *Removal of soiling*

A surgical vacuuming system (Bresciani S.r.l., Milano, IT) was used for controlled micro-suction and removal of the incoherent soiling. For cohesive deposits an experimentation including a preliminary laser irradiation testing phase on laboratory prepared samples was undertaken. Therefore, four flax fabric fragments were selected for the preparation of laboratory simulations. The soil material was first finely sieved, diluted in water at 50 wt% and 25 wt%, respectively, and then applied by brush on the fabric samples. In order to accelerate the degradation process of the fibres, soiled and unsoiled samples were exposed to hydrothermal ageing cycles following the UNI ISO 5630-3 standard (Rosati et al. 2014).

The laser irradiation parameters for cleaning tests were obtained by first considering the scientific literature concerning investigations on ancient and modern textile artefacts and then by conducting systematic laboratory tests. Two laser systems were tested: Q-Switched Nd:YAG (1064, 532 nm) laser (QS, 10 ns) and Long-Q-Switched Nd:YAG(1064 nm) laser (LQS, 120 ns).

2.3 *Removal of copper corrosion products*

For the removal of copper corrosion products several preliminary tests were carried out, based on: i) laser ablation; ii) complexing solution of Citric Acid and Sodium Citrate (ACCS) using four percentage ratios, 100% (pure), 50%, 35%, and 20%; iii) EDTA (Ethylene-Diamino Tetra Acetic) complexing solution combined with Peggy Nanogel 5. A 5% EDTA solution was prepared in 100 ml demineralised water in which small portions of Peggy 5 were immersed for 24 hours.

3 RESULTS AND DISCUSSION

3.1 *Diagnostics*

In order to understand the nature of the materials and the corresponding level of degradations, a characterization campaign was carried out.

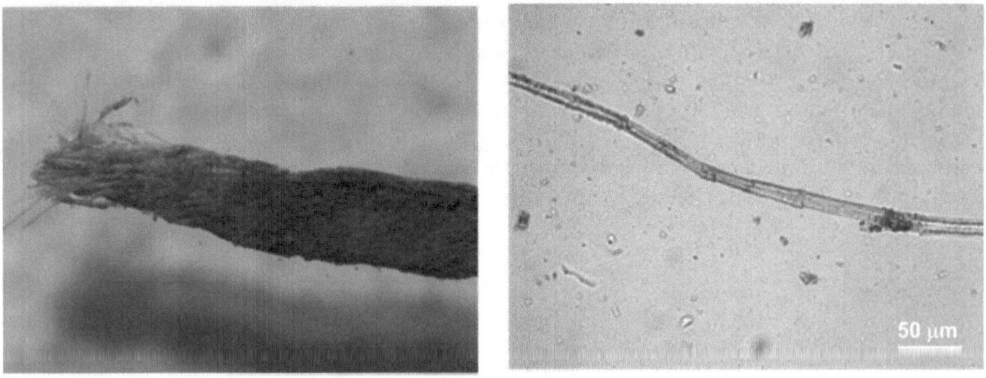

Figure 3. Fibre of linen: same fibre sample at OM and polarized transmitted visible light, 10× magnification.

OM observations revealed the fibrous material was a fine linen (Figure 3). Under IR false-colour examination, the corrosion products showed an intense blue colour, while UV induced fluorescence returned a completely black image of the object. The suppression of the fluorescence emission was reasonably due to the presence of earthy deposits that covered the surface.

A few micro-fragments of the soil deposits leaking from the fabric were taken and characterized through SEM-EDS analyses. Calcium (Ca), Silicon (Si) and Aluminium (Al) (i.e. calcium aluminosilicates) were identified as the major contaminating elements (Figure 4).

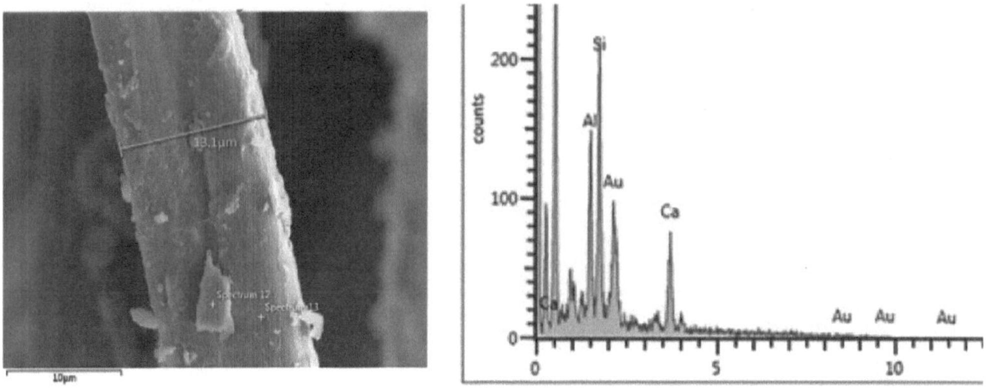

Figure 4. SEM-EDS analysis showing a fibre sample (7000× magnification) and the EDS spectrum of the soiling.

Some deteriorated fibres were taken and observed in cross-section under optical microscope: the degradation by contact with the metallic components completely altered their internal structure. SEM analysis identified the type of metal as Cu-Zn alloy, i.e. *orichalcum*, which was used for minting since the 1st century AD (Schirripa, 2014). Investigation by means of FTIR spectrophotometry was performed on the residual bright green material identified as malachite.

The radiographic investigations allowed to inspect the content of the present fabric container and to reconstruct the spatial arrangement of the objects within it. The front view of Figure 5 shows two separate metal objects of roundish shape and multi-angled inspections allowed to detect up to five similar objects of varying thickness that were identified as coins. These were arranged in two piles, one of two and one of three coins. Four of these coins were approximately 32-34 mm in diameter with thicknesses varying between 2.9 and 3.8 mm, at the thickest point. The smallest coin had a diameter of 25 mm and thickness of about 2-2.6 mm

(Figure 5). Thus, the composition of the corrosion products and the sizes of the larger coins could be likely associated with *orichalcum sestertii*, whose typical weight, diameter, and thickness were 25-28 g, 32-34 mm, and 3-4 mm, respectively. In contrast, the smallest coin certainly belongs to another group that could correspond more closely in size to the so-called *dupondius* (valued 1/2 of a sestertius or 1/8 of a denarius at the time of Augustus) or to other coins, like *asses*, *semisses*, or *quadrantes* in *orichalcum*, which were introduced during the monetary reform by Nero (63-64 AD) (Di Fazio 2019).

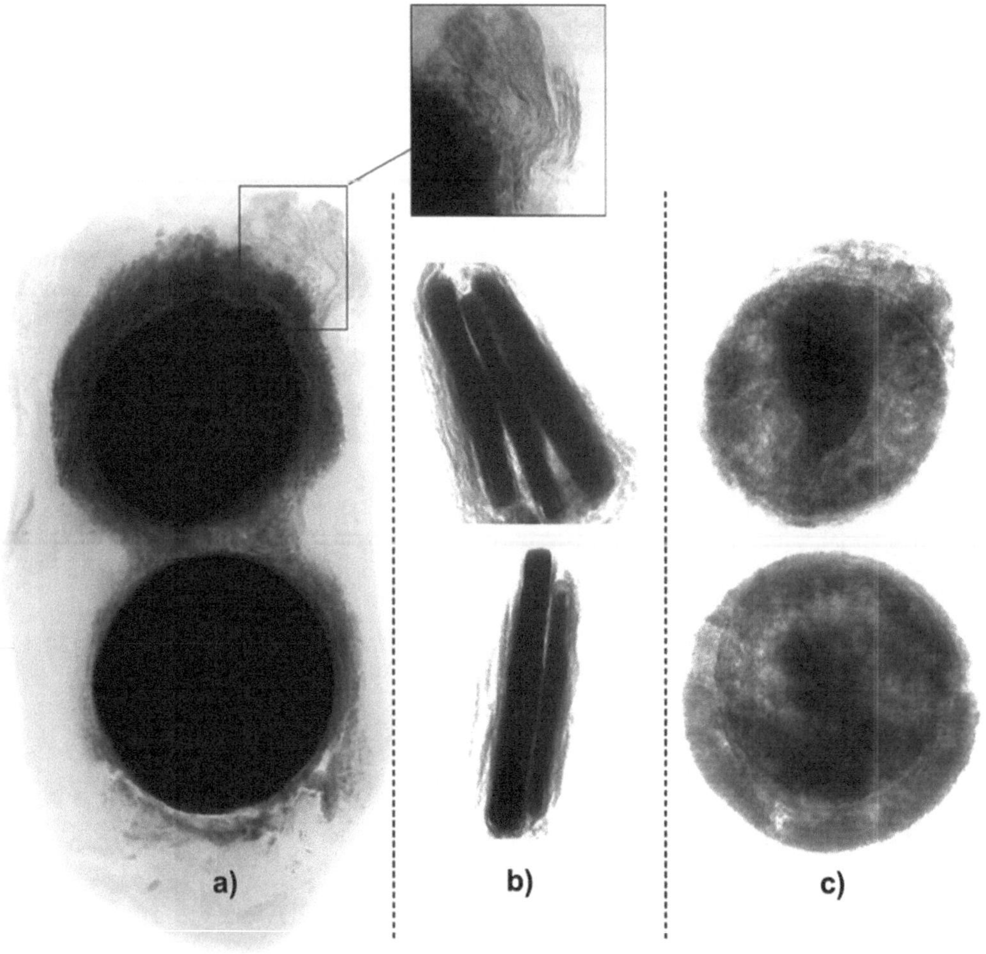

Figure 5. X-ray radiography of the 'coin purse': a) front view showing two separate cores of coins and in the inset a zoomed detail of the closing drawstring and the textile around the coins; b) lateral view where the number of coins, their edges and the spatial arrangement can be recognized; c) high-voltage acquisitions highlighting some structural features of the coins, i.e., coin's rims, superposed reliefs and shaded legends.

From the conservation point of view, the coins appeared morphologically altered by advanced corrosion. As well known, copper alloys undergo several corrosion processes over time, which can produce deep mineralization and associated significant volume increase. This likely prevented the possibility of reading the inscriptions and recognise the respective figures in the central parts of the coins (Figure 5).

Moreover, it was possible to discern a string that wasn't visible from the outside. It was placed at one end of the purse and may have originally been used as a closure (see inset of

Figure 5a). It was therefore assumed that this string was the termination of another fabric inside the visible one, i.e. an early coin wrapper.

3.2 *Conservation treatments*

It was decided to avoid any unfolding of the fabric in order to preserve the historical evidence of its presumed function of use, as well as to avoid the loss of the fibrous material altered by the serious degradation of the textile structure. At the same time, cleaning operations were planned.

We proceeded with the dry removal of superficial deposits in order to prevent undesired effects of humidity on the organic and metal components. Therefore, the removal of earthy deposits and of copper corrosion products was carried out by applying two different treatment methods, as described in the following.

3.2.1 *Removal of soiling deposits*

A preliminary controlled micro-suction was performed, which provided excellent results as it allowed the fibres to relax and soften, by removing deposits that were less adherent to the surface. Controlled micro-suction appears the main method for safeguarding archaeological textile artefacts and an excellent method for taking material samples for analysis (Conti 1996).

On the areas in which the soiling was more tenaciously adhering to the fibres, micro-suction did not return satisfactory results. Then, to remove the cohesive deposits dry laser removal was considered. The preliminary cleaning tests carried out on artificially aged and soiled fabric samples allowed to identify the most suitable type of laser and irradiation parameters.

Preliminary tests using the 2nd harmonic of Q-Switched Nd:YAG(532 nm) laser showed the use of this wavelength was unviable, as during the removal treatment the irradiation of the earthy deposit produced brown-grey surface stains. Instead, the fundamental wavelength of QS (Q-Switched) and LQS (Long Q-Switched) Nd:YAG(1064 nm) lasers did not induced such and undesired effect and were very effective in soiling removal. At the end of testing session, LQS was preferred as it offered the best results in terms of removal effectiveness and operative fluence ranges, especially for removing the more tightly packed earthy deposits. Optimal irradiation conditions found for LQS laser irradiation were fluences ranging between 0.7-1.0 J/cm^2 and a pulse repetition frequency of 2 Hz. Furthermore, the risk of possible thermal and mechanical effects on the most structurally altered and exposed fibres was reduced by choosing the LQS laser regime, which was safer than QS one (Ciofini 2022).

The parameters selected through the experimentation were then applied to the artefact by dividing the lateral side shown in Figure 6 in different areas. Laser trials were performed under stereomicroscopy in combination with micro-aspiration, in order to prevent re-deposition of the ablated products. This allowed to achieve a satisfactory removal result without any relevant undesired effect (Figure 6).

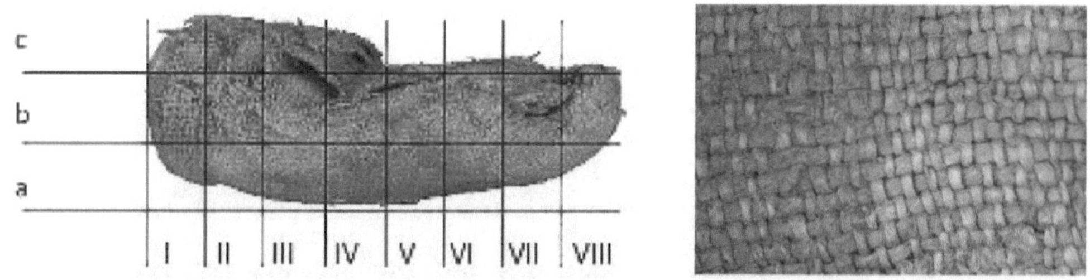

Figure 6. Grid showing areas where laser removal tests were performed. On the right image, soiled area before the treatment (left) and after LQS dry laser removal of earthy deposits (right) setting 0.75 J/cm^2, 2 Hz.

In the cleaning tests on samples, it was observed that the vertical laser irradiation of the fabric surface set on the horizontal plane led to partial removal and residual earthy deposit within the fibres. Thus, in order to increase the cleaning effectiveness and minimize possible substrate alterations horizontal laser cleaning was carried out. Basically, the fabric surface was set vertically and the laser irradiation was carried out perpendicularly to the latter (i.e. laser handpiece kept along the horizontal plane). In this way, slow re-deposition of ablated soiling was prevented. To this goal, a special plywood structure consisting of three sides and a synthetic fabric net was designed and constructed to support the present object, thus facilitating its handling and holding it in the desired position. A container underneath the net allowed collecting the ablated products.

Furthermore, in order to mitigate lateral sputtering of earthy particulate during the laser treatment, microaspiration assists was also used, while the overall intervention was carried out under microscopic observation. It should be noted, as already observed in previous works (Bracaloni 2013a, p. 25; Bracaloni 2013b, pp.17-28), the laser technique was effective in removing soil encrustations when not too thick, while some application difficulties were encountered in regions characterized by thicker deposits, especially when intimately bond to seriously altered portions of tissue. However, after repeated laser applications the deposits were completely removed from the surface without affecting the fibres and the final result was very satisfactory (Figure 7). Once the earthy residues were removed the artefact was appreciably fluorescence under UV illumination (Figure 7c) and the texture of the fabric was much more legible (Figure 7d).

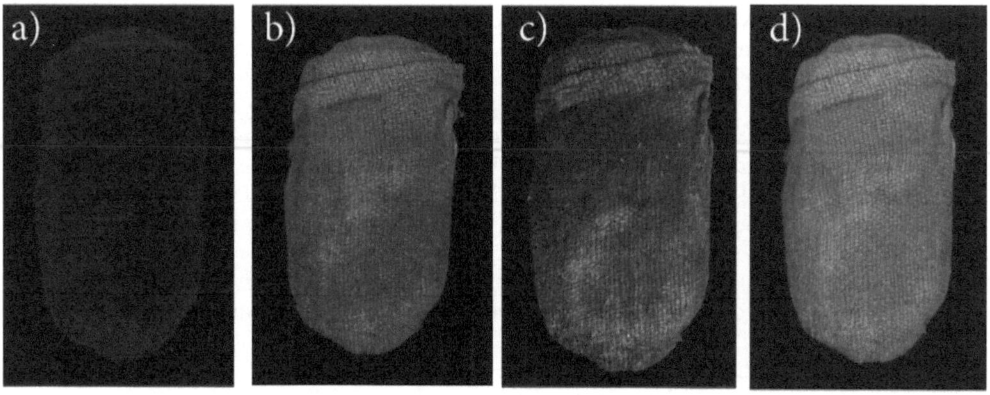

Figure 7. UV-induced Vis fluorescence (a and c) and Vis images (b and d) of side C before and after the conservation treatment using LQS laser.

3.2.2 *Removal of copper corrosion products*

Regarding the removal of copper corrosion products, another experiment was carried out using linen and cotton samples soiled with malachite. The results of the tests were the following.

1) Direct laser irradiation: it damaged the bonds between the underlying fibres, leading to a colour change from green to black due to the phototermal reduction of the copper compounds;
2) ACCS complexing solution: no one of the four percentage concentrations (100%, 50%, 35%, 20%) was suitable for the removal of copper alteration products, as poor cleaning results and a risk of irreversible colour change were pointed out.
3) EDTA complexing solution combined with Peggy Nanogel 5: after comparing the results of several laboratory tests, this solution proved to be the more appropriate and effective. EDTA acted as an intermediary between the corrosion product and the gel, i.e. by promoting the transfer of malachite from the textile surface to the gel, turning the gel to a bluish-green colour. On the other hand, EDTA is a Carboxylic acid that acts as a chelator for many stable complexes due to the presence of four carboxyl groups -COOH and two

nitrogen atoms that easily bind to calcium, magnesium, copper and zinc complexes. The Peggy Nanogel 5 is an opalescent hydrogel that, being flexible and elastic, also adheres to uneven surfaces; it acts as a controlled release agent and is excellent as a carrier. This transparent gel, which has the advantage of being cut to the desired size, has the characteristic of acting superficially on the tissue, while the EDTA load provides its chemical action.

The treatment was carried out by placing a Melinex sheet over the artefact to reduce the evaporation of the liquid contained within the gel (Figure 8b). Subsequently, the impregnated gel was replaced with portions of new gel for one more hour, while not covering the area with Melinex in order to allow them to dry so as to better remove the exogenous material. The ultraviolet fluorescence image of the coin purse showed that the fabric was uncovered and its texture became completely discernible (Figure 8c).

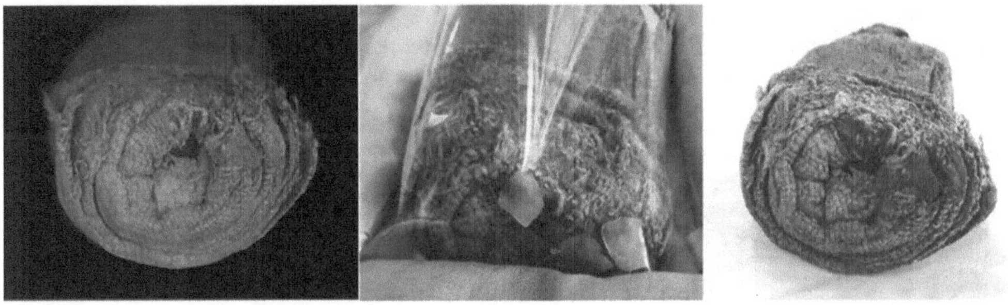

Figure 8. "Coin pursue" side E: before the treatment a); treatment with EDTA + Peggy's Nanogel b); after the treatment c).

4 CONCLUSIONS

The present work allowed to point out that the unique object investigated and subjected to cleaning treatments is a Roman coin purse made up of linen fibres, which contains five coins likely coined using a Cu-Zn alloy (i.e., orichalcum). Combined laser, chemical, and micro-suction conservation treatments were tested, optimized and eventually applied in order to remove dangerous deposits and corrosion products. These allowed to set free the fibres from the earthy concretions and copper minerals that formerly completely covered them and to recover the texture of the fabric.

ACKNOWLEDGEMENTS

The authors would like to thank Dr. Andrea Cagnini (OPD Scientific Department) for preliminary radiographic investigations and discussion and Ottaviano Caruso for visible and UV-Induced Vis Fluorescence photography.

REFERENCES

Bracaloni, E. Mascalchi, M. Cacciari, I. Conti, S. Osticioli, I. & Siano, S. 2013a. Sviluppo di nuove metodologie per lo studio e la conservazione dei manufatti tessili. In A. Brunetto (ed) *APLAR 4, Applicazioni laser nel restauro*, Roma, 14–15 giugno 2012: 267–280. Padova: Il Prato Editore.

Bracaloni, E. 2013b. In Patti M. & Siano S. (eds), *I tessili. Applicazioni laser e altre indagini per i materiali fibrosi.* (3): 17–27. Città di Castello: Nardini Editore.

Brun, E., Cotte, M., Wright, J., Ruat, M., Tack, P., Vincze, L., Ferrero, C., Delattre, D. & Mocella V. 2016. Revealing metallic ink in Herculaneum papyri. In R. Janko (ed.), *Proceedings of the National Academy of Sciences of the United States of America* 113(14): 3751–3754. Ann Arbor, Michigan: University of Michigan Press.

Ciofini, D., Rossignoli, G., Tosini, I., Lanterna, G., & Siano, S. (2022). Laser ablation treatment of soiled featherworks: the first validation study. *Journal of Cultural Heritage* 56: 118–129.

Conti S. 1996. Pulitura dei tessili antichi. Microaspirazione controllata, presentazione di un nuovo apparecchio, *OPD Restauro* 8:145–149. Firenze: CentroDi.

Di Fazio, M., Felici, A.C., Catalli, F. & De Vito, C. Microstructure and chemical composition of Roman orichalcum coins emitted after the monetary reform of Augustus (23 B.C.). Sci. Rep. 9: 12668. doi:10.1038/s41598-019-48941-4.

Fiorelli G. 1873. *Gli scavi di Pompei dal 1861 al 1872*, Napoli.

Galli, M., Coletti, F., Lemorini, C. & Mitschke, S. 2018. In Busana M.S., Gleba M., Meo F. & Tricomi A.R. (eds) The Textile Culture at Pompeii Project, *PURPUREAE VESTES VI. Textiles and Dyes in the Mediterranean Economy and Society. Proceeding of the VI International Symposium (Padova-Este-Altino, October 17–20, 2016)*: 276–277. Valencia.

Gualandris E. 2019. Il restauro di due reperti tessili archeologici provenienti dall'area vesuviana, ora al Museo Archeologico Nazionale di Napoli (MANN), MA dissertation A.A. 2017–2018 Opificio delle Pietre Dure, Firenze (unpublished).

Melillo, L. 2020. I reperti tessili dall'area vesuviana nel Museo Archeologico Nazionale di Napoli. In Giulierini, P., Coralini, A. & Calandra, E. (eds), *Miniere della Memoria. Scavi in archivi, depositi e biblioteche*, Firenze: 51–61.

Pappalardo, U. 1990. L'eruzione pliniana del Vesuvio nel 79 d.C.: Ercolano. In Livadie, A.C. & Widemann F. (eds), *Volcanology and Archaeology Volcanologie et archéologie. European Workshops Proceedings (Ravello, November 19–27, 1987, March 30–31, 1989). Strasbourg*: 197–215.

Rosati, C., Ciofini, D., Osticioli, I., Giorgi, R., Tegli, S. & Siano, S. (2014). Laser removal of mold growth from paper. *Applied Physics A* 117: 253–259.

Schirripa Spagnolo, G. Della Ventura, G. Bellatreccia, F. 2014. Analisi storico-iconografica e accertamenti tecnici su un falso sesterzio di Vespasiano con il tipo del Colosseo nella collezione Maruffi. In Calcani G. & Molinari M.C. (eds) *Terre, antichità, memorie. La raccolta numismatica Maruffi*: 253–259. Roma: TrE-Press Editore.

Lasers in the Conservation of Artworks XIII – Siano & Ciofini (eds)

Luca Trevisani's printed feathers: A laser case study

A. Di Matteo, G. De Cesare & A. Giovagnoli
Accademia di Belle Arti L'Aquila, L'Aquila, Italy

ABSTRACT: Contemporary artists have no limits for their new creations. Luca Trevisani in Wireless Fidelity, part of the MAXXI Collection, has used printed peacock feathers, mounted in a sort of antenna, with two orthogonal iron elements. He refers to the symbolic value of an ancient writing instrument in the era of speed communication, focalizing the modification of the nature by human high technology. This artwork appears destined to a difficult conservation for its fragile and ephemeral constituents. Thanks to the interview granted by the artist, it was possible to replicate a series of mock up, to test the various stages of conservation treatments and realize a conservation manual of recommendations. The application of laser cleaning was tested too, with irradiation at different wavelengths and pulse duration, to evaluate the risk threshold value and the possibility of offering a valid and safe instrument for future cleaning.

1 INTRODUCTION

Conservators are faced with difficult challenges to overcome, in order to transmit works of art to the future generations, even when the artists thought them as ephemeral.

Wireless Fidelity by Luca Trevisani, created in 2018, is a very interesting case study due to its particular technique and its imaginative project. The artist has chosen feathers and leaves in his most successful works, modifying them through digital printing technology. The fragile organic matter is prepared with a layer of alkyd enamel, a layer of acrylic varnish with a UV filter, to receive a wallpapers motif printed with acrylic emulsion inks catalysed with UV light. In the case of this work, the peacock feathers, some painted others left natural, are structurally strung and fixed with a heat-shrinking sheath in two orthogonal iron rods, to form a television antenna. The feathers were used to write and today they have been transformed to chase the speed of communication. Trevisani's focus is also man's manipulation of nature, and the desire to bend the natural vocation of materials to something artificially transformed. This work began to present mechanical conservation problems, induced by excessive bending during the exhibition in progress at the MAXXI L'Aquila. The first aid prompted to study the materials and the technique, interviewing the artist and reproducing his technique. On this information some mocks up were reproduced (De Cesare 2010) and the various solutions necessary for the restoration were sought in order to develop a manual with useful recommendations for conservation and restoration. For the cleaning after mechanical tests and very limited chemical results, it was decided to test the lasers system. The removal of compacted powder trapped in the three-dimensional structure of the organic material; turned out to be really complicated especially in some areas difficult to reach using flat brushes. The laser as non-contact technique appears to be one of the most interesting systems (Brunetto 2004). It should be noted that this technology has never been tested on such fragile and complex works, that combines the vulnerability of acrylic painting and the natural feathers, and therefore the planned experimentation can constitute an important preliminary study both from a scientific and a practical point of view. The few works currently available in the literature essentially concern the laser cleaning of feathers without superimposed layers (Pandozy et al. 2014, Ciofini et al. 2022) or the acrylic painting on different supports (De Cesare 2013, 2014).

The solubility tests were carried out with water solutions and organic solvents in a preliminary way, with the same limited possibility typical of the acrylic paints (De Cesare

DOI: 10.1201/9781003386872-21

2008, 2015). The dry cleaning with sponge, clothes and brushes have brought good outcomes. To test the effectiveness and non-interference of the laser instrumentation on this particular and innovative technique, different lasers by El. En. S.p.A. (Calenzano, Italy) were tested.

2 RESISTANCE OF THE PRINTING MATERIALS

Some mocks up have been realized, following the artist technique description:

- an anti-UV spray on a peacock feather, specifically the Universal Varnish Aero spray® by the company Schmincke type 50 594;
- several coats of a white alkyd spray, the Smalto 2000 Spray® from the company Talken Color, type 5748;
- inkjet printing with inks catalyzed by UV lamps. For the presence of acrylic emulsions, these formulations are difficult to treat for their sensitivity to the most part of organic solvents.

The preliminary experimentation had the aim to verify the resistance of the printing material to irradiations on the feather support and to identify the critical threshold value before the appearance of the damage. Different commercial lasers were tested on the mocks up, including: Q-Switched (QS) and Long Q-Switched (LQS) Nd:YAG (1064 nm) lasers (EOS 1000 series by El.En.) with pulse duration of 15 and 120 ns, respectively, and Free Running (FR) Er: YAG (2940 nm) laser (Light Brush 2 by El.En.) set with a pulse duration of 150 µs.

The irradiation started with a low fluence using large spots at minimum energy, to operate in total safety, then according to the results the operating parameters were adjusted.

Table 1 shows the irradiation conditions tested and the phenomena observed by visual examination of the treated area.

Table 1. Testing the resistance of the printing materials on the feather to laser irradiation.

Fluence J/cm^2	Phenomena observed
LQS Nd:YAG (1064 nm, 120 ns)	
0.54	4-5 laser pulses totally ablated the pink, while the purple color exhibited a bit more resistance, and was totally ablated using 11-12 pulses.
0.72	2 laser pulses totally ablated the pink color, while more than 12 pulses were needed to remove the purple features.
1.08	2-3 laser pulses were sufficient to totally ablate pink and purple features.
1.44	1-2 laser pulses produce total ablation of pink and purple features.
QS Nd:YAG (1064 nm, 15 ns)	
0.14	10 and 5 laser pulses for partial ablation of pink and purple areas, respectively. 14 laser pulses for total ablation of both colors.
0.28	10 and 3 laser pulses for partial ablation of pink and purple, respectively.
0.28	Natural feather shows great resistance to laser action with 20 pulses.
FR ER:YAG (2940 nm, 150 MS)	
1.59	Pink color showed partial ablation with 8 pulses while purple appeared very resistant up to 20 pulses, showing no signs of ablation.
3.20	The purple appeared very resistant up to 20 pulses.
4.78	Purple showed signs of ablation starting from 3 pulses.

As shown Table 1, the FR Er:YAG laser (2940 nm, 150 µs) provided encouraging results. For this reason, it was decided to work with more pulses at the lowest tested fluence. The results are reported in Table 2.

To evaluate the possible benefits of the water assists, the ink surface was wet with a damp swab of deionized water. However, irradiation tests using LQS Nd:YAG laser at 0.38 J/cm^2 and 2 Hz did not evidence any improvement.

At all the fluences tested, both the LQS and QS Nd: YAG lasers damaged the ink, causing its removal, while the natural feather remained unchanged. The pink was more fragile than purple under LQS, and the contrary occurred when using the QS laser.

The best results without damages were obtained using the FR Er:YAG (2940 nm, 150 μs), allowing to work safely up to 7 pulses with a fluence of 1.59 J/cm^2 for both colors, while operating at a fluence of 0.52 J/cm^2 and a frequency of 3 Hz it was possible to work without damaging the printing ink, up to a maximum of 5 series of pulses. Thus, the Er: YAG laser could represent a valid cleaning system of the original products for the removal of unwanted material.

Table 2. FR Er:YAG (2940 nm, 150 μs) irradiation tests of printing paints.

Fluence Frequency J/cm^2, Hz	Phenomena observed
1.59, 1	Up to 7 pulses it was possible to operate safely on both colors of the printing ink.
1.59, 2	Pink shows partial ablation with 1 pulse, while purple shows a good resistance.
1.59, 3	Purple shows the same result with 2 series of pulses.
0.52, 3	Partial ablation is observed for both colors using 6 series of pulses.

3 LASER CLEANING TESTS ON MOCK UP

The results obtained in the first phase showed the best result of Erbium laser on the printed feather.

Therefore, experimentation continued with this laser system. The same mock up was soiled by spreading real black dust and mapping the areas with a digital microscope before and after the laser tests. It was decided to irradiate both the printed and the natural feather in order to investigate the laser action.

Working at minimum fluence (0.04 J/cm^2) with a long series of pulses, the dirt was not removed and some printing damages were observed. By increasing the fluence at 0.52 J/cm^2, good satisfactory removal results were obtained using fewer pulses.

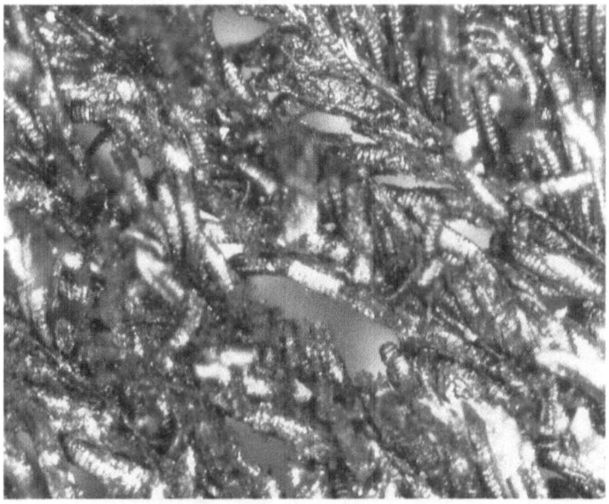

Figure 1. Carbonization of the natural feather (0.04 J/cm^2, 10 Hz) observed under microscope (65×).

Observing the treated mock up under digital microscope, the good macroscopic results obtained with lower fluence and higher number of pulses appear inappropriate. In fact, the Dino-lite in a microscopic level, showed the removal of the ink in some pixels, visible above

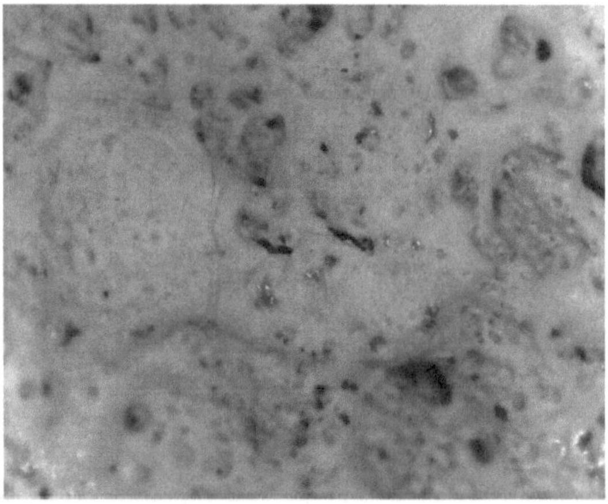

Figure 2. Melting of the printing ink with the underlying layers (0.04 J/cm^2 10 Hz) observed under microscope (185×).

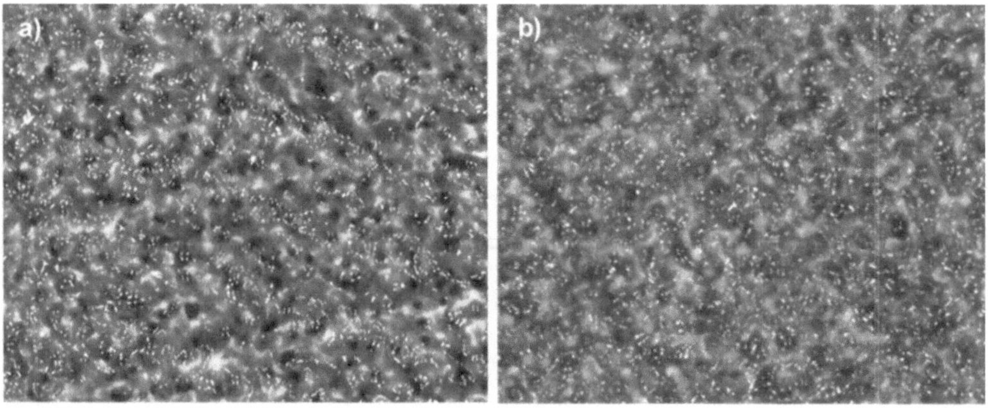

Figure 3. a) The soiled mock up detail with some black particles; before treatment observed under microscope (65×). b) Removal of the black particles but producing some white spots (0,04 J/cm^2 10 Hz) after treatment observed under microscope (65×).

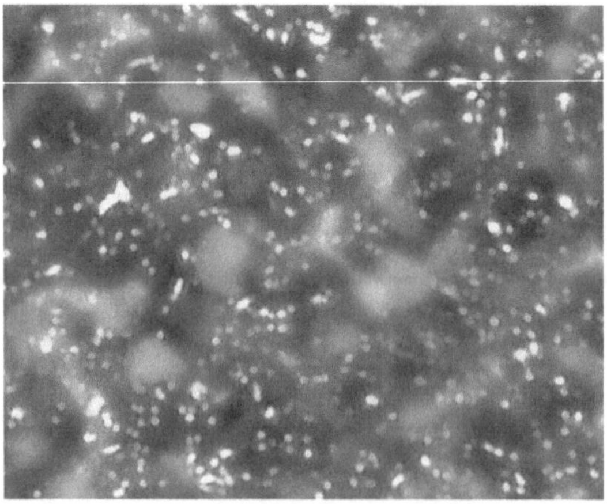

Figure 4. Detail of the white and opaque spots left by the laser linked to the use of inappropriate parameters (0,04 J/cm^2 10 Hz) observed under microscope (185×).

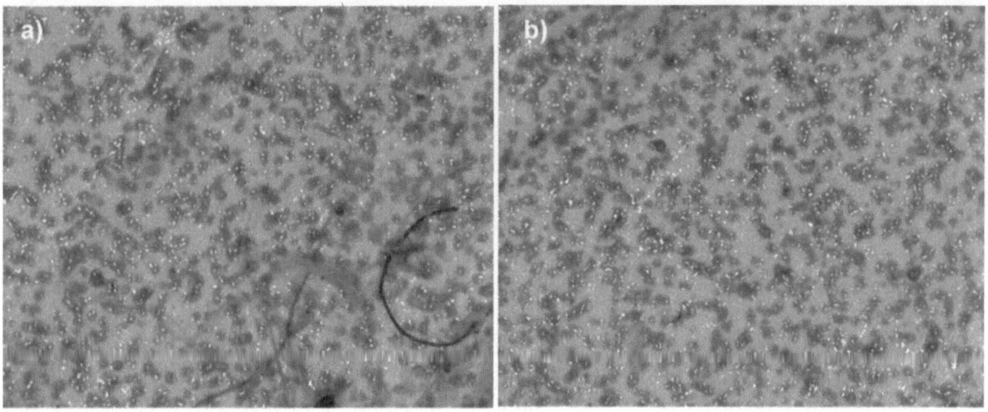

Figure 5. a) Pink area before laser cleaning (65×). b) Pink area after laser cleaning (0,52 J/cm^2 5 Hz) without damages (65×).

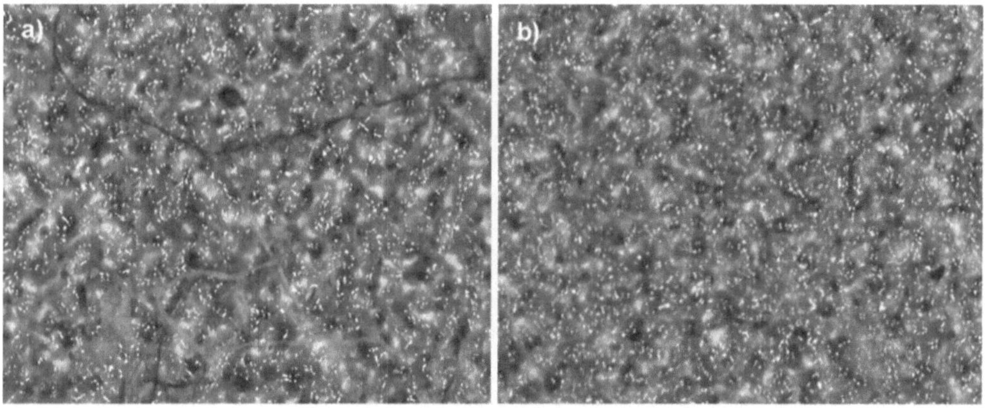

Figure 6. a) Purple area before laser cleaning (65X). b) Purple area after good laser cleaning (F = 0,520 J/cm^2 5 Hz) (65×).

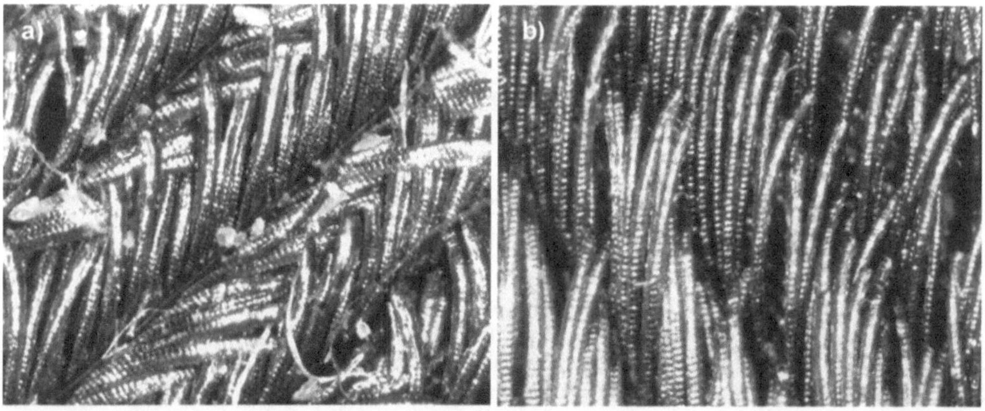

Figure 7. a) Natural feather before laser cleaning (185×). b) Natural feather after laser cleaning (F = 0,52 J/cm^2 5 Hz) (185×).

all on the purple areas (Figure 3). The ink has been modified in consistency and appearance, or completely eliminated. The laser left some areas white and opaque (Figure 4). Repeating the tests with a higher fluence with fewer pulses it is possible to reach a better result (Figures 5-6).

4 CONCLUSIONS

To validate a laser cleaning method, it is fundamental to preliminary test the chemical-physical resistance to the irradiation of the materials that must be safeguarded.

Therefore, we proceeded through the testing of different wavelengths and pulse durations, varying the fluences and the number of pulses. When a good result was macroscopically achieved, sometimes it was contradicted by a more accurate assessment with the microscope. In general, it is important to identify the "damage threshold" and the parameters considered "safe" for the artefact.

In our case study, the use of criteria deemed inappropriate led to the carbonization of the natural feather (Figure 1) and the melting of the printing ink (Figure 2) with the underlying layers. The LQS and QS Nd:YAG resulted incompatible with the cleaning of the digital printing layers. On the contrary the Erbium laser was applicable by suitably selecting the fluence and the number of pulses.

Finally, the satisfactory cleaning performed on this interesting artistic technique has highlighted a further aspect that needs mentioning: the passage of the laser beam on the natural feather, not stiff like the printed part has led to a disorder of the barbules, which was also visible to the naked eye (Figure 8); this aspect occurred using a frequency of 10 Hz.

The situation could be solved following the laser treatment with a soft bristle.

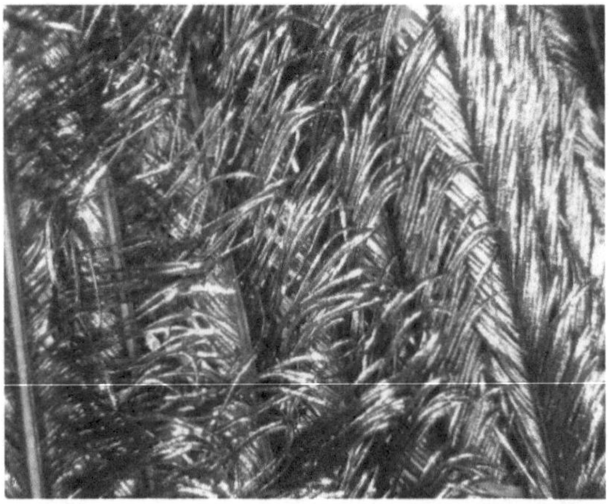

Figure 8. Disorder of the barbules following the passage of the laser beam.

In conclusion, laser instrumentation has proved to be a very effective cleaning method on the printed peacock feather, especially in the area free of coatings (Figures 7). The natural feather, whose three-dimensional structure tenaciously incorporates dirt, was very hard to clean. Although there are various studies carried out on the present topic, some peculiarities discussed in this experimentation have not been reported. In particular, the purple ink a little treated material and its laser cleaning was approached here for the first time.

REFERENCES

Brunetto, A. 2004. *L'utilizzo della strumentazione laser per la pulitura delle superfici nei manufatti artistici.* Padova: il Prato.

Brunetto, A. 2019. Laser cleaning on the mut temple wall paintings Gebel Barkal (Sudan). *APLAR 6 (Atti del convegno Firenze, 14-16 settembre 2017)*: 203–229.

Cappelloni, G., Tranquilli, G., Fumelli, F., Giovagnoli, A., Priori, G., Aramini, F., Conti, L., Talarico F. & Germinario, G. 2022. Applicazione della radiazione laser a diversa lunghezza d'onda per la pulitura delle campiture di azzurrite a tempera. *Aplar (Atti del convegno Venezia 7-8 novembre 2019)* 7: 325–339.

Croveri, P., Demmelbauer, M., Poli, T., Cavaleri, T., Giovagnoli, A. & Chiantore, O. 2013. La pulitura laser di leghe metalliche di interesse artistico: modificazioni morfologiche e chimiche di superfici patinate e non. *APLAR 4 (Atti del convegno Roma 14-15 giugno 2012)*: 295–305.

De Cesare, G. & Coladonato, M. 2008. Pitture acriliche-viniliche in emulsione: problemi di conservazione e proposte di restauro in un caso studio. *Lo Stato dell'Arte 6*, IGIIC, Spoleto 2008: 209–216.

De Cesare, G. & Iazurlo, P. 2010. Il laboratorio di restauro dei materiali dell'arte contemporanea dell'ISCR. XVII Salone dell'arte del restauro e della conservazione dei Beni Culturali, Ferrara 24–27 marzo 2010, Mibac – Direzione generale del Patrimonio Culturale: approccio di metodo: 61–62.

De Cesare, G., Melessanaki, K., Pouli, P., Domingues, J., Rosi, F., Miliani, C. & Fotakis C. 2013. Laser Cleaning Applied to Contemporary Paintings: Optimization of Working Parameters. In *Proceedings from the Cleaning 2010 International Conference Universidad Politécnica de Valencia and Museum Conservation Institute, Washington D.C. 2013, 2*: 237–238.

De Cesare, G., Iazurlo, P. & Biocca, P. 2014. La pulitura laser di vernici sintetiche su una tavolozza acrilico-vinilica: rimozione, resistenza e alterazioni. *APLAR 5 (Atti del convegno Musei Vaticani 18-20 settembre 2014)*: 383–393.

De Cesare, G. 2015. La conservazione delle pitture industriali: letteratura e approccio di metodo conservativo. In V. Bonino (a cura di) *Dall'olio all'acrilico, dall'impressionismo all'arte contemporanea. Studi, ricerche, indagini scientifiche ed interventi conservativi. Atti del VII Congresso Internazionale Colore e Conservazione. Politecnico di Milano, 13–14 novembre 2015*: 201–209.

De Cesare, G., Russo, C., Melessanaki, K. & Pouli, P. 2015. Lasers application for the cleaning of plastic supports. *FUTURE TALKS 2015 Processes. The making of Design and Modern Art Materials, Technologies and Conservation Strategies. October 20/30 2015. Pinakothek of Modern, Munich, Germany*: 180–187.

Giovagnoli, A. 2008. Valutazione dell'efficacia del processo di ablazione superficiale mediante utitizzo di tecnologia laser per il trattamento e la pulitura di componenti di interesse artistico. *APLAR 2(Atti del convegno Siena 4 luglio 2008)* 2: 179–189.

Giovagnoli, A. 2009. *Il laser. Pulitura su materiali di interesse artistico. Attività sperimentale.* Torino: Nardini.

Pandozy, S., Rivière, C., Brunori, M., Nepote, F., Rivalta, A., Santamaria, U., Morresi, F., Fraticelli F., Colantonio, C. & Brunetto A. 2014. Sperimentazione sull'uso del laser per la pulitura delle piume presenti nella collezione etnologica dei Musei Vaticani. In *APLAR* 5 *(Atti del convegno Musei Vaticani 18–20 settembre 2014)*: 423–441.

Stone and metal artefacts

Lasers in the Conservation of Artworks XIII – Siano & Ciofini (eds)

Combined use of Er:YAG and Nd:YAG lasers for cleaning the stone surfaces of the Monumental Cemetery of Pisa

A. Sutter, C. Di Marco, M. Spada & A. Trinchetti
Opera della Primaziale Pisana (OPA), Pisa, Italy

M. Spampinato
Spampinato Dr. Marcello Freelance petrographer, Lucca, Italy

A. Manariti, A. Andreotti & M.P. Colombini
Dipartimento di Chimica e Chimica Industriale, Università di Pisa, Pisa, Italy

ABSTRACT: The stone surfaces of the inner east side of Cemetery of Pisa are affected by extensive deterioration phenomena connected with the lithotype and the environmental condition. Past consolidating and protective treatments have improved the flaking and powdering of the stone, which made impossible to use traditional cleaning methods. For the gradual removal of soil and altered film, the single or combined action of two mid-infrared lasers operating at different wavelength were tested: Nd:YAG laser (1064 nm, 6 ns) and Er:YAG laser (2940 nm, 150 μs, 250 μs, 400 μs). The optimal cleaning parameters have been settled up varying the fluence, pulse duration, and wetting conditions for each typology of material to be removed. Test evaluations have been carried out by comparing the marble surface before and after the laser treatment with different analytical techniques. The optimized laser cleaning methodology has been successfully applied for the complete and safe cleaning of the pillar and its carved capital.

1 INTRODUCTION

The Monumental Cemetery of Pisa known as "Camposanto" is one of the oldest medieval Christian building intended for the cult of the dead. It is located in the north side of the Piazza del Duomo, flanking the northern boundary and completing the square shape. It was founded in 1277 to accommodate the graves that, until then, were scattered all around the Cathedral. It was the last monument to be fabricated in the square after the Cathedral, the "leaning Tower" and the Baptistery. The building has a rather atypical shape for a cemetery: it is closed to the outside by a massive stone masonry while the interior is composed by a loggia with round arches, decorated with "quadriforas" (four-light windows) in Gothic style. The stone walls are characterized by the typical Pisan Romanesque two-color motif, composed by linear horizontal bands and arches with black and white alternated marble blocks.

There are three main lithotypes, both in original elements and in replacements. 'San Giuliano marble' is the most widely used stone for the masonry; it is a sub-metamorphic fine-grained limestone from a local cave (Pisa/Lucca), present in pink-white or slightly grey varieties with dolomitic veins. 'Filettole stone' is a fine-grained gray sedimentary limestone from a local cave with marly veins and 'Apuan marble', which is a saccharoid medium-grained marble with dolomitic veins, from Apuan Alps (north Tuscany).

Deterioration problems are closely related with environmental and historical events. Many factors contributed to the stone degradation. First of all, the microclimate of the peculiar semiconfined site, characterized by cyclical water condensation and evaporation phenomena.

DOI: 10.1201/9781003386872-22

The presence of water with pollutants causes the deterioration of calcium carbonate and the migration of soluble salts from the inside towards the stone surface. Furthermore, in 1945, at the end of the Second World War, a catastrophic event occurred: an American bombing destroyed the lead roof of the monument causing the consequent fire of the entire monument. The high temperatures reached caused the so called "Marble firing" phenomena, which produced the expansion of the calcite crystals of the stone elements and the following disintegration and detachment of grains. In addition, past conservation treatments, such as the application of hydrophobic materials, produced severe conservative problems. The massive use of synthetic products such as fluorinated resin, siloxane resin, acrylic resin etc., both for the replacement of the original damaged elements, or for protective and consolidating purpose, occluded the superficial pores, restricting the natural transpiration of the stone. This degraded condition is particularly evident in the inner side of the loggia, where rainwater infiltrations entered from a leak in the roof and the cracks in the masonry. In this area, the inorganic crystallization of salts, located below the thick consolidating layers, amplified the degradation inducing powdering and disintegration with almost 1 cm of thickness.

A diagnostic campaign was launched more than 10 years ago for the investigation of the crystalline marble phases, the stratigraphy, the determination of soluble salts and the chemical-physical composition of the superficial film. With this aim petrographic analysis, polarizing optical microscope of thin section, optical microscopy with VIS and UV, attenuated total reflection (ATR) infrared spectroscopy (IR), water repellence tests by means of capillary absorption measurements with contact sponge (UNI EN 15801:2010 standard) and static contact angle measurements (UNI EN 15802:2010 standard) and finally gas chromatographic mass spectrometric analysis were performed. Because of this degraded and untouchable condition of the marble, the use of traditional cleaning methods would have caused the detachment of the superficial stone surface.

Over the last few years laser cleaning of large stone artworks and building was improved by the application of new conception laser systems such as Long Q-Switched (LQS) mode at high repetition rate up to 100 KHz, which are more productive with respect to traditional LQS, QS or Short Free Running (SFR) Nd:YAG laser (Canepa et al. 2022, Licciardi, R. et al. 2017). Nevertheless, during more than 10 years of restoration work on the masterpieces of the present monumental complex, the great difficulty of cleaning the various inhomogeneous and uneven surfaces with laser techniques was highlighted (Andreotti et al. 2006). For the removal of the layering of altered conservative materials from the non-homogenous and non-cohesive stone surface, we decided to apply the contactless cleaning method by means of Nd:YAG and Er:YAG lasers, even in combination. The tests have been carried out with different fluence of each laser sources, or in their combinations, as well as a variable number of consecutive pulses and in wet or dry condition. The laser test evaluation has been carried out by comparing the marble surface before and after the laser treatment with multispectral investigations by VIS scattered, grazing light and UV light, and water repellency measurements. The optimal cleaning conditions for each type of marble have been settled-up by performing spectroscopic and chromatographic analysis on scraped samples collected before and after the laser irradiation.

2 MATERIAL AND METHODS

2.1 *Instrumentation*

2.1.1 *Laser systems*

For this work, we compared two kinds of laser sources. The Thunder Art laser (Quanta System, Milan, IT), and the Light Brush 2 (El.En. S.p.A, Calenzano, IT). The former is a Nd: YAG laser system that allows the removal of different types of materials thanks to the possibility of selecting three different wavelengths 1064 nm, 532 nm, 355 nm. The Light Brush 2 is an Er:YAG(2940 nm) laser system optimized mainly for cleaning paintings on canvas and wood but also used on wall paintings, thanks to the strong absorption of its wavelength by those materials containing -OH bonds. The laser parameters adopted for the preliminary test are reported in Table 1.

2.1.2 *Analytical instrumentation*
The following analytical instrumentation were used for the investigation of the surfaces and the inner degraded stone component.

The attenuated total reflection (ATR) infrared spectroscopy (IR) spectra were acquired with a Nicolet iS50 FT-IR from Thermo Fischer interfaced with an ATR ITX accessory equipped with a diamond crystal (radiation penetration approx. 2 μm at 1000 cm^{-1}). The spectra were recorded in the range between 4000 and 650 cm^{-1} using 16 scans and a resolution of 4 cm^{-1}.

The chromatographic analyses were performed with a Micro-furnace Multi-Shot Pyrolyzer EGA/Py-3030D (Frontier Lab) coupled to a gas chromatograph 6890 Agilent Technologies (USA) equipped with an HP-5MS fused silica capillary column (stationary phase 5% diphenyl–95% dimethyl-polysiloxane, 30 m × 0.25 mm i.d., Hewlett Packard, USA) and with a deactivated silica pre-column (2 m × 0.32 mm i.d., Agilent J&W, USA). The GC was coupled with an Agilent 5973 Mass Selective Detector operating in electron impact mode (EI) at 70 eV. The instrumental parameters are reported elsewhere (Orsini 2017).

Table 1. Laser sources and parameters adopted.

	Thunder ART	Light Brush 2
Wavelength	532 nm, 1064 nm	2940 nm
Energy	1000 mJ (@ 1064nm) 550 mJ (@ 532nm)	300 mJ (VERY SHORT mode) 500 mJ (SHORT mode) 500 mJ (LONG mode)
Pulse duration	6 ns at 1064 nm 5 ns at 532 nm	VERY SHORT 150 μs SHORT 250 μs LONG 400 μs
Frequency	10 Hz	1, 2, 3, 5 Hz

The stone surface has been subjected to multispectral investigations by VIS scattered and grazing light and UV light (Dino-Lite Premier, 465nm) and the photographic shots were made using UX UV filters and 85A corrective filter.

The hydrophobicity of the stone was tested with Capillary absorption measurements with contact sponge water absorption tests (UNI EN 15801:2010 standard) and water adsorption drop test by static angle (Krüss mobile surface analyzer, UNI EN 15802:2010 standard).

2.1.3 *Chemical-physical characterization of materials*
The diagnostic campaign aimed at the complete characterization of the structure of the stone material and the investigation of its main degradation problems, as well as of the stratigraphy of the materials and their chemical composition. Petrographic analysis was previously carried out for the investigation of building stones and mortars of the apsidal walls of the Cathedral, the Leaning Tower and the Baptistery (Lazzerini et al. 2019, Ramaciotti et al. 2015). The main local limestones and white marble used as building stones were characterized (Lazzerini et al. 2020). The degradation can vary according to the different types of stone employed (Figure 1). The 'San Giuliano marble' (Figure 1a-1) is strongly affected by differential erosion, a phenomenon linked to the compositional heterogeneity of the material, which creates a partial erosion of the carbonate matrix and leaves the veins in relief (Figure 1b). 'Filettole limestone' (Figure 1a-2) is mostly prone to flaking and exfoliation due to its sedimentary nature (Figure 1c). 'Apuan marble' (Figure 1a-3) presents a serious erosion from the typical saccharoid crystalline structure ('sugaring') and sometimes deep cavities are created especially on the thinnest decorative elements (Figure 1d).

As regards, the mineralogical characterization and the deterioration products before laser cleaning, scraped samples were taken from the inner side of the loggia on the east side of the Monumental Cemetery, which showed a poor state of conservation of the marble stone surfaces. Figure 2 shows the investigated area of the pillar of the loggia and a few details of the swelling morphology caused by the thrust of the salts at the interface between stone and yellowish protective layer (Figure 2b), and the capital in VIS scattered (Figure 2c) and grazing light with the chromatic alteration of the protective film (Figure 2d).

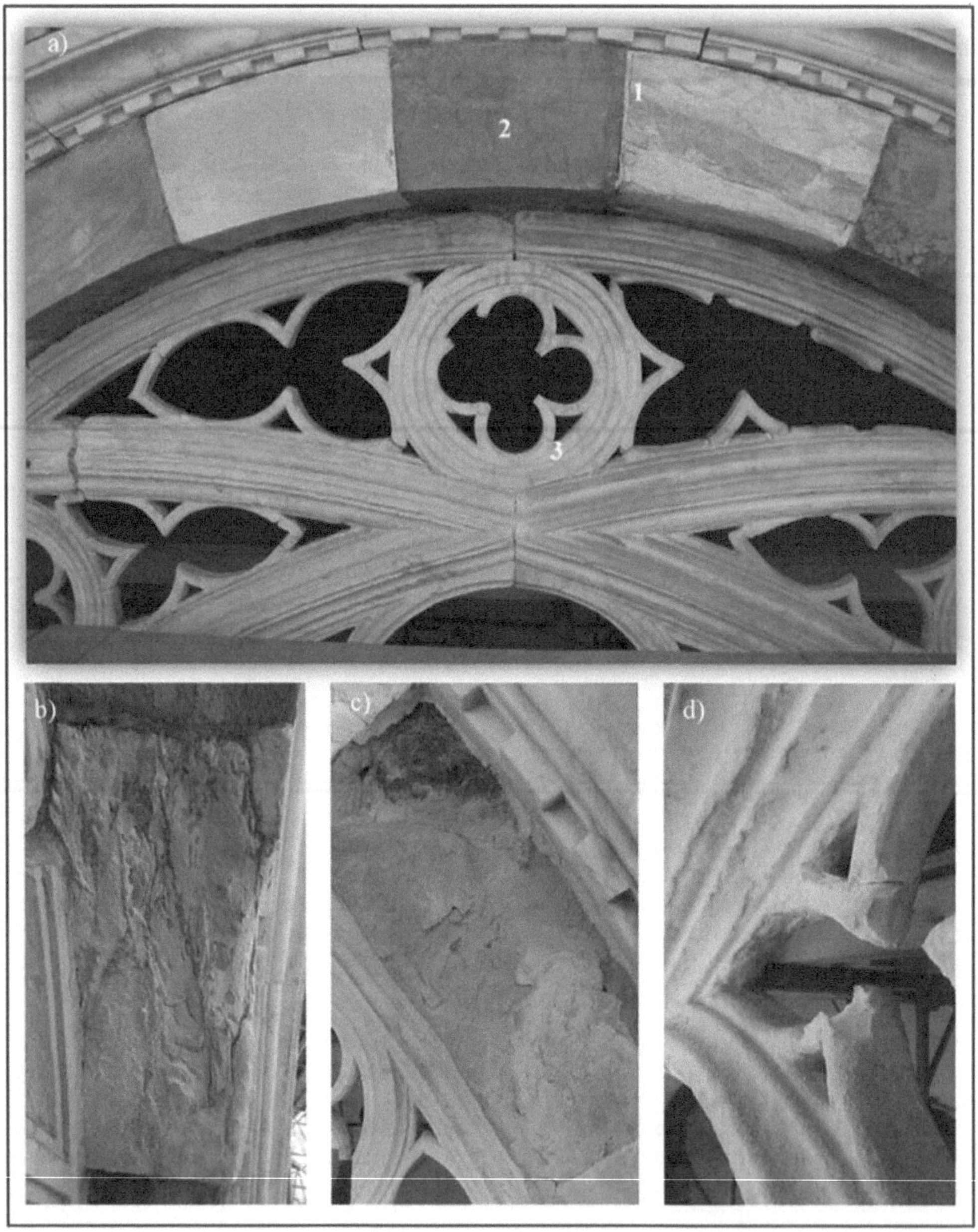

Figure 1. Details of the different lithotypes a) and their main deterioration problems: b) 'San Giuliano marble" (1) with differential erosion, c) 'Filettole limestone' (2) with flaking and exfoliation and d) 'Apuan marble' (3) with 'sugaring' effect.

The mineralogical and chemical investigation of the soluble salts were performed by means of the following analysis:

- petrographic analyses with a polarizing optical microscope on thin sections to investigate the mineralogical composition of the materials and the degradation products;
- mineralogical interpretation of the crystalline phases from x-ray diffraction analyses;
- chemical assays for the determination of soluble salts;

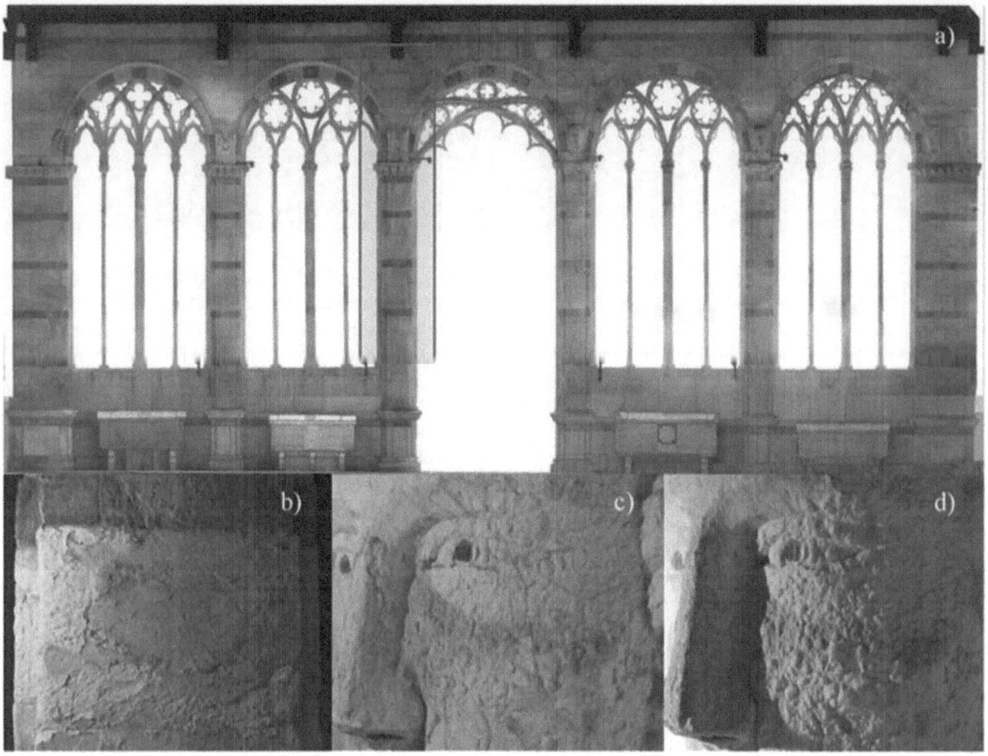

Figure 2. Pillar n°4 of the east side of the loggia selected for the laser cleaning test a); detail of the San Giuliano marble in VIS grazing light b); particular of the capital in VIS scattered c) and grazing light d).

The scraped samples from the pillar revealed the following crystalline phases: calcite and muscovite (hydrated potassium aluminosilicate) as marble component, gypsum (calcium sulphate dehydrate), whewellite (calcium oxalate monohydrate) and weddellite (calcium oxalate dehydrate). The chemical analysis confirmed a widespread presence of sulphates and evidenced others soluble salts such as clorides, nitrites and nitrates. The sulphates came mostly from the sulphation of the calcium carbonate of the stone, and it is the main problem of the Camposanto stone walls.

The high presence of sulphates and powdering phenomena induced by the "marble firing", were improved by the presence of elastomeric and hydrophobic products applied in past restoration works. Moreover, these resins, due to their elastic nature, tend to tear, in the form of films, and amplified the degradation processes causing peeling and swelling. The microphotographic documentation made it possible to understand the stratigraphy of the marble surface. The thin section observed in parallel nicols transmitted light of a sample including the stone surface and a film of resin mixed with marble dust (distinguished by its greater opacity to light due to the finer particle size fraction of calcite crystals), is shown in Figure 3a. Moreover, the crossed nicols transmitted light of the thin section (600x) evidenced the sulphate deposit (birefringent on grey) which is infiltrated between the decohesions of the marble calcite crystals, even in depth (Figure 3b). Finally, the crossed nicols transmitted light of the thin section showed that on the marble surface, mainly made up of very loose calcite crystals with a clear tendency to pulverize (1), a stucco with marble powder of various grain size (from a few microns up to about 0.2 mm in diameter) and an organic binder, likely a synthetic resin (Figure 3c) is present.

For the molecular characterization of organic binders of mortars, stucco and patina, attenuated total reflection infrared spectroscopy (ATR- IR) and Pyrolisis coupled with gas-chromatograpy-mass spectrometry (PY-GC-MS) have been used.

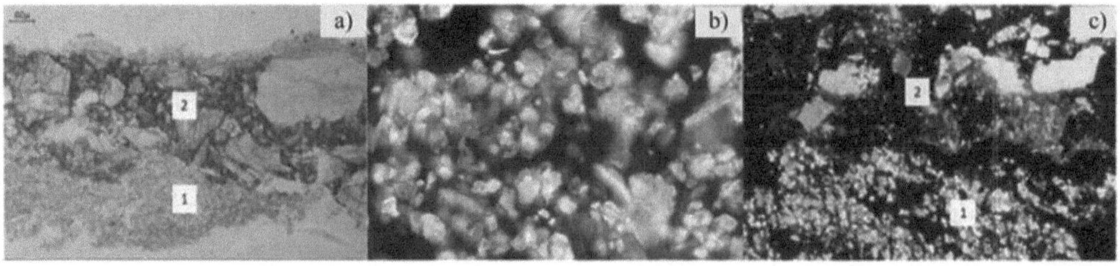

Figure 3. a) Parallel nicols transmitted light of the thin section including the marble surface (1) and a surface layer of a grouting with marble dust (2) (150×); b) crossed nicols transmitted light (600×) showing the sulphate infiltration (birefringent on grey) into the marble calcite crystals; c) crossed nicols transmitted light (150×) of very loose calcite crystals and sulphates into the marble surface (1), and superficial aggregate of marble powder of various granulometry with an organic binder (2).

The Py-GC-MS analysis showed the characteristic markers of perfluoropolyethers, and syloxanes, pointing to the use of both a fluorinated and sylossanic resin, together with trace of cholesterol and cholestanes, probably due to glycerolipid material contamination (Figure 4) (Tsuge et al. 2011). Markers of siccative oils, natural resins or waxes were absent.

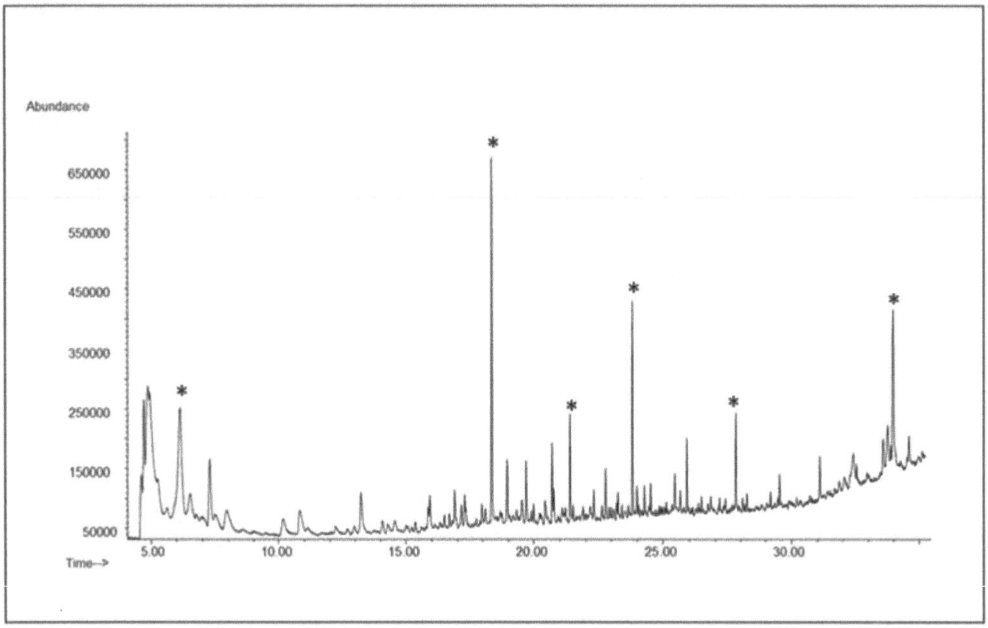

Figure 4. Pyrogramm of a scraped sample from the yellowish film from the pillar with the markers of perfluoropolyethers, and syloxanes (*).

The ATR difference spectra were obtained by subtracting the one recorded from the inner side of the marble sample, from the one recorded on the yellowish film (the difference spectrum is a line with few variations cantered on zero). The difference spectra show the absorption bands attributable to calcium carbonate, calcium oxalate and the intense band characteristic of the Si-O-Si stretching of silicone materials, and the CF_2 stretching of the fuoroelastomer (Derrick et al. 1999, Lluveras 2008). In particular, the intense two-point band at about 1100 and 1090 cm^{-1} is characteristic of the Si-O-Si stretching of silicone materials such as Rhodorsil RC-80. The absorption band which points to the presence of an fluoroelastomer such as akeogard BA plausible, is the CF_2 stretching at 1216-1218 cm^{-1} (Figure 5).

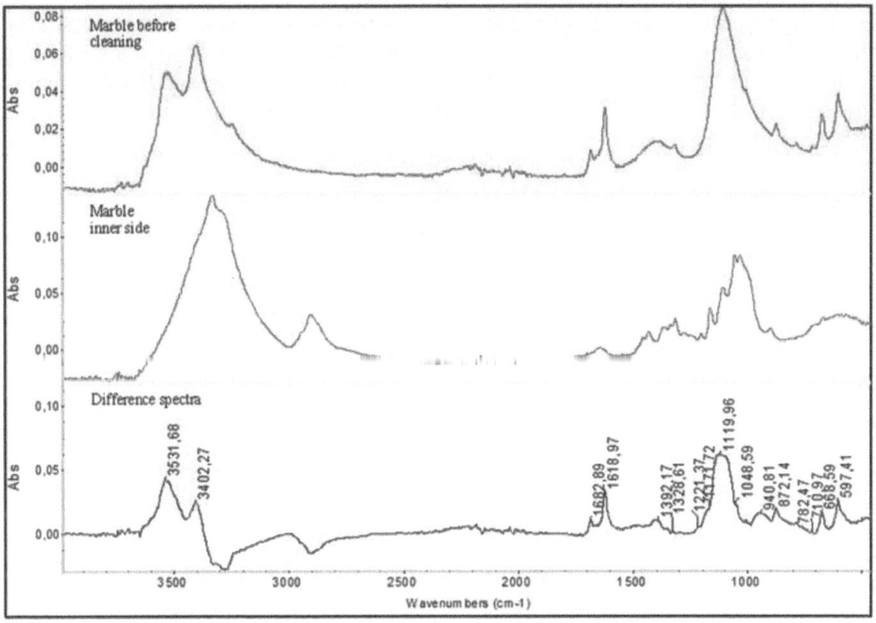

Figure 5. ATR spectra of the marble before cleaning, the one from the inner side of the marble and the relative difference spectra.

3 RESULTS

3.1 *Experimental*

The first aim of the cleaning process was to obtain a safe gradual cleaning of the surface through the removal of the soil and the reduction of the chromatic alteration of the protective film and secondly, to obtain the gradual thinning of the hydrophobic protective film. This operation could increase water permeability of the surface, allowing for a better migration of hygroscopic salts and a subsequent consolidation treatment. For these purposes, the single or combined action of two laser sources Nd: YAG(1064,532nm) laser and Er: YAG(2940 nm) laser at different pulse duration were tested. Laser tests have been carried out with different fluences as well as a variable number of pulses and in wet or dry condition. The laser test evaluation has been carried out by comparing the marble surface before and after the laser treatment.

We chose areas with different degradation characteristics. The first one consisted of a San Giuliano marble block, which had a bad conservative condition, with powdering, swelling and detachment of the superficial layers. The second one consisted of a San Giuliano marble block, which showed a better conservative condition, without detachment problems but nevertheless showed an evident yellowing and soil deposits. The third one consisted of a Filettole limestone block (grey sedimentary limestone).

We started experimenting by varying the different parameters. Initially, the Nd:YAG operating at the second harmonic (532 nm) was also tested, but it proved to be ineffective as the layers of superficial and consolidating deposits remained well cohesive to the surface.

We tried the laser irradiation on the altered film by pre-wetting the surface before each consecutive laser passages on the same area with ethanol, water or isopropyl alcohol. Likely given the complexity of the mixture of restoration and deterioration materials present on the stone, the partial solubilisation of the compounds made their removal difficult. It was therefore decided to operate dry to have a more controlled action during the 'thermal' ablation of synthetic materials which involves bubbling, splashing and swelling phenomena (Pouli et al., 2006).

The Er:YAG Light Brush 2 has the possibility to work with very short (150 μs), short (250 μs) or long pulse duration (400 μs). The long pulse duration proved to be ineffective, most probably because the high non-homogeneity of the surface in terms of thicknesses and materials, with its higher thermal effect induced in the areas surrounding the laser spot, generated a spotty cleaning.

The fluence range tested for the Thunder Art (1064 nm) was in between 0.2-0.3 J/cm^2, while the frequency and was 10 Hz, for the Light brush 2 (2940 nm) these were was 0.1-0.4 J/cm^2, and 1-5 Hz, respectively.

3.2 *Laser tests and control methods*

The multispectral investigation allowed to highlight the removal degree of the surface deposits and protective/consolidating agents (Figure 6). The UV showed the presence of at least two different substances on the stone surface: the first greyish one with no fluorescence, and a second one with an intense orange fluorescence, more homogeneously spread over all the pillar surface.

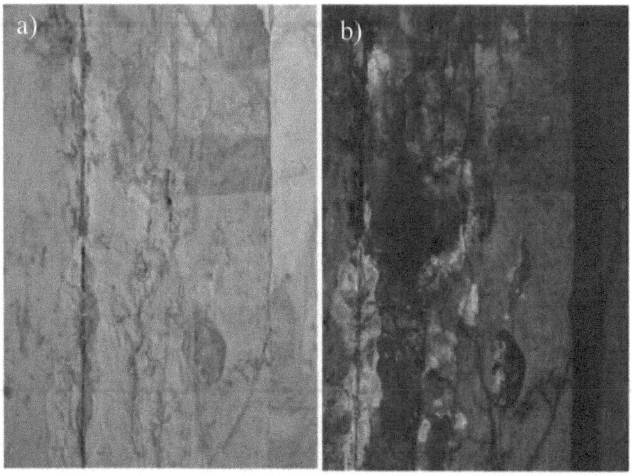

Figure 6. Vis a) and UV b) light images of the pillar with tested area.

Standalone Nd:YAG (at 1064 nm) removed soiling deposits very well. Thanks to the observation in ultraviolet fluorescence it was possible to note how the Nd:YAG at a low fluence (0.20 J/cm^2) was able to eliminate the semi-coherent surface deposit, thus uncovering the consolidant/ protective layer, which appeared more orange under UV rays. Anyway, it did not seem to sufficiently act on the thicker underlying organic layer, while the repeated action with consecutive laser passages (at least 3 passages) became difficult to manage because it tended to remove the layer unevenly, with disintegration phenomena (Figure 7a-b). The action of Er: YAG with 150 μs pulse duration, on the other hand, removed the yellowish degraded film but didn't remove the soil deposits layer even after 3 consecutive laser irradiation at 0.3 J/cm^2, in fact the surface remained rather grey (Figure 7c). The best laser treatment on the white marble included a first scan with Nd:YAG radiation and a consecutive one with Er:YAG (Figue 7d).

As for the grey 'Filettole stone', only Er: YAG was used with a fluence ranging between 0.1-0.4 J/cm^2 (150 μs, 15 Hz, 1-2 runs), because the Nd:YAG laser induced bleaching effect even at very low fluences. The test carried out on the Filettole stone gave a good result, the Er:YAG laser in fact managed to break the film, interacting more effectively thanks to its particular wavelength, without damages at the surface lithotype (Figure 8).

The ATR analysis performed on scraped superficial samples after a few laser irradiation tests confirmed the multispectral analysis evidence. It shows that the surface resins are more effectively removed by the combined action of the two lasers. The difference spectra were obtained by subtracting the one recorded from the samples before and after the Nd:YAG irradiation (Figure 9a) or the two laser combination tests (Figure 9b), and showed a few absorption bands. These spectra point to the presence of only a trace of other chemical species

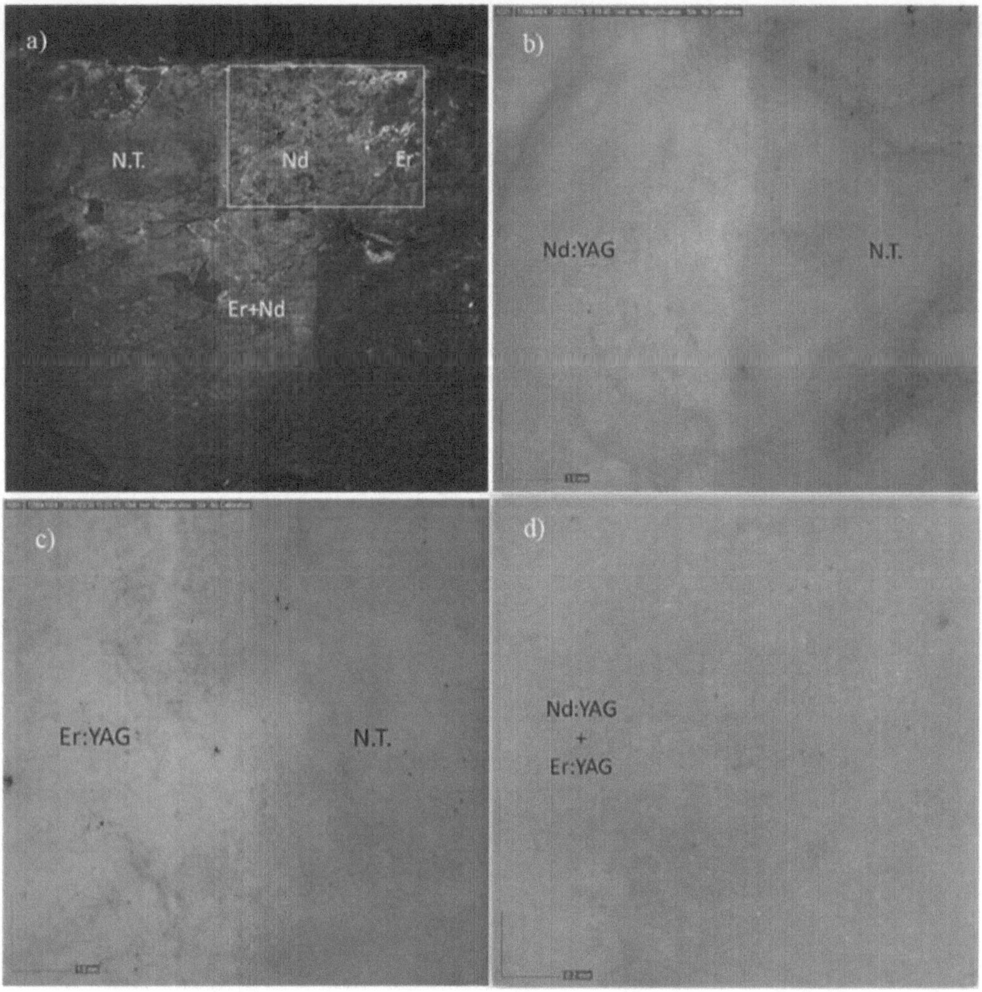

Figure 7. UV a) and magnification Vis light images b-d) on the tested area on the pillar: Nd:YAG (0.20 J/cm^2) vs not treated area (N.T.) b), Er:YAG (0.30 J/cm^2) vs not treated area b), cleaned area with the combination of the two laser sources c).

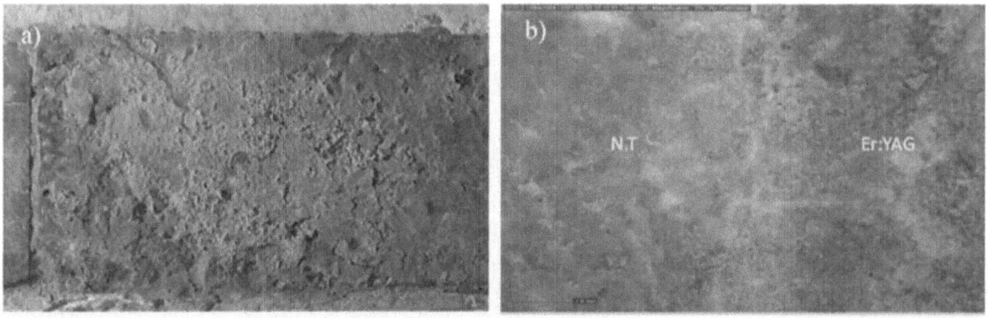

Figure 8. Image of the 'Filettole stone' block before the cleaning a) and magnification of the Vis light images of a detail before (N.T) and after Er:YAG irradiation (Er:YAG: 0.2J/cm^2 (150 μs, 15 HZ, 1 pass).

with respect to the ones of the marble itself. In particular, the absorption bands at about 1100-1000 cm^{-1} and that of 1216-18 cm^{-1} ascribable to the two consolidants, are almost absent after the combined laser action.

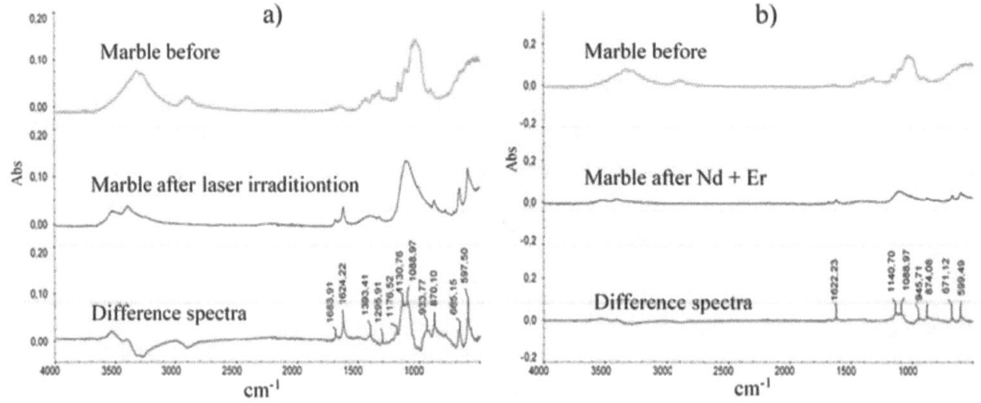

Figure 9. ATR spectra recorded before and after laser cleaning tests with Nd:YAG laser a) and with the combination of the two laser and b) their relative difference spectra.

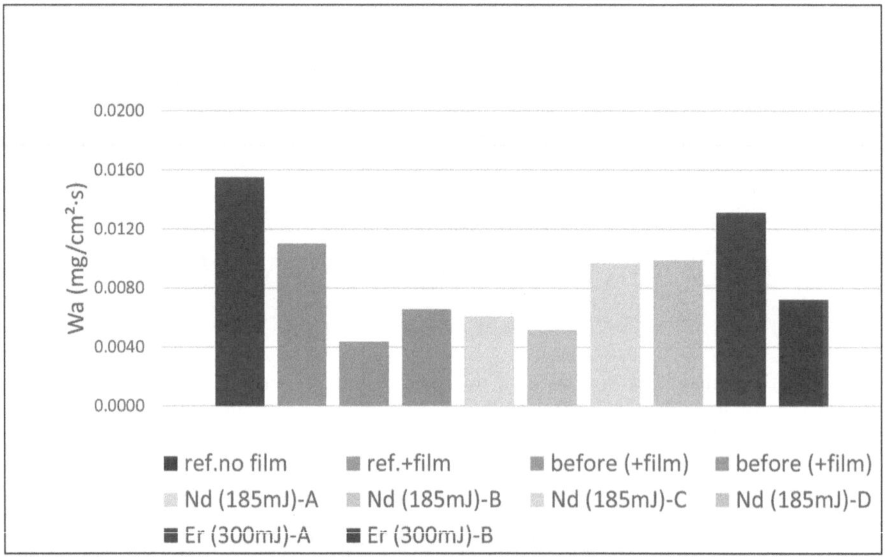

Figure 10. Absorption measurements using contact sponges carried out on untreated block of marble ('ref. no film'), in comparison with the ones performed on blocks with protective layer but in good condition ('ref.+film' or 'before + film') and after Nd:YAG or Er:YAG irradiation. The energy values reported in mJ are the ones referred to a diameter spot of 1 mm.

The absorption of water measurements using contact sponges were carried out on non-decomposed areas and in presence of protective, after dusting (to eliminate incoherent deposit and saline efflorescence, which would have distorted the results). Unfortunately, because of the bad state of conservation of the stone surface, a great variability in the water absorption (Wa) values in relation to the thermal hygrometric changes was registered.

The tested areas were analysed before and after laser cleaning. Comparison with an untreated block of marble was carried out in order to evaluate not only the change after the laser removal of the film-forming material, but also in relation to environmental thermohygrometric changes.

At least 6 measurements on each area have been performed (two different days with a different T and RH), holding the sponge in contact with the surface for 5 minutes. The areas treated with Er:YAG laser showed the highest Wa values similar to the standard ones, probably thanks to the partial removal of the coherent surface salts deposit and protective film. As regards Nd: YAG laser, even after two consecutive laser passes the Wa of the surface was almost unchanged, although on a visual observation there has been a significant improvement (Figure 10).

Contact angle drop test measurements for water repellence and wettability also showed an high variability, even on the stone before laser irradiation. However, considering only the measurement with a good standard deviation (between 5 and 10) a general trend line of increased hydrophobicity could be identified.

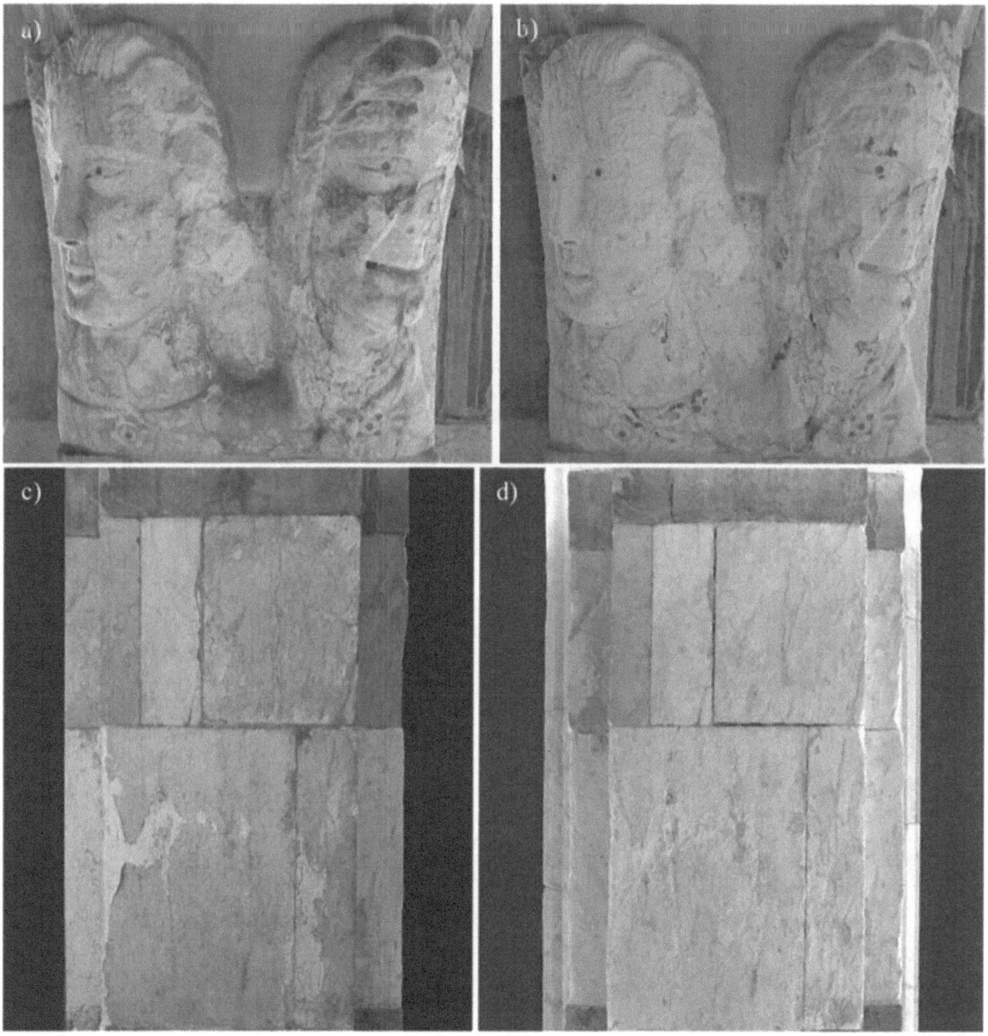

Figure 11. Carved capital and pillar before (a and c respectively) and after the laser cleaning (b and d respectively).

The water adsorption measurements by capillarity demonstrated that the Er:YAG laser reach a deeper effectiveness of the cleaning of the elastomer film, but on the other hand was able to create a superficial melting of the organic material into the porosity of the degraded marble, which resulted in an increase of hydrophobicity measured with the contact angle test.

It was interesting to note that the hydrophobicity on the surface increased after the Er:YAG laser cleaning. In literature a dental research on dentin treatment with Er:YAG and Nd:YAG sources, evidenced similar result. The Er:YAG laser removal of the superficial yellowing layer mainly through a thermal action, induced a contemporary melting of organic material just on the dentine surface (Birang et al. 2007). Another previous study on the comparison between Er:YAG and Nd: YAG laser for the removal of degraded synthetic protective from stone reference specimens, demonstrated that when using the Er:YAG laser, a deeper effectiveness of the cleaning is reached respect to the other laser, although a deeper closure of the inner porosity of the stone was found.

4 CONCLUSIONS

This research demonstrated the effectiveness and safety of optimized laser cleaning of very damaged and fragile stone surfaces of the Monumental Cemetery of Pisa. The best operating mode selected for the Nd:YAG laser demonstrated to be able to remove the semi-coherent surface deposit, unrevealing the consolidating/synthetic protective film. Anyway, the altered organic protective was removed only with consecutive laser passages in an uneven way, not in respect of the safeness of the stone surface.

As regard the efficacy of Er:YAG laser for the removal of the organic film, it has been proved where the soil deposit were less consistent and cohesive. However, in the presence of thicker surface layers, it acted only with a partial removal of the altered protective products whit a resulting non-homogeneous cleaning. This laser was able to remove the unwanted material from the dark coloured 'Filettole stone', without affecting the colour of the stone.

The combined action of Nd:YAG (1064 nm) for the first laser pass and Er:YAG for the cleaning of the altered consolidation materials, gave the best results for a safe and selective removal of the degraded film from the carved capital and pillar as visible in Figure 11.

REFERENCES

Andreotti, A, Colombini, M. P., Nevin, A., Melessanaki, K., Pouli, P. & Fotakis, C. 2006. Multi-analytical study of the Laser Pulse Duration Effect in the IR laser-cleaning of the Monumental Cemetery wall paintings of Pisa. *Laser Chemistry* 2006: Article ID 39046.

Birang, R., Poursamimi, J., Gutknecht, N., Lampert, F. & Mir M. 2007. Comparative evaluation of the effects of Nd:YAG and Er:YAG laser in dentin hypersensitivity treatment. *Lasers Med Sci* 22(1): 21–24.

Canepa, M.C., Zenucchini, F., Coco, R.M.A., Piccirillo, A., Manchini, L., Appolonia, L. & Cardinali, M. 2022. Applicazioni laser su superfici lapideee di grandi dimensioni: tempi di applicazione e risultati a confronto. In A. Andreotti, A. Brunetto, G. Lanterna, B. Mazzei (eds) *Atti del Convegno - Applicazioni laser nel restauro - APLAR 7*: 203–2019. Firenze: Nardini Editore.

Derrick, M.R., Stulik, D. & Landry, J.M. 1999. *Infrared Spectroscopy in Conservation Science, Scientific Tools for Conservation*, Getty Conservation Institute, Los Angeles.

Lezzerini, M., Spampinato, M., Sutter, A., Montevecchi, N., & Aquino, A. 2019. Petrographic characteristics of the mortars from the Pisa's cathedral apse. Paper presented at the 2019 IMEKO TC4 *International Conference on Metrology for Archaeology and Cultural Heritage*, MetroArchaeo, 459–463.

Licciardi, R., Mascalchi, M., Siano, S. & Megna, B. 2017. In A. Brunetto (ed) *Atti del Convegno - Applicazioni laser nel restauro - APLAR 5*: 15–30. Firenze: Nardini Editore

Lluveras, A., Boularand, S., Roqué, J., Cotte, M., Giráldez, P. & Vendrell-Saz, M. 2008. Weathering of gilding decorations investigated by SR: development and distribution of calcium oxalates in the case of Sant Benet de Bages (Barcelona, Spain), *Applied Physics A* 90(1): 23–33.

Orsini, S., Parlanti, F. & Bonaduce, I. 2017. Analytical pyrolysis of proteins in samples from artistic and archaeological objects. *Journal of Analytical and Applied Pyrolysis* 124: 643–657.

Pouli, P., Nevin, A., Andreotti, A., Colombini, M.P., Georgiou, S. & Fotakis, C. 2009. Laser assisted removal of synthetic painting-conservation materials using UV radiation of ns and fs pulse duration: Morphological studies on model samples. *Applied Surface Science* 255: 4955–4960.

Ramacciotti, M., Spampinato, M., Lezzerini, M.2015. *The building stones of the apsidal walls of the Pisa's Cathedral.* In *Atti Soc. Tosc. Sci. Nat., Mem.*, Serie A, 122, 55–62.

Tsuge, S., Ohtani, H., Watanabe, C. 2011. *Pyrolysis–GC/MS Data Book of Synthetic -Pyrograms*, Thermograms and MS of Pyrolyzates, Oxford, UK, Elsevier.

Lasers in the Conservation of Artworks XIII – Siano & Ciofini (eds)

Laser cleaning of graffiti and cleaning evaluation

F. Surma, P. Schloegel, V. Aguilar & L. Rosenbaum
Epitopos, Strasbourg, France

M. Labouré
Mescla, Strasbourg, France

ABSTRACT: Graffitis' removal is a current challenge for buildings conservation. The aim is to restore the affected area as close as possible to its initial state and to avoid any kind of additional degradation. Yet, the use of chemical products or sandblasting can alter the stone surface. This study focuses in the cleaning of painted surfaces by laser using two different power intensities. Eighteen sandstone and marble slabs have been sprayed using oil based and acrylic paint products or markers employed nowadays for street-art. A preliminary analysis using LIBS and IRTF-ATR has been performed to find mineral and organic pigments or bindings in the paints. Then, the cleaning has been carried out. For each sample, a no-cleaned surface and two surfaces cleaned at 50W and 75W were analyzed. A DRIFT acquisition has been performed to realize a quantitative cleaning evaluation.

1 INTRODUCTION

Graffiti is often found on monuments and removing it is not always easy. Indeed, classical chemical or sandblasting techniques can damage the surface, which is why alternatives such as laser cleaning are being considered (Giusti et al. 2020, Ricci et al. 2020). Therefore, it seems relevant to study the impact of laser cleaning of graffiti on typical stones of buildings and monuments in order to understand what some of the factors are that influence the ablation process to remove undesirable coatings. The present study seeks to evaluate the impact of a Nd-YAG-1064 nm cleaning laser for two types of graffiti paint products (acrylic and oil based) on marble and sandstone supports. We used paint products that are the most commonly used to produce street art (Bosi et al. 2020, Marazioti et al. 2022). Samples of graffiti on marble and sandstone slabs were prepared by the Mescla company and then studied in Epitopos' laboratory. One of the parameters that must be considered when it comes to laser ablation is the power of the laser. Indeed, one of the effects of the laser is also the degradation of the support. Thus, one of the intentions of this study is to compare the effects of the laser using two different powers and to observe what differences exist between both.

2 MATERIALS AND METHODS

The objective is to characterize the effect of laser cleaning on different types of graffiti applied to marble and sandstone slabs. A paint product or a marker product had been applied to sandstone and marble slabs of dimensions 10 x 10 x 2 cm. The painted side of the slab was then divided into three areas which were given different powers of laser cleaning (Z0: no cleaning, Z50: 50W and Z75: 75W) using a Nd-YAG (1064 nm) cleaning laser. It should be noted that the back of the plates has no paint and is used as a control sample.

DOI: 10.1201/9781003386872-23

2.1 *Sampling*

Different paint products were applied on marble and sandstone slabs (Figure 1), we used oil-based paint (green, red, blue and yellow Montana Matt), acrylic paint (black, green, red, blue and yellow Gold Acrylic), permanent (oil based) markers (black, blue and red Maxiflo markers) and water based markers (green, black, blue and red Uni Procket markers). The paint products have been applied by aerosol to the surfaces of sandstone and marble slabs to be cleaned. We have used gray Niderwiller sandstone slabs (open porosity 20% and apparent density 1835 kg/m^3) and white Carrara marble slabs (open porosity 0.2% and apparent density 2790 kg/m^3).A total of 18 painted slabs (5 acrylic and 4 oil-based paints) and 14 marked slabs (4 water based and 3 permanent markers) were studied.

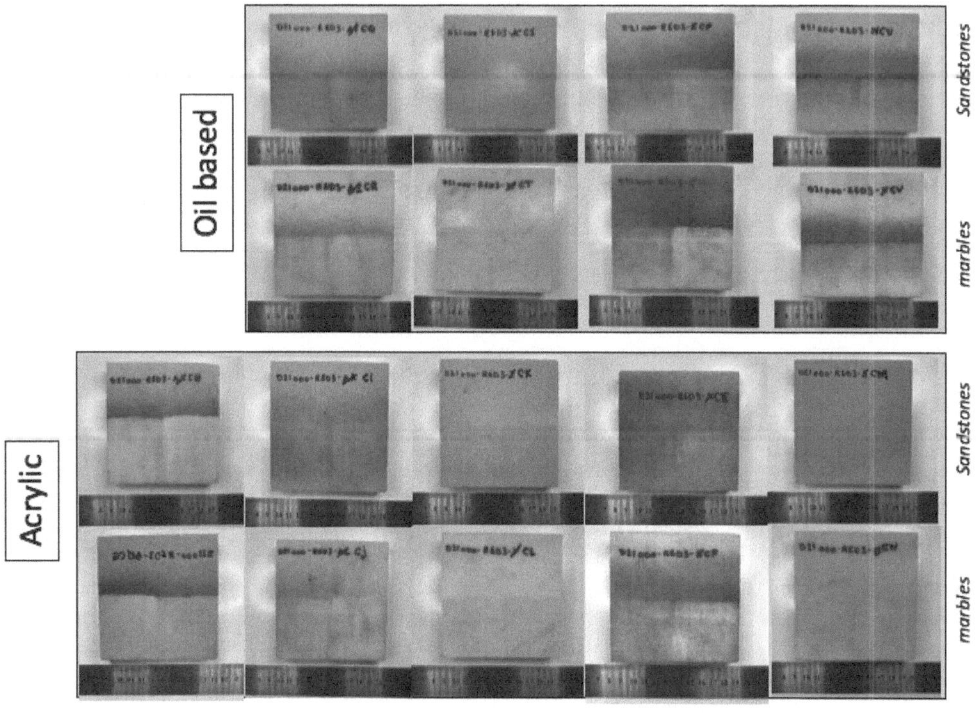

Figure 1. Sandstone and marble slabs painted and cleaned.

The raw paint products were prepared to be collected for infrared analysis in ATR mode in order to identify their components. A LIBS acquisition was also performed to complete the characterization. The plates were then analyzed before (control sample and Z0 area) and after laser cleaning (Z50 and Z75 areas). Surface observations (appearance and color measurement) were first made, and then infrared analyses were performed.

Laser cleaning: the principle of laser cleaning is based on the removal of unwanted layers or contamination on a solid surface. Laser cleaning has great potential to replace chemical methods or other types of cleaning such as sandblasting. The action of the laser on the surface to be cleaned involves mechanisms of thermal ablation, mechanical effect, or a combination of both. The proportion of these mechanisms depends on the chosen characteristics of the laser beam and the properties of the material being ablated.

In this study, we will evaluate the effect of a Nd-YAG (1064 nm) which is a laser used for the cleaning and preservation of stones (Siano et al. 2012). The 'Infinito' cleaning laser (from El.En. company), set to 2 different powers, on the sandstone and marble slabs that we have prepared with four different paint products. The parameters chosen for the laser cleaning are the following: frequency set to 100 kHz, pulse energy going from 0.50 mJ to 0.75 mJ, giving a power from 50 W to 75 W, spot size of 15 mm and scan speed of 10 m/s.

2.2 *LIBS analysis (Laser Induced Breakdown Spectroscopy)*

LIBS is a measurement technique used to identify the elements in a material (Siano & Agresti 2018). The principle is simple: a 1064 nm pulsed Nd-YAG laser (a high focalization laser) excites during 10 ns the surface of a sample to be analyzed. The power density at the material surface induces a laser ablation. Moreover, the high-power density generates the formation of a plasma at the material surface. Plasma emits optical radiation when coming back to a lower energy state, this optical radiation is detected by optical fiber and analyzed by a spectrograph. The spectra interpretation allows to determine elements composing the analyzed material.

We have used this technique to analyze the raw products of paint, the results of this analysis allowed us to characterize pigments of mineral nature.

2.3 *FTIR analysis (Fourier Transform Infrared Spectrometry)*

FTIR is a portable technique used to determine the composition of certain organic and/or mineral materials (Paris 2012). It measures the result of the interaction of controlled infrared radiation with a material. The mode of absorption, transmission and reflection of the light spectrum will vary according to the molecules passed through.

It is a fast and accurate method that has a lower signal-to-noise ratio than other methods. Infrared light is guided through an interferometer and then through the sample. The signal is then collected and processed with Fourier Transform functions to obtain an absorbance spectrum as a function of wavelength (Servant et al. 2011).

Two modes of acquisition have been used in this study. The first one is ATR mode or Attenuated Total Reflectance, for this mode of analysis the sample is placed on a diamond cell and the measurement is performed with a small press that maximizes the contact surface. ATR is used here to the identification of pigments and binders of raw products. The second mode of acquisition of FTIR is DRIFT mode or Diffuse Reflectance Infrared Fourier Transform. This modality allows to obtain an absorbance spectrum from a surface that reflects IR light. There is no need to sample and the measurement can be made directly. The DRIFT module is used to measure on a known surface in a repeatable manner, allowing comparable quantification analyses. We use a DRIFT module to perform the analysis of painted slabs.

The spectra obtained in DRIFT mode differ from the spectra available in the databases, which are often made in ATR mode. Indeed, several types of reflections come into play depending on the surface condition: if the surface is smooth, a specular reflection is favored whereas if the surface is rough, a diffuse reflection is rather observed. It is therefore necessary to perform a Kramers-Kronig transform in certain cases.

2.4 *Color measurement*

The total surface color change E can be measured using the color space Lab. This is calculated through the parameters L, a and b using two different measurements. The parameter L corresponds to the lightness (0 = black, 100 = white), a to the ratio red/green (with a+ indicating a red trend and a- indicating a green trend) and b to the ratio yellow/blue (with b+ indicating a yellow trend and b- indicating a blue trend).

A ColorCatch Nano colorimeter was used to obtain these parameters. An average value is established per sample after three measurements. ΔE (1) is calculated for sample areas cleaned to 50W and 75W of power. The following equation shows the calculation of ΔE:

$$\Delta E = \sqrt{(\Delta L^2 + \Delta a^2 + \Delta b^2)} \tag{1}$$

where ΔL, Δa, Δb are the differences of the respective parameters between two measurements, a state before cleaning (Z0) and after cleaning (Z50 or Z75).

3 RESULTS

Firstly, the aerosols used to create the graffiti were studied by infrared (FTIR in ATR mode) and/or laser (LIBS) in order to identify the nature of the pigments, charges and binders and thus improve the monitoring of the impact of laser cleaning. The prepared marble and sandstone slabs were then observed before and after cleaning. Colorimetry measurements also completed the study. Finally, infrared analyses in DRIFT mode were carried out and the data was processed by computing.

3.1 *Identification of raw products*

The nine types of aerosol paintings used were studied by infrared (FTIR in ATR mode) and laser (LIBS) to determine the nature of the pigments, binders and charges.

Using LIBS we were able to identify elements such as carbone, copper, iron and titanium which are the main elements presents in some pigments (carbon black, phtalocyanine, oxyde iron and white titanium). FTIR allowed us to identify organic compounds and their wavelength bands to keep for the analysis of paint abnd the evaluation of paint removal on slabs (Daher 2012,Daurelio et al.2018, Nodari et al. 2019,Wiesinger et al.2018). We have found an alkyd resin present in all oil based paints (1741 cm^{-1}) and acrylic styrene in acrylic paints (1740 cm^{-1}).

We have also analyzed the control samples of rock and kept the value of wavelength bands (Jozanikohan & Abarghooei 2022) for marble (calcite at 1801 cm^{-1}) and sandstone (quartz at 1875 cm^{-1}) slabs.

3.2 *Evaluation of laser cleaning*

The evaluation of laser cleaning is first carried out visually and using a field microscope, then color measurements and infrared analyses (DRIFT mode) are performed. A computer processing of the spectra and values of colorimetry is realized for a quantitative evaluation of cleaned areas.

3.2.1 *Observations of slabs after cleaning*

Each sample has been examined macro and microscopically. The observations of all samples are the following:

- Control samples: the sandstone slab has a heterogeneous composition and a gray beige hue. Minerals such as quartz and micas are present and noticeable. The marble slab is heterogeneous in composition and white in color. Some oxides are visible.
- Black painted samples: Laser cleaning of black acrylic paint works very well: an important difference is visible between Z0 and cleaned areas. Visually, laser cleaning at 75W seems to work better (Figure 2), indeed it seems that there are few traces of paint left. The black acrylic paint seems to clean up just as well on the sandstone as on the marble.
- Green painted samples: Laser cleaning of green oil and acrylic paints works relatively well: some differences are visible between Z0 and cleaned areas, but there is still paint left at Z75. The 75W laser cleaning seems to work slightly better than the 50W cleaning. Green oil paint clean better than acrylics on both surfaces.
- Yellow painted samples: Laser cleaning of yellow oil and acrylic paints does not work very well:a very little difference is visible between Z0 and cleaned areas. Laser cleaning at 75W seems to work slightly better than cleaning at 50W. Yellow acrylic paint seems to clean slightly better than oil paint on both surfaces. Nevertheless, according to our observations there is a possible redeposition of the paint after cleaning tentatives.
- Blue painted samples: Laser cleaning of blue oil and acrylic paints works relatively well: some differences are visible between Z0 and cleaned areas, there is still paint left at Z75 (Figure 3). Laser cleaning at 75W works better than cleaning at 50W. Blue oil paint seems to clean better than acrylic paint on both surfaces.
- Red painted samples: Laser cleaning of red oil and acrylic paints works relatively well: some differences are visible between Z0 and cleaned areas, there is still paint left at Z75. The oil paint cleaning seems to work better on the sandstone. Acrylic paint cleaning seems to work better on marble.

3.2.2 *Apparent cleaning using colorimetry*

Colorimetry measurements were carried out to characterize the quality of the cleaning beside the power of laser applied to treat the surfaces. These measurements have been used to compute the total surface color change ΔE for three different cases :

- an uncleaned area (Z0) and a cleaned area at 50W (Z50) to compute a ΔE50 value
- an uncleaned area (Z0) and a cleaned area at 75W (Z75) to compute a ΔE75 value
- an uncleaned area (Z0) and a surface without paint (control sample of sandstone or marble which would represent a 100% perfect cleaning) in order to compute a ΔEMarble or ΔESandstone value.

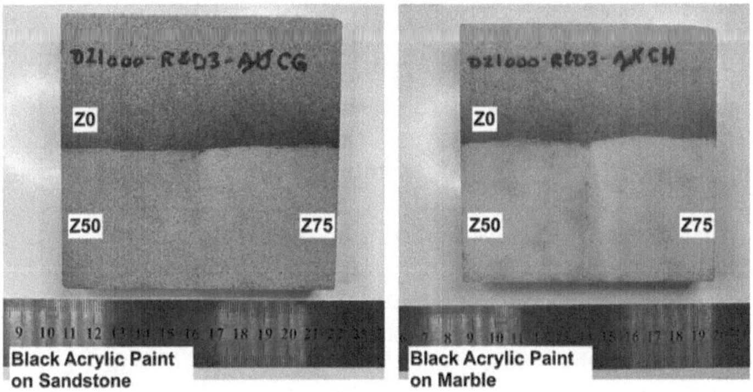

Figure 2. Black painted samples (Z0: no cleaning, Z50: 50W and Z75: 75W).

Table 1. Apparent cleaning from color measurement.

Oil based

Sandstone				Marbles			
Sample/Color		Cleaning 50W	Cleaning 75W	Sample/Color		Cleaning 50W	Cleaning 75W
CQ	Green	70%	83%	CR	Green	54%	63%
CS	Yellow	15%	18%	CT	Yellow	14%	29%
CP	Blue	53%	74%	CO	Blue	57%	68%
CU	Red	80%	89%	CV	Red	65%	68%
Mean		55%	66%	Mean		48%	57%

Cleaning average at 50 W(oil based paint) = 51%
Without yellow sample = 63%
Cleaning average at 75 W(oil based paint) = 62%
Without yellow sample = 74%

Acrylic

Sandstone				Marbles			
Sample /Color		Cleaning 50W	Cleaning 75W	Sample /Color		Cleaning 50W	Cleaning 75W
CG	Black	81%	85%	CH	Black	87%	88%
CI	Green	20%	50%	CJ	Green	34%	42%
CK	Yellow	2%	14%	CL	Yellow	19%	31%
CE	Blue	29%	58%	CF	Blue	48%	46%
CM	Red	17%	39%	CN	Red	45%	59%
Mean		17%	39%	Mean		37%	45%

Cleaning average at 50 W(acrylic paint) = 27%
Without yellow sample = 32%
Cleaning average at 75 W(acrylic paint) = 42%

A calculation of three values of ΔE for each sample what corresponds to three different states of the slabs: a perfectly cleaned slab (control samples), a 50W cleaned slab and a 75W cleaned slab. A scale for each sample is then computed to determine a percentage value of the cleaning that we name here «Apparent cleaning».

According to the calculation results (Table 1), in most cases, cleaning at 75W seems to be more effective. The laser seems to provide a better cleaning average (> 80%) on black acrylic paint (sandstone and marble) and on red oil paint (sandstone).On the other hand, for yellow paint, oil or acrylic, the cleaning rate is the lowest (< 30%) for both type of surfaces.

A cleaning average has been calculated from all apparent cleaning percentages and without taking into account values of yellow paints. The cleaning average for oil paint using 50W is 63% and 74% using 75W. In the same way, the cleaning average for acrylic paint using 50W is 32% and 48% using 75W.

3.2.3 *Semi-quantification of cleaning using FTIR*

The three zones defined as Z0, Z50 and Z75 of each sample had been analyzed using DRIFT mode from an infrared FTIR device. All acquisitions have been realized using an area of 5,25cm^2 in order to keep repeatable measurements and to be able to carry a semi-quantitative evaluation of cleaning. The spectra obtained were all normalized and a calculation of absorbance ratios has been performed using the following wavelength bands: 1741 cm^{-1} for alkyd resin, 1740 cm^{-1} for acrylic styrene, 1875 cm^{-1} for quartz and 1801 cm^{-1} for calcite. In this way, alkyd resin and acrylic styrene wavelength bands are the representative bands for paint products and quartz and calcite bands are the mineral bands related to pure surfaces of slabs.

Four spectrograms were obtained for each slab (Z0, Z50, Z75 and control sample), each spectrogram is build from an average of 128 FTIR acquisitions. When spectrograms are displayed to visualization a sequence in intensity is noticeable, which remains coherent with the acquisition area. The absorbance value of paint bands decrease while mineral bands increase in accordance to the intensity of cleaning. Nevertheless in certain cases the absorbance value ofmineral bands overpass the maxima of the control sample, more visible on the marble spectra (calcite) which is strongly linked to the state of the surface of the samples (higher values of absorbance for mineral bands could be maybe interpreted as a consequence of the surface degradation).

The calculation of the absorbance ratios is achieved using an algorithm that allows us to choose the exact wavelength value where the band reaches its maximum point, this, in order to take into consideration possible offsets due to acquisition in DRIFT mode. The reference wavelengths related to paint products are derived from the spectra without cleaning (Z0) and from the spectra of control slabs for the mineral bands. These wavelengths are then used for all 4 areas of an analyzed sample: Z0, Z50, Z75 and marble or sandstone control sample. For each representative spectrum two absorbance values are collected, one associated to the paint band (noted P) and the other associated to the mineral band (noted R). The P/R ratio (2) is then calculated for each zone: Z0, Z50, Z75 and the control slab. A level of noise was subtracted in each case.

$$\frac{P}{R} = \frac{(Absorbance\ of\ paintband - noise\ associated)}{(Absorbance\ of\ rockband - noise\ associated)} \tag{2}$$

Three P/R values are then obtained for each sample, one value for each zone, which are compared to a P/R value of a pure sandstone or marble zone. This calculation is performed and the higher the P/R value is, the more paint is present in the sample. An absorbance ratio was calculated on the control samples in order to define a reference absorbance ratio value for a 100% paint free sample. This reference value should be the lower value to be obtained. A first observation of the results of this computation is that oil paint is the easier to remove from the samples. Secondly, black and red paints are the easier to remove on marble and sandstone, also cleaning at 75W seems to be more effective.

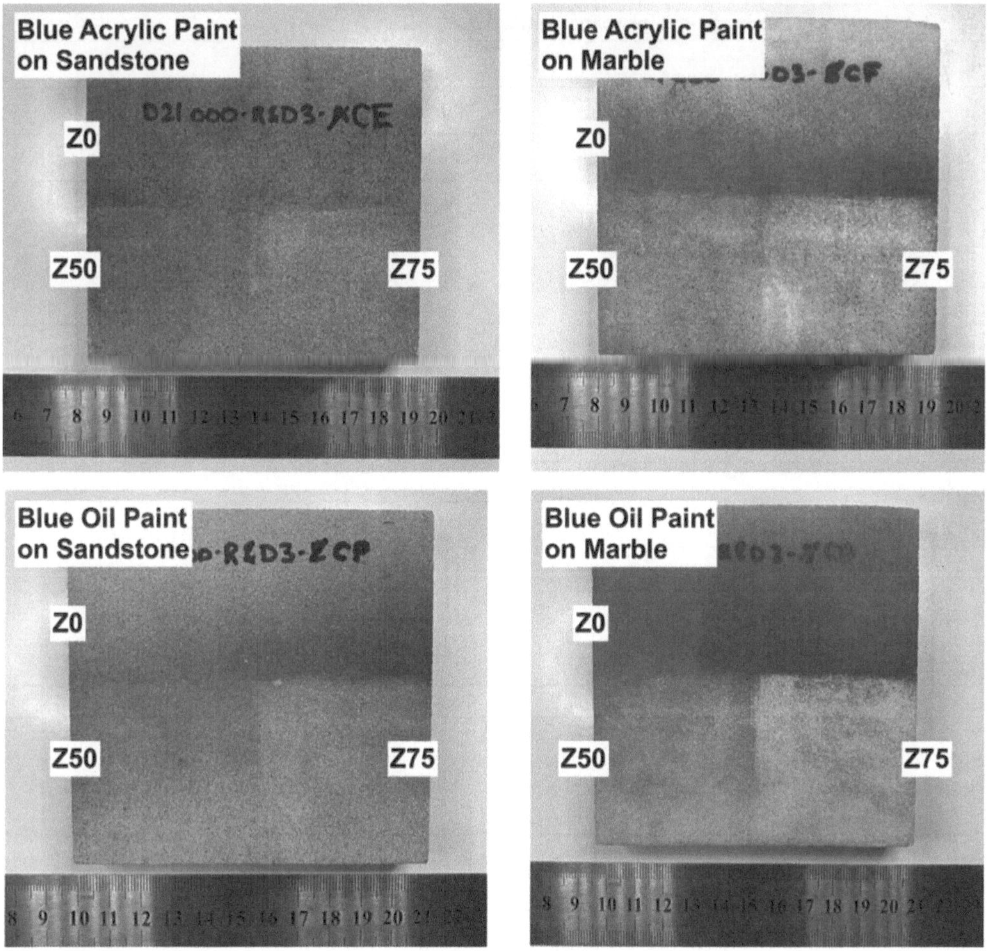

Figure 3. Blue painted samples (Z0: no cleaning, Z50: 50W and Z75: 75W).

Therefore, the same approach as for the colorimetric values has been used herein to define apparent cleaning percentages. The percentages will allow us to compare different natures of support and different colors of paint. The calculation of apparent cleaning or percentage of paint removed was carried out under the following conditions: The P/R ratio of the Z0 area is considered to be 100% paint and the P/R ratio of the sandstone or marble is 0%. The difference obtained from Z0 (P/R ratio) - control sample (P/R ratio) is interpreted as 100% of paint from which the Z50 or Z75 ratios are subtracted to calculate the proportion of paint removed. The results of this calculation for all paints tested for 50W and 75W and apparent cleaning averages are given below (Table 2).

It can be observed that for some samples the proportion of paint removed is higher at 50W cleaning than at 75W, this is the case for samples with similar cleaning effectivity. In addition, some values are close or above 100%, but it is visually demonstrated that this is not the case. This means that P/R ratios of samples are lower than the reference ratio (control samples). A correction of data has been performed but the results of apparent cleaning remain at the same magnitudes. It seems that the diffused reflection of minerals of slabs is strongly intensified by the surface state. In other words, the degradation resulted of the laser ablation intensifies the response of mineral bands in DRIFT mode.

Table 2. Apparent cleaning from absorbance ratios P/R.

Oil based

Sandstone				Marbles			
Sample/Color		Cleaning 50W	Cleaning 75W	Sample/Color		Cleaning 50W	Cleaning 75W
CQ	Green	88%	97%	CR	Green	106%	110%
CS	Yellow	48%	46%	CT	Yellow	52%	89%
CP	Blue	67%	99%	CO	Blue	79%	84%
CU	Red	103%	101%	CV	Red	104%	102%
Mean		77%	86%	Mean		85%	96%

Cleaning average at 50 W(oil based paint) = 80.875 %
Without yellow sample = 91%
Cleaning average at 75 W(oil based paint) = 91%
Without yellow sample = 99%

Acrylic

Sandstone				Marbles			
Sample /Color		Cleaning 50W	Cleaning 75W	Sample /Color		Cleaning 50W	Cleaning 75W
CG	Black	33%	88%	CH	Black	82%	43%
CI	Green	0%	61%	CJ	Green	66%	59%
CK	Yellow	1%	26%	CL	Yellow	24%	26%
CE	Blue	19%	62%	CF	Blue	63%	97%
CM	Red	7%	12%	CN	Red	68%	74%
Mean		7%	40%	Mean		55%	64%

Cleaning average at 50 W(acrylic paint) = 31 %
Without yellow sample = 37 %
Cleaning average at 75 W(acrylic paint) = 52 %
Without yellow sample = 61%

Although, the values are in line with the results of apparent cleaning obtained by colorimetry. Thus, cleaning averages were calculated in order to compare with the colorimetry percentages of apparent cleaning. The results indicate that cleaning at 75W is indeed the more effective, also, oil paint is the easiest to remove on marble or sandstone, 82% cleaning average for oil paint and 60% for acrylic paint. Among oil paints, red color is the better cleaned with 90% of apparent cleaning average, then green and blue from 70 to 80% and finally yellow paint is the most complicated to remove (45%). For acrylic products, black shows an apparent cleaning average of 85%, green and blue between 50% and 60% these on marble or sandstone slabs. However, red acrylic paint has a different cleaning average depending on the nature of the surface cleaned, indeed when red paint is cleaned from marble slabs the cleaning average at 75W is about 70% but when it is removed form sandstone slabs the cleaning average is about 25% just as yellow paint on both type of surfaces.

3.2.4 *Markers cleaning evaluation*

The set of marble and sandstone samples that were marked with permanent (oil based) and water-based markers were analyzed and evaluated using same procedure as on the previous samples. A cleaning at 50W and 75 was carried out and keeping the areas Z0, Z50 and Z75. Colorimetric measurements were carried out and a calculation of apparent cleaning was made. For these samples, DRIFT acquisition was not carried out because the painted surfaces are not large enough to allow measurements. The values obtained are given below (Table 3).

As before, we can see that the 75W cleaning seems to give the best results. Also, the permanent (oil based) type markers seem to be easier to remove, however the difference is very small compared to the water-based markers. As for paint, the black marker is the easiest to be cleaned.

Table 3. Apparent cleaning from colorimetry applied to markers.

Sample	Support	Marker type	50W	75W
BQ	Sandstone	Green Water based	54%	54%
BR	Marble	Green Water based	36%	68%
BS	Sandstone	Red Water based	37%	52%
BT	Marble	Red Water based	25%	74%
BU	Sandstone	Blue Water based	30%	69%
BV	Marble	Blue Water based	40%	54%
BY	Sandstone	Red permanent	43%	48%
BZ	Marble	Red permanent	29%	60%
CA	Sandstone	Blue permanent	31%	50%
CB	Marble	Blue permanent	18 %	52%
CC	Sandstone	Black permanent	54%	63%
CD	Sandstone	Black permanent	61%	71%

4 CONCLUSIONS

This study shows the effects of the laser on two different types of facings. A laser cleaning has been performed using two levels of power. The analysis of colorimetry and the data acquired from FTIR allowed us to realize a cleaning evaluation following some parameters such as the nature of paint and its color and also the nature of the facings being ablated.

The first observation is that the cleaning feasibility depends on the nature of the binding, indeed, a difference between oil and acrylic paint is evidenced, with oil being an easier product to remove using a cleaning laser method. Similarly, there is a difference coming from the color of the paint to remove. It can be concluded that black and red paints are the easier to remove, in contrast, yellow color remains the more difficult to clean from marble and sandstone with a supposed re-deposition of particles after the action of the laser. The most relevant information that we can acquire from the colorimetric and FTIR cleaning evaluation is that in all cases a 75W power has been the more effective. Black and red colors are the preferred when it comes to laser clean. This applies for paint products but also markers for which permanent ink was the easiest to remove.

The degradation induced by the laser has an impact on the DRIFT acquisition (amplification of the mineral bands of sandstone and marble). This is why apparent cleaning of FTIR data is slightly overestimated. However, FTIR results are in line with results obtained from colorimetric measurements. A combination of both could improve the accuracy of the cleaning evaluation and it is necessary to consider a new parameter which is surface degradation.

A roughness test was carried out on two samples to highlight the intensity of the degradation but additional study is still needed. Thus, the future parts of this study will connect the present results with the degradation state of slabs after laser cleaning. Further tests on the rest of the samples as well as possible scanning electron microscope observations are envisaged.

ACKNOWLEDGMENTS

The financial support of this work has been provided by Mescla and Epitopos. The authors would like to thank Meazza society in Mundolsheim for providing the marble and sandstone slabs used for the investigation. We would like to thank Gilles Morvan, from the EOST, for his contribution to the study with interferometric analysis (roughness tests).

REFERENCES

Bosi, A. and al. 2020. Street art graffiti: Discovering their composition and alteration by FTIR and micro-Raman spectroscopy. *Spectrochimica Acta Part A: Molecular and Biomolecular Spectr.* 225: 117474.

Daher, C. 2012. Analyse par spectroscopies Raman et infrarouge de matériaux naturels organiques issus d'objets du patrimoine: méthodologies et applications. *PhD thesis , Université Pierre et Marie Curie - Paris VI.*

Daurelio, G. et al. 2018. The removal of sprayed paints on Calcareous stones: A comparative study of the application of N-Mode, short Free Running and Q-Switch Nd-YAG beam. *RESTAURO Review Forum fur Restauratoren, Konservatoren und DenKmapflege*: 400–409. Munich: Verlag Callwey.

Giusti, C. and al. 2020. Graphic vandalism: Multi-analytical evaluation of laser and chemical methods for the removal of spray paints. *Journal of Cultural Heritage* 44: 260–274.

Jozanikohan, G., Abarghooei, M. 2022. The Fourier transform infrared spectroscopy (FTIR) analysis for the clay mineralogy studies in a clastic reservoir. *Journal of Petroleum Exploration and Production Technology* 12: 2093–2106.

Marazioti, A. et al. 2022. Contemporary murals: Chemical characterisation of artists' spray-paints. *Research Square*. https://doi.org/10.21203/rs.3.rs-1519449/v1.

Paris, C. 2012. Méthodologies spectroscopiques pour l'étude de matériaux: objets du patrimoine de la fin du 19ème au début du 20ème siècle. *PhD thesis, Université Pierre et Marie Curie - Paris VI.*

Ricci, C. and al. 2020. Developing new cleaning strategies of cultural heritage stones: Are synergistic combinations of a low toxic solvent ternary mixtures followed by laser the solution?. *Coatings* 10(466).

Servant, L. et al. 2011. Comprendre la spectroscopie infrarouge: principes et mise en oeuvre. *Photoniques* 53: 68–73.

Siano, S., Agresti, J., Cacciari, I., Ciofini, D., Mascalchi, M., Osticioli I., & Mencaglia A. A 2012. Laser cleaning in conservation of stone, metal, and painted artifacts: State of the art and new insights on the use of the Nd:YAG lasers. Applied Physics A. *Applied Physics A*, 106: 419–446.

Siano, S. & Agresti, J. 2018. Laser-Induced Breakdown Spectroscopy (LIBS). In S. L. López Varela (ed.) *Encyclopedia of Archaeological Sciences*, Wiley. DOI: 10.1002/9781119188230.

Wiesinger, R., Pagnin, L., Anghelone, M., Moretto, L.M., Orsega, E.F. & Schreiner M. 2018. Pigment and binder concentrations in modern paint samples determined by IR and Raman spectroscopy. *Angewandte Chemie International Edition* 57(25): 7401–7407.

Lasers in the Conservation of Artworks XIII – Siano & Ciofini (eds)

The Flight of Night: Laser cleaning coated plaster

E. Promise
Isabella Stewart Gardner Museum, Boston, USA

ABSTRACT: Conservators at the Isabella Stewart Gardner Museum have used a Q-Switched Nd:YAG(1064 nm) laser unit to clean marble surfaces in the collection with gratifying results. Success in cleaning stone surfaces has prompted occasional experiments with more delicate materials such as gilded frames. ISGM conservators have found that, in contrast with the parameters for working on marble, using a low energy pulse and increasing working distance to produce a diffuse spot size can, with care, effectively remove a dark layer of soiling from a vulnerable surface. The author used a similar approach to clean heavy grime and soot from a plaster cast with a darkened resin surface while preserving the resin coating. Results suggest that further studies of the potential for Nd:YAG lasers to safely clean coated surfaces should be undertaken.

1 INTRODUCTION

The conservation department of the Isabella Stewart Gardner Museum owns a Compact Phoenix Q-Switched Nd:YAG laser from Lynton Lasers, Ltd. Since 2007, this tool has been used to clean many collection objects, mostly stone sculptures, but also gilded surfaces. Examples of the latter include the 2008 cleaning of a water-gilt frame for the painting *King Philip IV of Spain* by Velázquez. This treatment provided a successful case study at the Gardner Museum of working at a low fluence to reduce soiling while preserving an object surface that could be damaged by the laser irradiation at higher energy density levels.

In late 2018, ISGM conservators began planning for the reinstallation of a small gallery called the Vatichino, a space that holds objects of personal significance to the museum's founder and namesake, Isabella Gardner. Turnaround on this project was tight, as the team needed to evaluate, and in many cases treat, more than 120 objects in a span of about six months.

One such object was a plaster cast entitled *The Horses of Anahita* or *The Flight of Night* by William Morris Hunt. The cast depicts a scene of three charging horses and a male nude figure in high relief. Prior to treatment, the entire surface was covered with a dark, sooty layer of dirt and grime. Underneath the soiling, a patchy amber coating was visible.

2 RESEARCH AND TESTING

Several other institutions in the United States own copies of The Horses of Anahita. Among these are the Metropolitan Museum of Art, Pennsylvania Academy of the Fine Arts, the Art Institute of Chicago, and the Boston Athenaeum. The plaster models are identical, but the state of the coatings varies across the copies from a thick brownish tone on the cast at the Met to nearly bare plaster at the Art Institute. In consultation with the curatorial department at the Gardner Museum, it was decided that the coating on the ISGM copy should be retained.

DOI: 10.1201/9781003386872-24

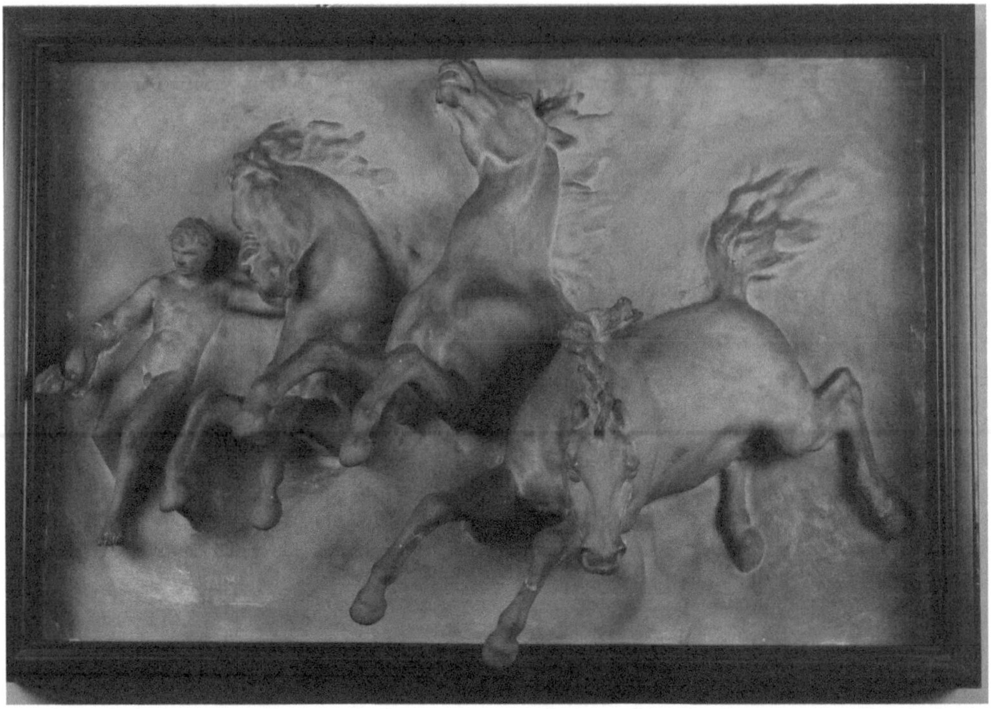

Figure 1. Before treatment image of *The Horses of Anahita* featuring heavy soiling and some cracking of the plaster in areas of high relief.

Dry cleaning methods such as cosmetic sponges and wet cleaning methods such as swab cleaning with saliva were explored unsuccessfully prior to testing the laser. Not only were these methods ineffective at removing the heavily entrenched layer of dirt, but the plaster surface, which was cracking in some areas, could not sustain the application of pressure required for mechanical or solvent cleaning.

The *Port Sunlight War Memorial* featured in *Laser Cleaning in Conservation* (Cooper & Larson 1998) offered an example of a successful laser cleaning of a plaster object with a coated surface. Although no precise working parameters were documented for this treatment, the author was inspired by this successful treatment to carry out testing on *The Flight of Night.*

3 WORKING PARAMETERS

Spot testing on the side edge of the object demonstrated that the laser could be used to remove the dark soiling without disturbing the underlying coating. The laser unit was operated at an energy set corresponding to 85 mJ/pulse and a repetition rate of 5 to 10 Hz. The working distance was approximately 8" from the surface with a spot size of 6 mm, resulting in a fluence of 0.30 J/cm^2 to create a diffuse beam of energy to clean the plaster.

4 FINDINGS

Several challenges occurred during the cleaning process. The greatest challenge was that the coating could only withstand a short dwell time before being disrupted by the laser. Undercuts and tight passages required more passes to clean, necessitating a tight balancing act in order to remove the grime without removing the coating.

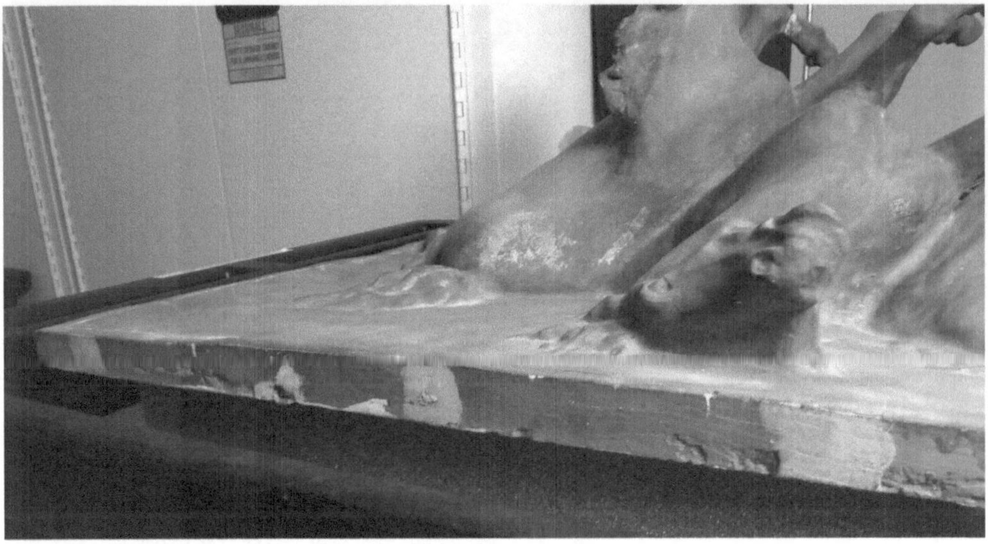

Figure 2. Spot testing on the margins of *The Horses of Anahita*.

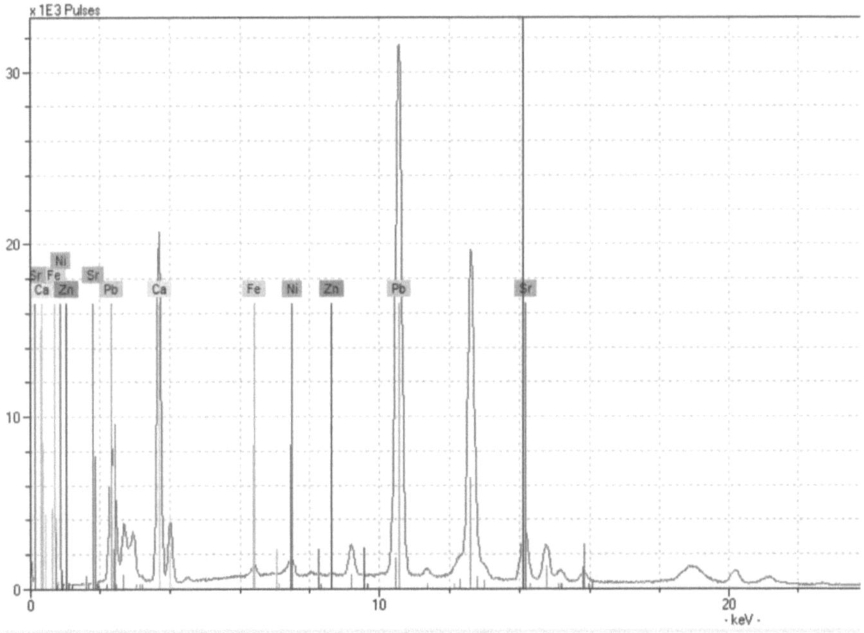

Figure 3. XRF spectrum results indicating the presence of lead in the plaster coating.

Additionally, after analyzing the object using X-ray fluorescence (XRF), the coating was found to contain lead, likely lead white, which discolored to a gray tonality in some spots after interaction with the laser energy. These results were presented to the curatorial team, who found the appearance achieved through laser cleaning to be acceptable.

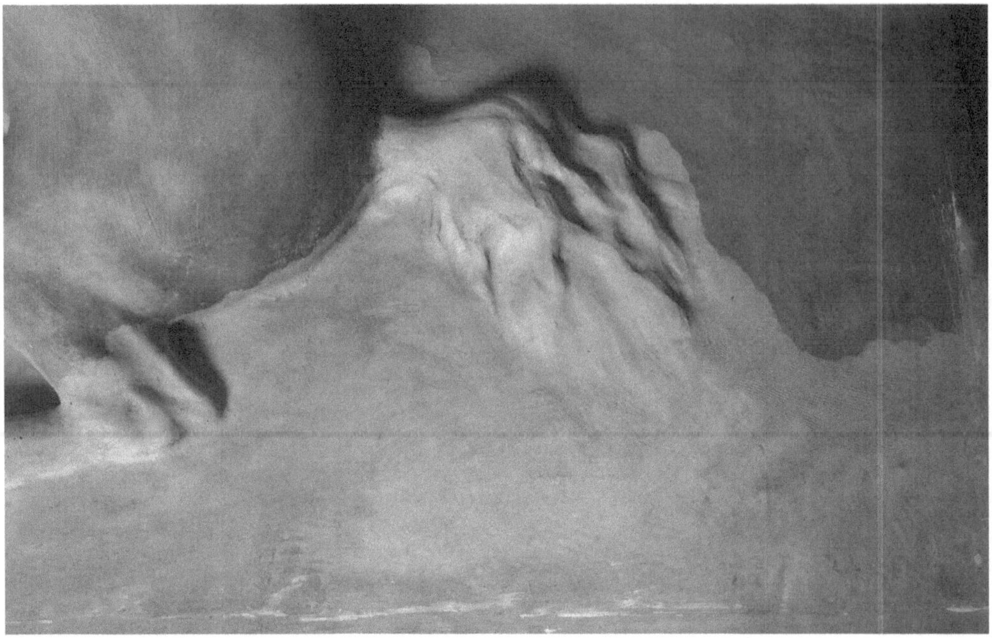

Figure 4. Cleaned portion of The Horses of Anahita showing some color shift in the lead component of the coating after interaction with the laser.

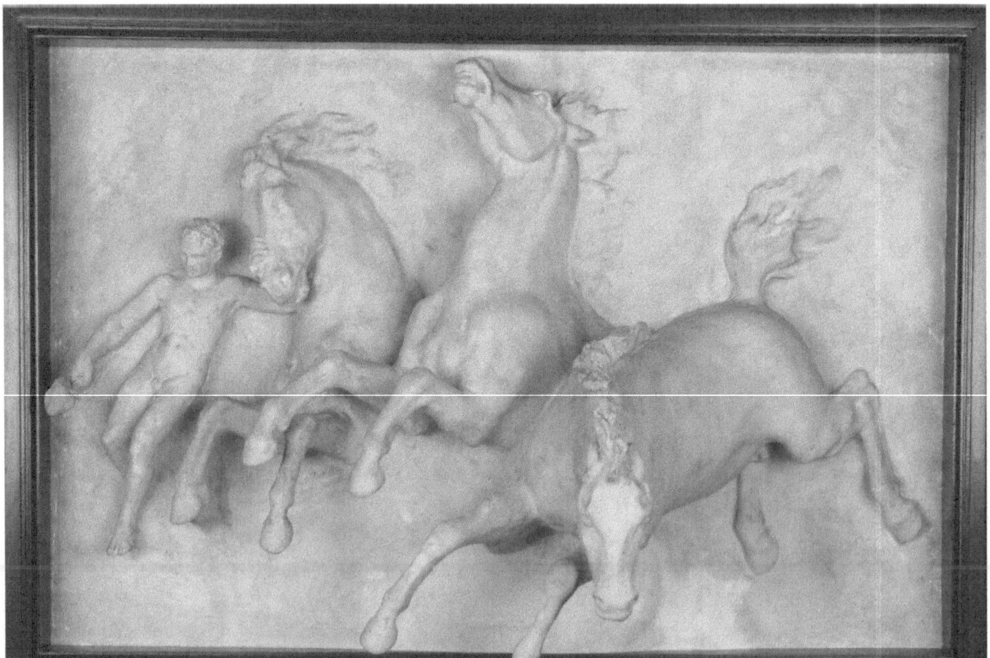

Figure 5. After treatment image of *The Horses of Anahita*.

5 CONCLUSIONS

In spite of the obstacles encountered in this treatment, the laser was used to achieve an even cleaned surface. Other cleaning methods proved ineffective or risky to the underlying plaster. Using the Nd:YAG laser, the plaster was undamaged, and the coating and its golden tonality were preserved.

This project would have benefited from access to Nd:YAG lasers with alternative properties to compare their efficacy with the ISGM's Q-Switched 1064 nm unit. Because the damage threshold of the coating was very close to the fluence required to remove the dirt layer, this treatment protocol was difficult. Nonetheless, the available Nd:YAG laser proved to be the most effective tool that was tested on this object to achieve the project's aims.

REFERENCES

Cooper, M. and Larson, J. 1998. *Laser Cleaning in Conservation*. Oxford: Butterworth-Heinemann.

Lasers in the Conservation of Artworks XIII – Siano & Ciofini (eds)

Evaluation of femtosecond laser texturing on carbonate heritage stones

A.J. López Díaz & A. Ramil
Laboratorio de Aplicacións Industriais do Láser, Campus Industrial de Ferrol, Universidade da Coruña, Ferrol, Spain

D.M. Freire-Lista
Universidade de Trás-os-Montes e Alto Douro, UTAD, Escola de Ciências da Vida e do Ambiente, Quinta dos Prados, Vila Real, Portugal
Centro de Geociências, Universidade de Coimbra

ABSTRACT: The effect of femtosecond laser texturing on Carrara marble, Montesclaros marble, and Redueña dolostone was evaluated. Mineralogical characterisation, laser structuring of thin sections and hand samples of the three heritage stones were carried out. The topography of the textured surfaces was obtained by means of interferometric microscopy, and a petrographic study of laser-generated craters in thin sections was performed to elucidate the effect of the crystalline structure of calcite on the morphology of the laser-generated structures. The differences observed in thin sections allowed to understand the differences observed in laser-structured bulk samples of the three heritage stones; which is key aspect when designing conservative strategies based on laser texturing.

1 INTRODUCTION

Carbonaceous stones have been used in the carving of sculptures throughout history (Cassar 2010, Freire-Lista 2021). The availability of these stones, particularly marble and dolostone, determined their widespread use. These building stones are found in much of the world's heritage, thanks primarily to its purity, homogeneity, colour, and polishability (Siegesmund et al. 2000). In addition, dolomite has been chosen as building stone for fountains due to lower solubility than marble (Freire-Lista & Fort 2017, 2019).

Paseo del Prado of Madrid (Spain) was promoted by King Carlos III in the 18th century. It has three aligned fountains: Cibeles at the northern end; Apollo (or Four Seasons), in the centre; and Neptuno, at the southern end. Cibeles and Neptuno Fountains are built mainly with Montesclaros marble and Apollo Fountain is built with Redueña dolomite (Freire-Lista 2020). Prado-Retiro axis was declared a UNESCO World Heritage Site in 2021 and it is one of the most touristic areas of Spain. The conservation of this urban heritage is a challenging task; it is conducted through a multiplicity of both theoretical and methodological perspectives that require an interdisciplinary vision (De Wever et al. 2017; Huerta-Murillo et al. 2019; Salvini et al. 2022).

Hydrophobic protective treatments, which decrease the water ingress in the stone surface, being developed. In this sense, chemicals with varying degrees of toxicity are often applied to create a surface layer. Laser texturing, both in nanosecond (ns) and femtosecond (fs) pulses, currently used in different industrial and technological fields has been investigated as an environmentally friendly alternative to reduce the wettability of stones (Fiorucci et al. 2014, Pou-Álvarez et al. 2021). Laser texturing of marble has resulted in an increase in the contact angle

DOI: 10.1201/9781003386872-25

above 90º, giving the surface a hydrophobic, and even superhydrophobic, character (Ariza et al. 2022, Carrascosa et al. 2022, López et al. 2019); however, in other stones such as slate, quartzite or granite, laser texturing caused a decrease in the contact angle, thus increasing the hydrophilic nature of the surface (López et al. 2022). These results highlight the need further research on laser texturing in natural stones; given that the challenge is their heterogeneity that causes different interaction mechanisms of the laser pulses with each type of grain, leading to irregular surface absorption. Moreover, the petrographic properties of heritage stones largely condition their durability and forms of deterioration (Fehér & Török 2022; Martínez-Martínez et al 2011; Weber et al. 2011), and thus, it is necessary a petrographic study of the laser-generated structures in thin sections, not only in bulk samples, owing that this knowledge is essential to know the morphology of the ablated area in each of the constituent minerals of heritage stones.

The aim of this study is to assess, at the petrographic microscope scale, the effect of femtosecond laser texturing of the calcareous heritage stones used in the main fountains of Paseo del Prado (Madrid, Spain); that is: Carrara marble (CM), Montesclaros marble (MM) and Redueña Dolostone (RD). The methods used, and data obtained are useful for heritage conservation purposes. The femtosecond laser effect in each calcareous stone allows to solve the problem of the lack of information about the response to the femtosecond laser of each of the mineralogical textures present in different carbonate stones used in the Prado-Retiro axis UNESCO World Heritage Site 2021, and to establish a conservation and restoration plan (Salvini et al 2023).

2 MATERIALS AND METHODS

2.1 *Carrara Marble (CM)*

CM was used since pre-historic times and found in many Roman patrician structures such as the Pantheon at Agrippa and Trajan's Column, it was exported from the port at Luni to the entire Roman Empire. CM, specifically 'Statuario Michelangelo' marble is quarried at Monte Altissimo in the Apuan Alps around Carrara, Tuscany (Italy) (44 0300800 N, 10 1400400 E). It bellows to the Apuane Unit, the lowermost of the Tuscan units. This marble is characterised by a uniform composition and colour (white, locally streaked with thin brownish veins and dark sub-millimetric laminates) and exhibited no cracking visible to the naked eye (Murru et al. 2028). The two putti located at the back of the carriage of Cibeles Fountain are carved in this marble (López de Azcona et al. 2002).

2.2 *Montesclaros Marble (MM)*

MM outcrops in the Gredos Complex, in a band approximately 11 km long and up to 1 km wide, between the villages of Hontanares and Montesclaros, where these are high grade regional metamorphic materials. MM is quarried approximately 140 km from Madrid and approximately 15 km north of Talavera de la Reina (Toledo, Spain). It is predominantly white, white-bluish in colour. In addition, there are grey, blue grey, cream, pink and cream marbles. This dolomitic marble has crystals visible to the naked eye. The fountains of Cibeles and Neptuno are built with this marble (Freire-Lista 2020).

2.3 *Redueña Dolostone (RD)*

Due to difficulties in financing of CM for the fountains, the architect Ventura Rodríguez proposed the use of Redueña dolostone (RD) for the construction of Apollo fountain. Redueña stone includes limestone and dolomicrites, normally of cream colour tones. Its historical quarries were very dispersed, notably the villages of Redueña, Guadalix de la Sierra, El Molar, Venturada and Torrelaguna, as well as other towns in Guadalajara, Spain (Fort et al. 2013, Sanz et al. 2015). The dolostone correspond to the Montejo member that forms part of the

Castrojimeno Formation (Santonian). This member has a level of Trochactaeon Lamarcki specie gastropods and it outcrops in the around Redueña Village (Freire-Lista & Fort 2017). This member has been used for the construction of Apollo fountain, in the Paseo del Prado of Madrid and it is the one that has been tested in this work.

2.4 *Sampling*

Samples of CM are of 'Statuario Michelangelo' marble, a variety quarried at Monte Altissimo in the Apuan Alps around Carrara, Tuscany (Italy), in the quarry located at the coordinates: 44.05183618, 10.24049532. Samples of MD were extracted from a historical quarry at Montesclaros (40.078169, -4.933098). Samples of RD (Castrojimeno Formation) was extracted from Redueña village, located approximately 50 km North of Madrid, in a historical quarry located at the coordinates 40.80208, -3.59141.

All samples were cut and polished to make cuboids of 2 cm× 5 cm× 5 cm (bulk samples) and thin sections (2 cm × 3 cm × 30 μm). After cutting samples with a low speed saw, they were polished with sandpaper and progressively smaller diameter size of diamond spray (from 6, 3 to 1 μm of grain diameter) to get polished thin sections. The thin sections were observed under a Leica DM4500 P polarisation microscope, equipped with a digital FireWire Camera Leica DFC 290 HD that worked with the Leica application suite software LAS 4. To understand the effect of laser texturing, two micromosaics (parallel and crossed Nicols) were made with more than 200 microscopic images each, covering an approximate area of 4 cm^2. To facilitate the understanding of the results, a colour code has been established, assigning blue to CM, green to MM, and yellow to RD.

2.5 *Laser texturing*

The laser used was the Spirit system from Spectra Physics with emission wavelength 1040 nm and pulse width < 400 fs. The intensity profile at the laser output was near- Gaussian (M2 < 1.2) and the beam diameter at the exit of the laser head was 1.5 mm. The laser beam presents horizontal polarization (> 100:1). Pulse rate can be selected from single shot to 1 MHz, with maximum pulse energy of 40 μJ at 100 kHz. The maximum mean power output is > 4 W. A two-mirror galvanometric scanner (Raylase SuperscanIII-15) was used and scanned the laser beam in X-Y directions. The beam was focused by means of a F-theta objective lens, 160 mm focal length, up to a diameter of 30 μm. At the working plane, the beam polarisation was parallel to Y direction. All the processing experiments were performed in ambient air and off-line control of the ablation process was accomplished by optical microscopy (Nikon Eclipse L150) and digital image processing software.

Once the most suitable laser parameters for each stone were selected, structuring patterns were generated on both thin sections and bulk samples. In the case of thin sections, 20 horizontal series of craters 200 μm apart were made as it is shown in Figure 1; the number of laser shots per crater was increased from 1 shot (bottom) to 20 shots (top). In bulk samples, two types of texturing patterns were used; a matrix of holes and parallel grooves, and 40 μm, 60 μm and 80 μm separation between holes or grooves.

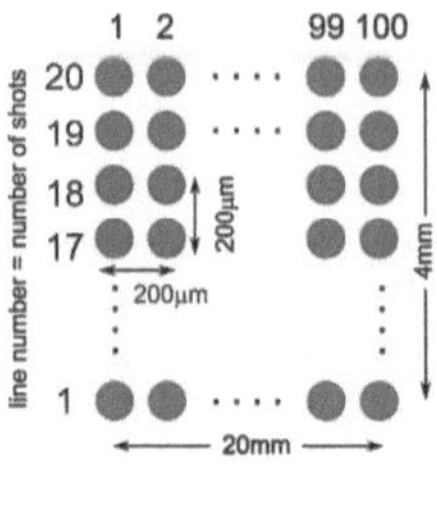

Figure 1. Scheme of laser texturing on the three thin sections (MM, CM and RD).

2.6 *Surface roughness*

A scanning white light interferometer (SWLI) microscope Zygo NewView 600 mounted in a vibration isolation table was used to evaluate the results of the texturing process. The vertical scan range was 150 µm with a vertical resolution of 0.1 nm. The image resolution was 640 × 480 pixels that gives a field view of 349.1 µm × 261.9 µm.

Data processing, analysis and visualisation were implemented by using Scientific Phyton. The same equipment was used to characterise the surface roughness of laser processed bulk samples in accordance with standard UNI-EN ISO 25178–2 (Int. Org. Standardization 2012).

3 RESULTS AND DISCUSSION

3.1 *Petrography of thin section mosaics*

Each carbonate stone has a different petrography. Figure 2 shows the mosaics of the texturing of the three thin sections (CM, MM and RD) using both parallel Nicols (a, b, c) and crossed Nicols (a', b', c'). The different colours of calcite crystals in Figures 2a', 2b', and 2c' indicate different crystallographic orientations. The extinction angle of calcite is symmetric or inclined with respect to the exfoliation traces.

CM is equigranular monocrystalline (Calcite), with a mean crystal size of 300 µm. Lamellar twins of calcite are common under {01-12}, and single twins are common under {0001}. Some crystals show perfect (rhombohedral) cleavage at {10-11}. The edges of the crystals are often straight. Intercrystallite microcracks border the edges of calcite crystals (Figure 2a). Rhombohedral exfoliation in the calcite contributed to the formation of normally straight intracrystalline microcracks.

MM is very coarse to coarse-grained (crystal size up to 5 mm) and it has granoblastic texture. The crystal boundaries are very sinuous, although few straight intercrystalline fractures are observed (Figures 2b and 2b').

RD is a massive dolostone, formed by micritic cement (particles ranging in diameter up to 4 µm formed by the recrystallisation of lime mud) with few mottled colours and poikilotopic and blocky mosaic cements predominate. RD has fossils filled with randomly oriented calcite that can reach sizes of 1 mm. The fossils are mainly composed of gastropods and fragments of bivalves. These fossils are found as molds filled with calcite crystals. The thickness of their shells can reach 3 mm thick. Remnants of their original depositional texture are preserved, generally disposed fossils aligned longitudinally to the lamination with few mottled colours and poikilotopic, and blocky mosaic cements predominate (Figures 2c and 2c').

An increase in diameter of the holes in relation to the number of laser shots delivered is appreciated in Figure 2. In this way, CM shows complete lines of craters perceptible under an optical microscope starting from shot number 6 (Figure 2a).

MM shows complete lines of craters perceptible from shot number 1 (Figure 2b). In case of RD thin section, different behaviour can be seen depending on the laser irradiated material; micritic cement shows complete lines of craters starting at line number 10 and recrystallized calcite shows craters at line 1.

Figure 3 depicts a scope of the areas highlighted in Figure 2, which allow us to observe the morphology of the laser generated structures: The craters on CM present an irregular morphology and, conversely, craters on MM present the most circular and regular morphology; with regards to RD, the morphology of the craters depends on the mineralogy on which the laser strikes. Differences in morphology can be intuited depending on the crystallographic direction of the calcite crystal on which the laser is incident. In this way, RD1 presents rectilinear fractures bordering the craters, while RD3 presents a rectangular morphology with the characteristic angles of the calcite structure. When the laser hits the micritic cement, the craters are more diffuse, maintaining a pseudo-circular morphology.

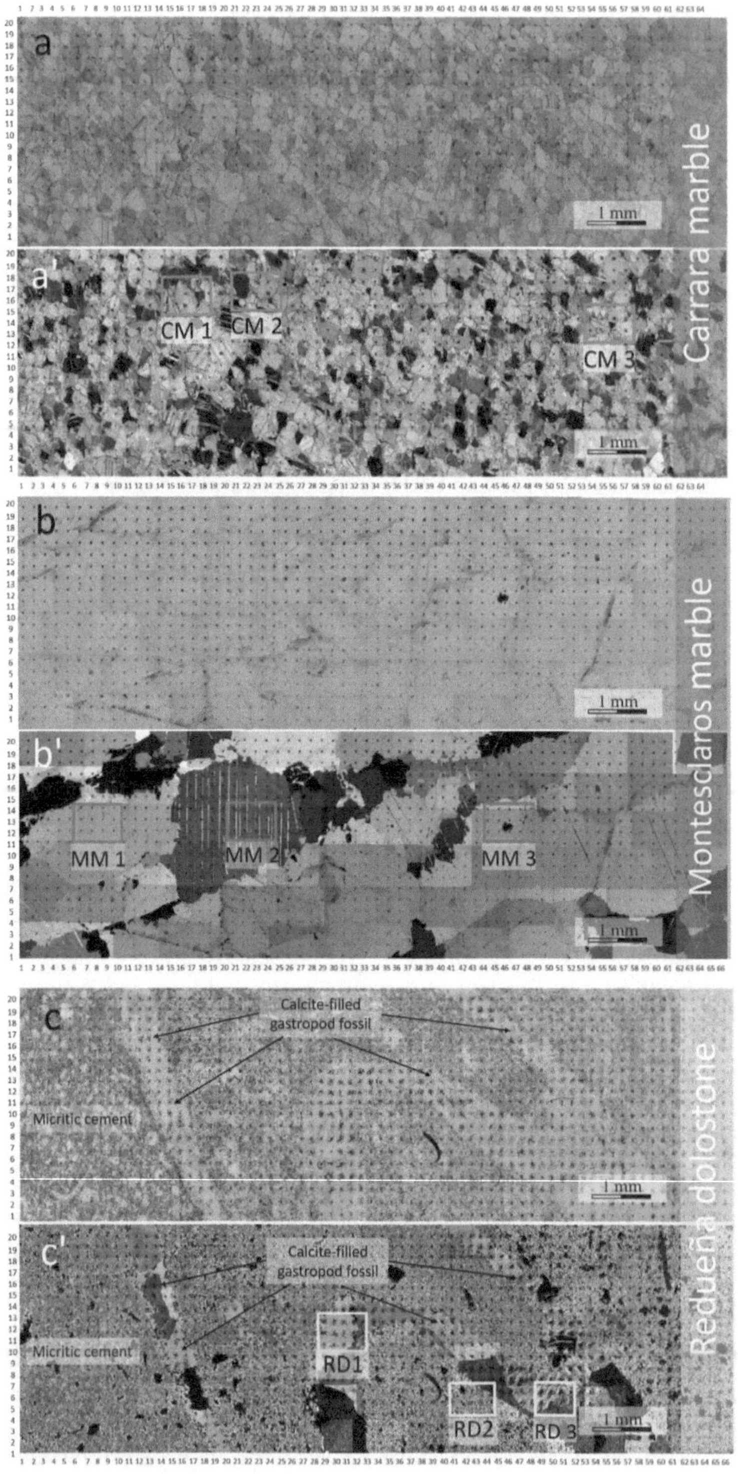

Figure 2. Petrographic mosaics of laser textured thin sections of Carrara marble (CM), Montesclaros marble (MM) and Redueña dolomite (RD).

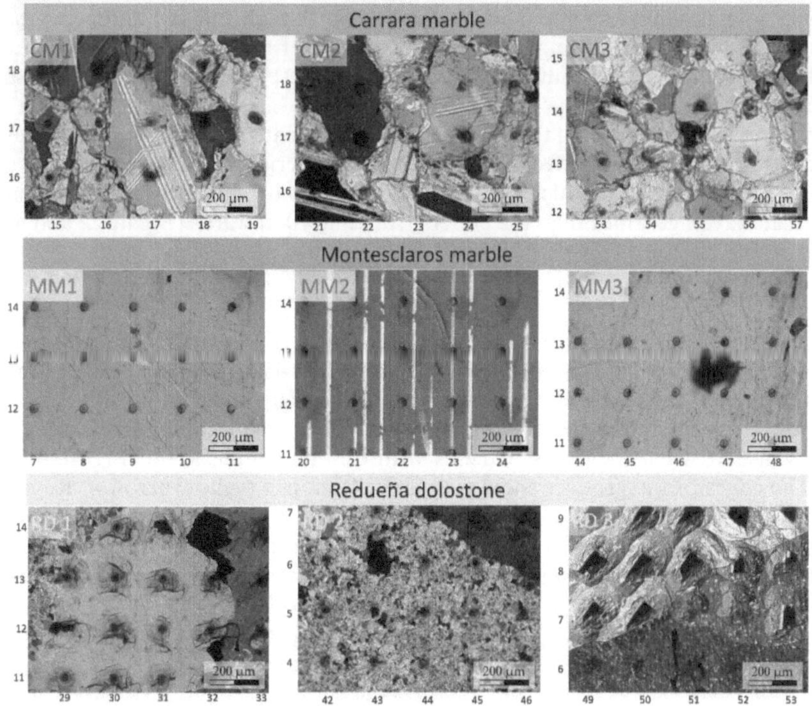

Figure 3. Enlargement of the rectangles indicated in the thin sections of Figure 2. Craters produced by laser texturing in Carrara marble (CM), Montesclaros marble (MM) and Redueña dolomite (RD) are shown in more detail.

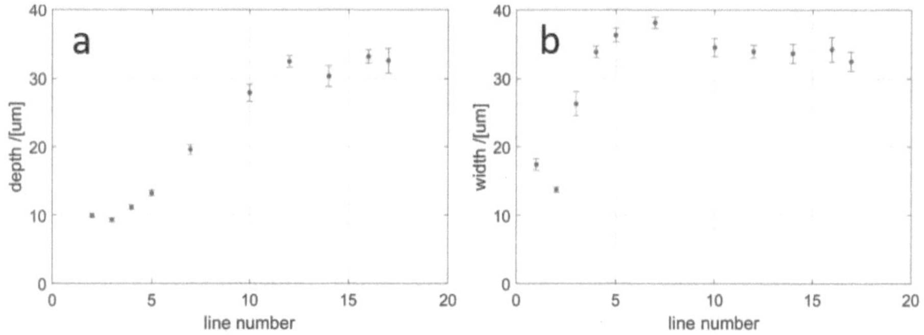

Figure 4. Crater size (μm) produced by laser texturing in Redueña Dolomite (RD) thin section as a function of the number of laser shots: a) depth, b) width.

3.2 *Thin section craters: Depth and width*

Topographies, obtained by fitting crater profiles to Gaussian bidimensional curves, allowed us to obtain the geometric parameters (depth and width) of the laser-generated structures. Figure 4 illustrates the dependence of the hole depth and width, on the number of laser shots in RD. Values were calculated by averaging 20 holes in each line.

In case of depth, Figure 4a), after the first five laser shots the dependence is nicely linear up to approximately 10-11 laser shots where saturation occurs for a hole depth of 30 μm.

This behaviour is similar in CM and MM marbles, though the saturation depth is lower than RD. Regarding width, a rapid increase is observed for the first five laser shots up to saturation that occurs for a hole width of 35 μm, as it was been expected given the diameter of the laser beam.

Note that starting from line 10, the average depth and width values present a greater dispersion and in this area the laser affects both calcite and micritic cement crystals. On the other hand, the lines that cross more calcite are from 1 to 8, 13 and 14, and the craters of lines 5 and 7 have a greater average width than the craters from line 10 onwards (Figures 2 and 4).

3.3 *Bulk samples*

Bulk samples were laser structured with two different patterns, a matrix of holes and parallel grooves, with different separation between them, "pitch": 40 μm, 60 μm and 80 μm, respectively. Figure 5 shows the topographies of both patterns at 60 μm pitch.

Images of the textured samples of CM, MM and RD, show how the grooves change their dimensions in CM when they cross calcites with different orientation (Figure 6 CM2, red arrows). The edges of the grooves show triangular fractures (yellow arrows). Regarding the hole matrix in CM, hole sizes vary according to the orientation of the calcite (Figure 6 CM3). In relation to MM, it has a larger crystal size, and it shows a more homogeneous hole and groove size. The image of RD textured with grooves shows a less uniform appearance.

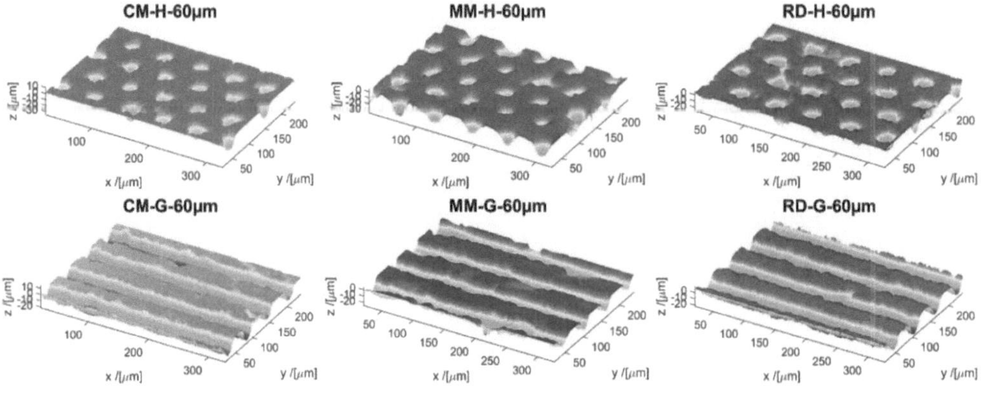

Figure 5. Topographies of the textured samples of Carrara marble (CM), Montesclaros marble (MM) and Redueña dolomite (RD). Texturing patterns are H (matrix of holes) and G (parallel grooves). Separation between structures 60 μm.

The areal roughness parameters of the textured surfaces were calculated from the topographies. Figure 7 depicts the values of the root mean square height, Sq, as a function of the "pitch" in Carrara marble (CM), Montesclaros marble (MM) and Redueña dolomite (RD). It can be seen that the roughness decreases with the pitch, in both holes and grooves; corresponding the maximum value to the lower pitch, 40 μm, given that increasing the separation between structures increases the surface that is not laser ablated. Moreover, values of the root mean square height, Sq, are higher for grooves than holes but diminish faster with the pitch in the case of holes than grooves. On the other hand, for the same pitch, CM and RD present higher roughness (above 5 μm) tan MM (≤5 μm). The same behaviour has been observed in the mean roughness, Sa, but with lower values than in Sq, because the latter is a more sensitive parameter than the arithmetic mean to large localized irregularities, as it is the case of the laser texturing regular patterns used in this work.

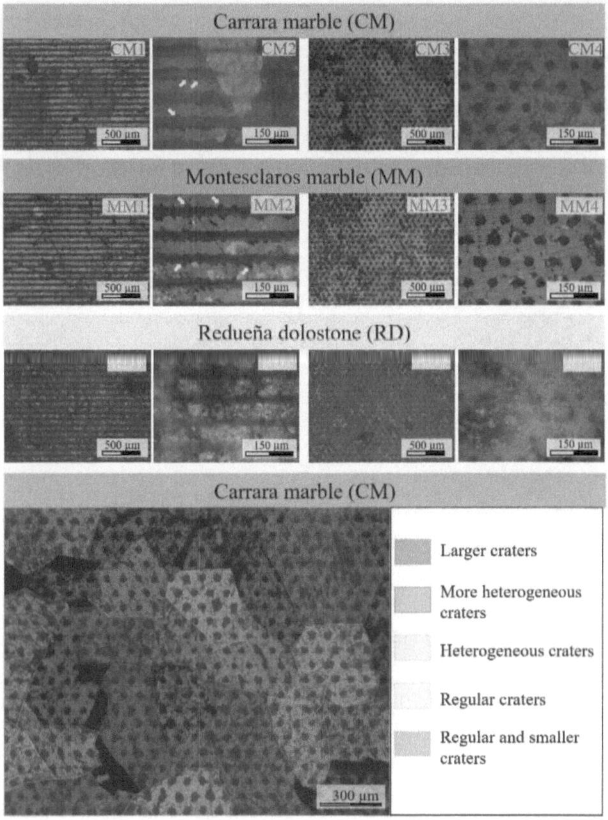

Figure 6. Bulk samples texturing (parallel grooves and hole matrix) of Carrara marble (CM), Montesclaros marble (MM) and Redueña dolomite (RD). Bottom figure: Enlargement of the MM3 bulk sample with the division of the calcite crystals according to the morphology of the craters.

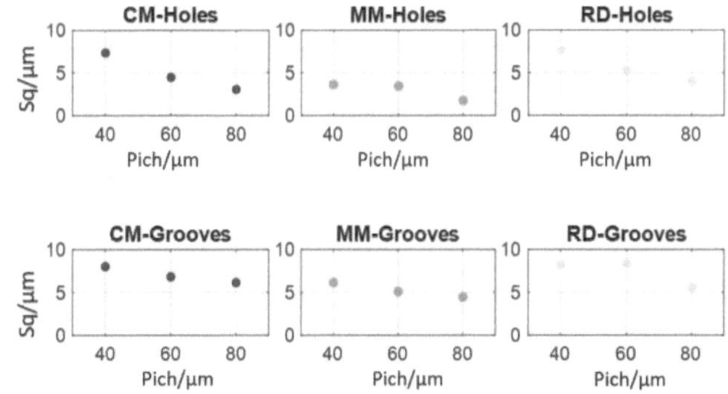

Figure 7. Root mean square height Sq vs. pitch, of the textured samples of Carrara marble (CM), Montesclaros marble (MM) and Redueña dolomite (RD).

To explain the different responses under femtosecond laser texturing of these carbonate stones; Carrara marble (CM), Montesclaros marble (MM) and Redueña dolostone (RD); the structure of calcite must be considered (Figure 8): A calcite crystal is shown in Figure 8a, and the acute rhombohedral unit cell of calcite indicating the cleavage rhombohedron setup in Figure 8b. The true unit cell includes 2 $CaCO_3$ with calcium ions at the corners of the rhombohedrons and CO_3 groups, each of which consists of a carbon ion at the centre of a planar group of oxygen atoms whose centres define an equilateral triangle (Figure 8c). This atomic arrangement conditions the fracture produced by laser texturing in calcite crystals, especially when this calcite has not undergone a metamorphism process, as is the case with marbles. For this reason, the craters of the marbles show a more circular morphology than the craters produced in the neoformation calcite included in the fill of RD fossils. These craters are triangular (Figure 3 RD3), depending on the direction of incidence of the laser with respect to the structure of calcite.

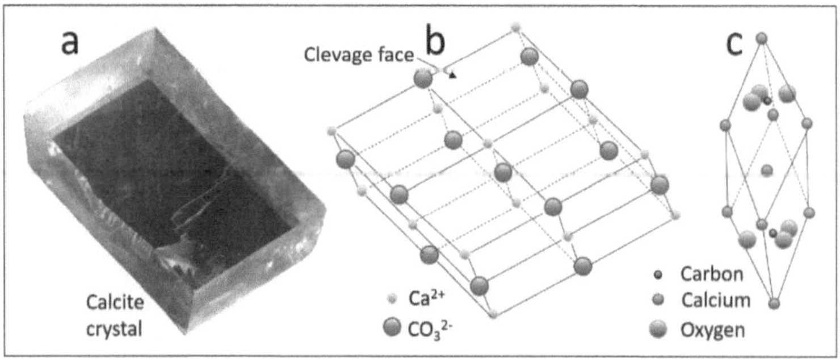

Figure 8. Calcite structure. a) Calcite crystal, b) Calcite unit cell, c) Atomic arrangement. Modified of Encyclopedia Britannica, 1994.

Crystal size and microcracks also influence crater morphology. In this sense, CM has crystal size of 300 μm and these crystals have intercrystalline microcracks. This disposition of the crystals generates more heterogeneous craters than MM (average crystal size of 1 mm), because there are more planes of weakness. The greater size of craters in the crystal boundaries is due to crater spreads along the crystal boundary. In addition, a greater deformation in this zone due to the grain boundary sliding must be considered (Hou et al., 2022). In the case of the micritic cement present in RD, its low crystallinity does not produce this effect, for this reason the craters appear circular and diffuse.

The combination of these characteristics means that, despite being three carbonate stones, their responses to laser texturing give different morphology of the generated structures and, consequently, different surface roughness (Figure 7). Moreover, with regards to the stones, CM and RD present values of Sq above 5 μm; while MM is always below this value. This indicates that a greater heterogeneity of the stones in terms of the type of calcite (RD), but also in terms of number of intercrystalline limits (CM), results in a greater surface roughness than in the case of stones like MM with large crystal size.

4 CONCLUSIONS

Calcite present in the marbles (CM and MM) has generated more regular craters than the neoformation calcite that fills the fossils in Redueña dolostone (RD). The craters generated in the neoformation calcite are more conditioned by microcracks due to the microstructure of the calcite. In the case of the micritic cement present in the RD, its low crystallinity does not produce this effect and the craters appear circular and with fuzzy edges. The differences observed in thin sections allowed us to understand the differences observed in laser-structured bulk samples and, specifically, in the roughness values achieved by each of the heritage calcareous stones as a function of the textured pattern. This is a crucial parameter when designing conservative strategies based on laser texturing.

Finally, policies to maintain and restore built heritage must be enacted in consultation with specialists in the science of stone conservation to follow an appropriate strategy for preservation. European directives and the scientific community advocate preventive conservation rather than interventional conservation (UNESCO 2011). Therefore, the causes of building stone decay must be identified, and the means needed to mitigate them must be utilized.

ACKNOWLEDGMENTS

This work was supported by the Portuguese Foundation for Science and Technology (FCT) in the framework of the Strategic Funding, with UIDB/00073/2020 and CEECIND/03568/2017 projects. Authors acknowledge Erasmus+ program: HERDADE Consortium (2021-1-ES01-KA130-HED-360 000007519). This paper has been written during a scientific stay at the Centro de Ciencias Sociales y Humanas (CSIC) in Madrid within the framework of the project entitled. Use and function of the peninsular granary caves: an approximation based on archaeobotany (PID2021-127936NB-I00).

REFERENCES

Ariza R., Alvarez-Alegria, M., Costas, G., Tribaldo, L., R. Gonzalez-Elipe, A. & Siegel, J. 2022. Multiscale ultrafast laser texturing of marble for reduced surface wetting. *Applied Surface Science* 577: 151850. https://doi.org/10.1016/j.apsusc.2021.151850.

Carrascosa, L.A.M., Zarzuela, R., Botana-Galvín, M., Botana, F.J. & Mosquera, M.J. 2022. Achieving superhydrophobic surfaces with tunable roughness on building materials via nanosecond laser texturing of silane/siloxane coatings. *Journal of Building Engineering* 58: 104979. https://doi.org/10.1016/j.jobe.2022.104979.

Cassar J. 2010. The use of limestone in a historic context – the experience of Malta. *Geological Society, London, Special Publications* 331: 13–25. https://doi.org/10.1144/SP331.2.

De Wever, P., Baudin, F., Pereira, D., Cornée, A., Egoroff, G. & Page, K. 2017. The Importance of Geosites and Heritage Stones in Cities—a Review. *Geoheritage* 9: 561–75. https://doi.org/10.1007/s12371-016-0210-3.

Fehér, K. & Török, Á. 2022. Detecting short-term weathering of stone monuments by 3D laser scanning: lithology, wall orientation, material loss. *Journal of Cultural Heritage* 58:245–55. https://doi.org/10.1016/j.culher.2022.10.012.

Fiorucci, M.P., López, A.J. & Ramil, A. 2014. Comparative study of surface structuring of biometals by UV nanosecond Nd:YVO4 laser. The International Journal of Advanced Manufacturing 394 *Technology*; 75: 515–521. https://doi.org/10.1007/s00170-014-6164-1.

Fort, R., Alvarez de Buergo, M., Perez-Monserrat, E.M., Gomez-Heras, M., Jose Varas- Muriel, M. & Freire D.M. 2013. Evolution in the use of natural building stone in Madrid, Spain. *Quarterly Journal of Engineering Geology and Hydrogeology* 46:421–9. https://doi.org/10.1144/qjegh2012-041.

Freire-Lista, D.M. & Fort, R. 2017. Historical Quarries, Decay and Petrophysical Properties of Carbonate Stones Used in the Historical Center of Madrid (Spain). *AIMS Geosciences* 3:284–303. https://doi.org/10.3934/geosci.2017.2.284.

Freire-Lista, D.M. & Fort, R. 2019. Historical City Centres and Traditional Building Stones as Heritage: Barrio de las Letras, Madrid (Spain). *Geoheritage* 11: 71–85. https://doi.org/10.1007/s12371-018-0314-z.

Freire-Lista, D.M. 2020. Geotourism from fuente de cibeles of Madrid. History, building stones and quarries. *Cadernos Do Laboratorio Xeoloxico de Laxe* 42:69–94. https://doi.org/10.17979/CADLAXE.2020.42.0.7286.

Freire-Lista, D.M. 2021. The Forerunners on Heritage Stones Investigation: Historical Synthesis and Evolution. *Heritage*; 4:1228–68. https://doi.org/10.3390/heritage4030068.

Hou, C., Liu, J., Zheng, Y., Sun, Y., Zhou, B. & Fan, W. 2022. Prolonged grain boundary sliding in naturally deformed calcite marble at the middle crustal level. *Journal of Structural Geology* 161:104658. https://doi.org/10.1016/J.JSG.2022.104658.

Huerta-Murillo, D., García-Girón, A., Romano, J.M., Cardoso, J.T., Cordovilla, F., Walker, M., Dimov S.S. & Ocaña, J.L. 2019. Wettability modification of laser-fabricated hierarchical Surface structures in Ti-6Al-4V titanium alloy. *Applied Surface Science* 463:838–46. https://doi.org/10.1016/J.APSUSC.2018.09.012.

López, A.J., Ramil, A., Pozo-Antonio, J.S., Rivas, T. & Pereira, D. 2019. Ultrafast laser Surface texturing: A sustainable tool to modify wettability properties of marble. *Sustainability* 11: 4079. https://doi.org/10.3390/su11154079.

López, A.J., Pozo-Antonio, J.S., Moreno, A., Rivas, T., Pereira, D. & Ramil, A. 2022. Femtosecond laser texturing as a tool to increase the hydrophobicity of ornamental stone: The influence of lithology and texture. *Journal of Building Engineering* 51: 104176. https://doi.org/10.1016/j.jobe.2022.104176.
López de Azcona, M.C., Fort González, R. & Mingarro Martín, F. 2002. La conservación de los materiales pétreos en la Fuente de Cibeles, Madrid (España). *Materiales de Construcción* 52: 65–75. https://doi.org/10.3989/mc.2002.v52.i265.345.
Martínez-Martínez, J., Benavente, D. & García-del-Cura, M.A. 2011. Spatial attenuation: The most sensitive ultrasonic parameter for detecting petrographic features and decay processes in carbonate rocks. *Engineering Geology* 119: 84–95. https://doi.org/10.1016/j.enggeo.2011.02.002.
Murru, A., Freire-Lista, D.M., Fort, R., Varas-Muriel, M.J. & Meloni, P. 2018. Evaluation of post440 thermal shock effects in Carrara marble and Santa Caterina di Pittinuri limestone. *Construction and Building Materials* 186: 1200–1211. https://doi.org/10.1016/j.conbuildmat.2018.08.034.
Pou-Álvarez, P., Riveiro, A., Nóvoa, XR., Fernández-Arias, M., del Val, J., Comesaña, R., Boutinguiza, M., Lusquiños, F. & Pou, J. 2021. Nanosecond, picosecond and femtosecond laser surface treatment of magnesium alloy: role of pulse length. *Surface and Coatings Technology* 427:127802. https://doi.org/10.1016/j.surfcoat.
Salvini, S., Bertoncello, R., Coletti, C., Germinario, L., Maritan, L., Massironi, M., Pozzobon, R. & Mazzoli C. 2022. Recession rate of carbonate rocks used in cultural heritage: Textural control assessed by accelerated ageing tests. *Journal of Cultural Heritage* 57: 154–64. https://doi.org/10.1016/j.culher.2022.08.010.
Salvini, S., Coletti, C., Maritan, L., Massironi, M., Pieropan, A., Spiess, R. & Mazzoli, C. 2023. Petrographic characterization and durability of carbonate stones used in UNESCO World Heritage Sites in northeastern Italy. *Environmental Earth Sciences* 82: 49. https://doi.org/10.1007/s12665-022-10732-y.
Sanz, M., Oujja, M., Ascaso C., de los Ríos, A., Pérez-Ortega, S., Souza-Egipsy, V., Wierzchos, J., Speranza, M., Cañamares, M.V. & Castillejo M. 2015. Infrared and ultraviolet laser removal of crustose lichens on dolomite heritage stone. *Applied Surface Science* 346:248–55. https://doi.org/10.1016/j.apsusc.2015.04.013.
Siegesmund, S, Ullemeyer, K, Weiss, T & Tschegg, E.K. 2000. Physical weathering of marbles caused by anisotropic thermal expansion. *International Journal of Earth Sciences* 89:170–82. https://doi.org/10.1007/s005310050324.
Taelman, D, Delpino, C & Antonelli, F. 2019. Marble decoration of the Roman theatre of Urvinum Mataurense (Urbino, Marche region, Italy): An archaeological and archaeometric multi-method provenance study. *Journal of Cultural Heritage* 39: 238–50. https://doi.org/10.1016/j.culher.2019.03.009.
Weber, J, Beseler, S & Sterflinger, K. 2007. Thin-section microscopy of decayed crystalline marble from the garden sculptures of Schoenbrunn Palace in Vienna. *Materials Characterization* 58:1042–51. https://doi.org/10.1016/j.matchar.2007.04.014.

Lasers in the Conservation of Artworks XIII – Siano & Ciofini (eds)

The restoration of the amalgam gilded bronze elements from the Baptismal Font of Siena: Remarks on laser ablation in combination with other cleaning methods

M. Baruffetti, S. Agnoletti, P. Belluzzo, A. Brini, A. Cagnini, S. Porcinai, E.C. Ortolani & M. Galeotti
Opificio delle Pietre Dure di Firenze, Laboratori di restauro e scientifici, Florence, Italy

L. Bartoli
El.En. SpA, Calenzano, Italy

S. Tartaglia, S. Casu, E. Della Schiava, A. Mignemi & M. Nesi
Individual conservation-restoration enterprises

ABSTRACT: This paper refers to the conservation work in progress in Florence, at the Opificio delle Pietre Dure, on the bronze elements from the Baptismal Font of Siena (1417-1430). In the bronze conservation community, it is well known that the laser cleaning of gilded bronzes was developed during the restoration of the *Gates of Paradise* by Lorenzo Ghiberti and it was subsequently adopted for the restoration of the other doors of the Florentine Baptistery. Since then, laser ablation has been used for various cases and materials, whether involving or not surface gilding. In this new case study, the focus of the cleaning of large amalgam gilded surfaces was on indoor stored artworks that underwent various maintenance procedures with different materials. The consequent conservation issue concerns the interaction of metal substrates with altered organic coatings (such as waxes, oils and varnishes) and the effects of some drastic cleaning methods operated in the past. The method selected by the conservation team was based on a combined approach to remove altered materials from the surface. Particularly effective was a preliminary step based on the treatment with steam and the use of "oil in water" emulsions, which allowed to partially remove the organic layers and to proceed with lower fluence laser cleaning. The selection of the laser type and the optimization of the operating parameters, along with scientific investigations aimed at the characterization of surface materials and monitoring of ablation, have led to a deeper awareness of some operating solutions.

1 INTRODUCTION

The metals artifacts of the Baptismal Font of the Siena Cathedral (1417-1430 A.D.), a well-known masterpiece of the Italian Renaissance manufactured by some of the leading sculptors of that time, is currently under restoration at the Opificio delle Pietre Dure in Florence, Italy. It consists of marble pieces carved by Jacopo della Quercia and other artists, copper alloy elements and enamelled stripes (Figure 1).

Copper alloy elements were cast by lost wax technique and fire-gilded (mercury amalgam technique) by Giovanni di Turino (Siena, 1385? – post 1455), Goro di Ser Neroccio (Siena 1386 – 1444), Donatello (Firenze, 1386 – 1466), Jacopo Della Quercia (Siena, 1374 – 1438), Lorenzo Ghiberti (Pelago, 1378 – Firenze, 1455). They include six reliefs (approx. 60×60 cm^2) depicting episodes of the Baptist's Life, six sculptures representing the Virtues (h around 50 cm), and four out of the six putti (h around 41 cm) originally decorating the marble tabernacle on top of the baptismal basin. The door of the tabernacle (around 47×23 cm^2) and twelve enamelled and gilded

DOI: 10.1201/9781003386872-26

perimetral stripes (around 100×4 cm^2) decorating the edges of the marble frames are also included in the conservation project. The latter started in 2020 and it is involving three Restoration Departments of OPD (Stone, Bronze, and Jewellery), along with freelance restorers (for the bronze panels and sculptures and for the enamelled stripes), and OPD's Scientific Laboratory.

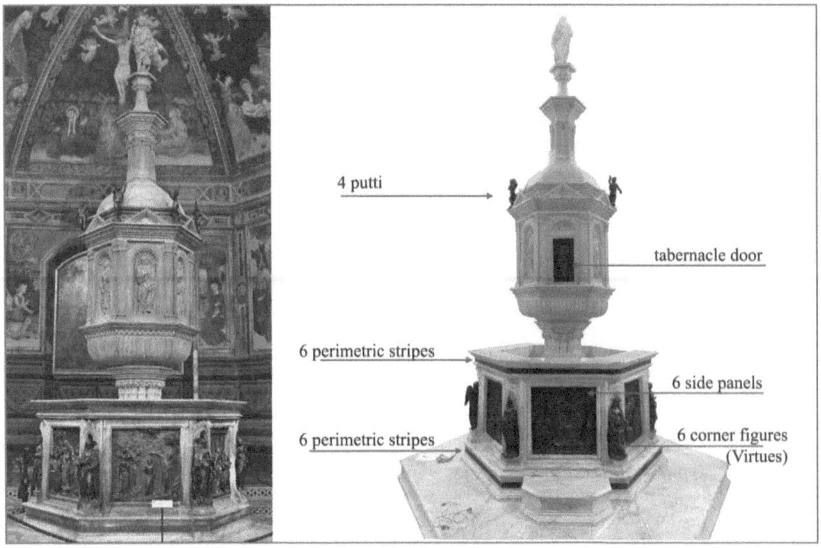

Figure 1. The Baptismal Font of Siena.

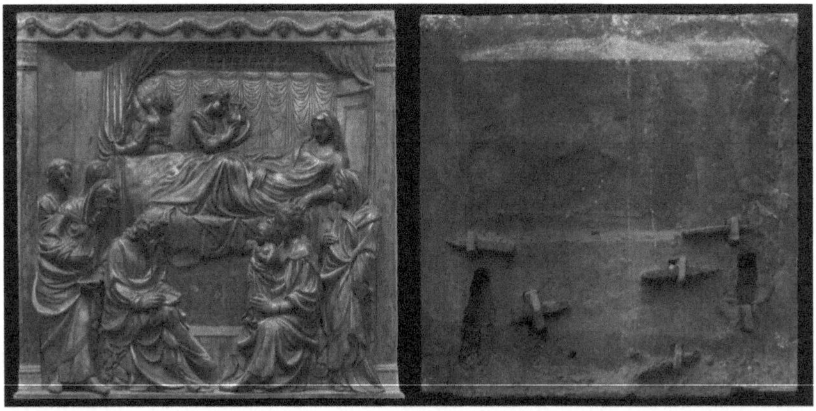

Figure 2. Giovanni di Turino and Turino di Sano, *Birth of the Baptist*, gilded bronze, lost wax casting and mercury amalgam techniques, front and backside of the panel before restoration.

Despite the many studies on this artwork, some technical aspects remain unknown; for instance, many sources seem to suggest that the fire gilding of Donatello's cast bronzes was carried out by different workshops (e.g., Bacci 1929; Mather 1937; Fattorini 2005). The choice to take all the bronze elements apart (four putti, the tabernacle door, six reliefs, six statuettes, six top and six bottom perimetral stripes) gave the chance to study the materials and the manufacturing techniques and to inspect and treat even hidden parts. For example, the backside of Turino di Sano e Giovanni di Turino's *Birth of the Baptist* panel (Figure 2), revealed a smart solution to join each cast figure to the background of the panel (similar iron spines, as mechanical joints for separated casting, came out behind the *Saint John the Baptist Preaching*, another panel made by the same author).

At the time of LACONA XIII (September 2022), all the enamelled perimetric stripes, three reliefs, six small sculptures representing the Virtues, the four putti, and the small tabernacle door had been transferred to Florence. Among these elements, it is worth mentioning the statue *Hope* and the well-known panel of *Herod's Feast* by Donatello, both put on display at the latest Florentine exhibition on Donatello (Palazzo Strozzi, 19th March – 31st July 2022). The third relief under restoration as of September 2022 was the Lorenzo Ghiberti's and Giuliano di Ser Andrea's scene, the *Arrest of John the Baptist* (Figure 3).

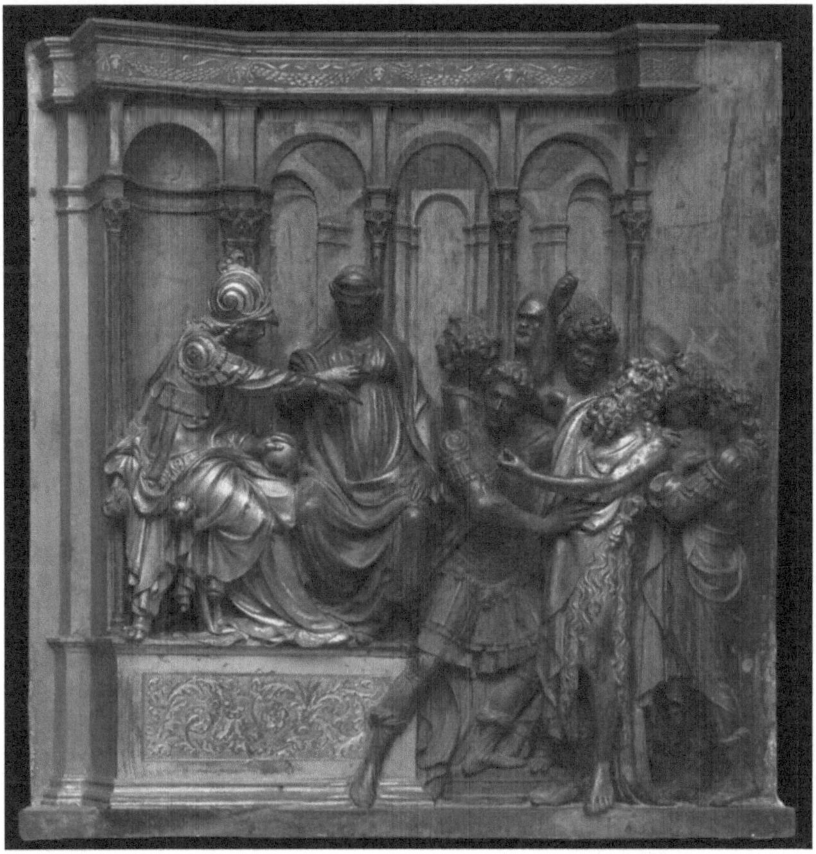

Figure 3. Lorenzo Ghiberti, *Arrest of John the Baptist*, gilded bronze, lost wax casting and mercury amalgam techniques, during restoration treatment.

Starting from the laser technique developed years ago for cleaning the *Gates of Paradise* of the Florentine Baptistery (Siano & Salimbeni 2001), our aim was to adapt it to the specific needs of the pieces of the Baptismal Font of Siena. Unlike the *Gates of Paradise*, the gilded elements of Siena are preserved indoors and show a different state of alteration. For this reason, it was necessary to test different cleaning options, changing the type of laser and combining the laser with other methodologies. The results of these tests and the protocol adopted to evaluate them will be illustrated in the following sections.

2 THE CASE-STUDY: DESCRIPTION OF THE STATE OF PRESERVATION

The Baptistery of Siena has specific microclimatic conditions that can be correlated to the atmospheric deposits, the encrustations, and the chemical compounds (such as copper salts) found on the gilded bronzes. Further alteration may be due to the use of the Font in religious rituals, to old treatments with organic substances to enhance the surface appearance, and to plaster left in the undercuts after casting to make gypsum copies. The organic and inorganic materials on the surfaces were carefully analysed with invasive and microinvasive techniques which included portable and bench FTIR (Fourier Transformed Infrared) spectroscopy, GC-MS (Gas-Chromatography-Mass Spectrometry) and SEM-EDS (Scanning Electron Microscopy-Energy Dispersion Spectrometry). The alloy composition was determined with a combined approach using SEM –EDS analysis on microfragments taken either from hidden parts or back sides and portable X-Ray Fluorescence (pXRF) on shavings drilled from 0.7 mm wide holes.

Data show that the panels and the statues are generally made of a ternary alloy of copper, tin and lead as major elements, with high content of copper (in the range 92-96%).

The gilding was applied with the mercury technique and the gold thickness ranges between 5 and 13 μm with a mean value of 10 μm. Preliminary results show that a significant amount of copper is intermixed with gold, with a CuAu phase present in the whole thickness of the gilding. The gilding shows many micropores related to the process of mercury evaporation from the amalgam. Other faults on the gold surface are due to scratches, abrasions, losses and microcracks, corrosion compounds from the underlying alloy, and deposits (Figure 4). The latter include gypsum, possibly due to old plaster cast, nitrates, sulphates, and chlorides. In some areas where UV imaging showed significant luminescence, we identified (with FTIR) oils, natural and synthetic waxes and copper carboxylates (oleates and palmitates or stearates), along with copper oxalates.

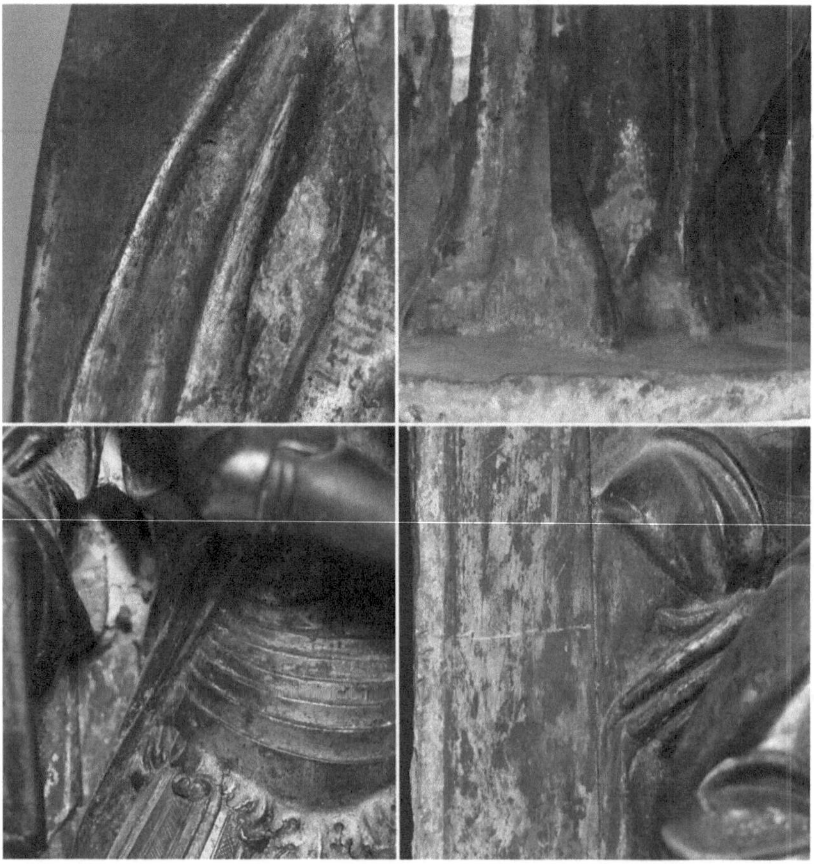

Figure 4. Gilded bronze elements of the Baptismal Font of Siena, some details of the state of preservation before restoration.

3 CLEANING TREATMENTS AND THEIR ASSESSMENT

Our cleaning methodology was based on various operations. After dusting the surfaces, steam was used to remove part of the wax applied during old treatments and the dust embedded therein. Cleaning tests were then arranged, following the Wolbers and Cremonesi protocol (e.g. Wolbers 2004; Cremonesi 2011; Cremonesi 2019) with the aim to select the best solvents for the preliminary step of removing old organic matter. Given the characteristics of the surfaces, the protocol indicated the mixture of isooctane and acetone as the most suitable. The solvents mixture was applied by an "oil-in-water" (O/W) emulsion designed to be kept over the surface and removed after 20 or 40 minutes with water only (Di Virgilio et al. 2023). The option of solvents in emulsion was compared with chelating gels, which are commonly used in gilded bronze cleaning. Visual inspection and images at different wavelengths before and after cleaning tests helped to compare the two approaches. In some cases, the chelating gels have led to an effective cleaning of the gilded layer, but in some areas, after the treatment, an unwanted light-yellow shiny appearance was found. The emulsions of organic solvents, on the other hand, allowed a selective and progressive cleaning of the surface coatings. Diffuse microporosity and several losses of the gold that reveal the underlying alloy required a thorough and immediate rinsing with water of both gels and emulsions; especially with chelates the washing had to be very deep to prevent the chelating agents from damaging the thin layer of the natural aging patina of the alloy. Therefore, emulsions turned out to be the most effective and safe way to remove the first few layers, so we decided to apply isooctane and acetone emulsion on three different portions of the panel *Arrest of John the Baptist* by Ghiberti for the second step of the cleaning treatment after steam. After the removal of large parts of the organic matter, each portion was divided into several smaller areas to test the effects of the laser cleaning for the next step, i.e. cleaning of the corrosion crust (Figure 5).

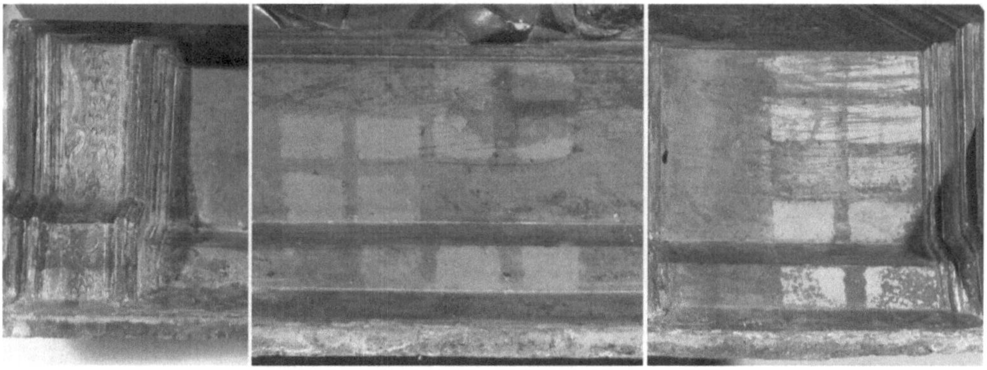

Figure 5. The left side of the gilded bronze panel by Ghiberti, with the three test areas after different treatments with laser cleaning after steam and oil-in-water emulsion (O/W) n. 2 (isooctane-acetone blend).

The employed device is an implementation of the first model (called VARIO) designed more than 20 years ago as a result of the collaboration between the Opificio delle Pietre Dure, IFAC-CNR and El.En. during the restoration project of the *Gates of Paradise* by Lorenzo Ghiberti (Siano & Salimbeni 2001; Siano et al. 2008; Siano & Salimbeni 2010; Siano et al. 2012). The energy rates (starting from 40 mJ) were matched with different spots (diameter between 5 – 3.2 mm, corresponding to notch on the handpiece 3 – 5) and frequencies (from 6 to 10 Hz) to reach the desired cleaning results. The fluence went approximately from 0.12 to 0.6 J/cm^2. Better and safer fluence rates were determined around 0.4 J/cm^2.

Nowadays, El.En. has introduced a new version of the Long Q-Switched laser, named EOS1000 LQS (https://www.lightforart.com/eos-1000-lqs/). The pulse duration has values similar to those of the VARIO prototype. The operative parameters were set up comparing the best result of the VARIO and EOS1000 LQS cleaning tests (Table 1).

Table 1. VARIO and EOS1000 LQS operative parameters ranges used for cleaning amalgam gilding.

Model	VARIO	EOS1000
λ	1064 nm	1064 nm
Pulse duration	70 ns (LQS)	100 ns (LQS)
E (mJ)	80-150	130
Optical filter	T25	T25
Ƒ (Hz)	2(min) – 7(max)	2(min) – 7(max)
Ø (mm)	ca 3.5–4 (notch 4-6)	ca 3.5-4 (notch 4-6)
F (J/cm^2)	0.12(min) – 0.62(max)	0.19(min) – 0.46(max)
Water	by means of brushes, water tank, sprays	by means of brushes, water tank, sprays

Localized spot corrosions were treated with a specific laser protocol. The passivation of the hydroxychlorides was induced following the same protocol used for the chlorides treatment of the *Winged Victory* from Brescia, recently restored by our Department (Pucci & Brini, 2021). The laser system used to this purpose was the Smart Clean 2, operating in Short Free Running pulse duration. The Smart Clean 2 handpiece was removed to reduce the beam area. Without the lens, the fiber induces a very punctual thermic effect that allows a sort of heat-sealing of the corrosion pits (E 200 mJ, F 1 Hz). The fiber must be handled carefully, and the tip needs to be cut away when it turns into the shape of a burned drop (Figure 6). Apart from the treatment for chlorides, the laser cleaning with the Short Free Running regime against encrustations and compounds resulting from the alteration of the alloy with waxes and oils has proven unsatisfactory in some cases. Due to the presence of a very hard encrustation, it was necessary to increase the laser fluence, but side effects often occurred. These include a broad purpling of the underlying bronze alloy and the appearance of stains in scratches and areas where the gold was abraded in the past and is now thinner.

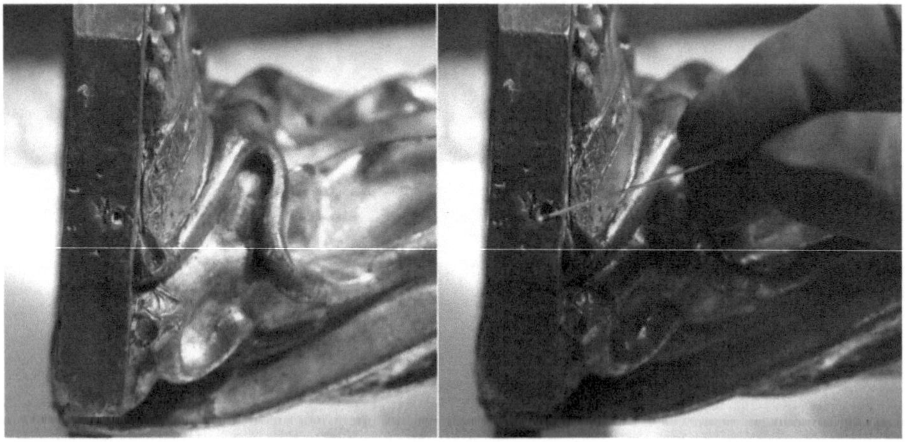

Figure 6. Using Smart Clean 2 for localized treatments against chlorides.

To reinforce our assessment of EOS1000 LQS laser as the best device for cleaning this kind of gilding (even abraded or coated by thick layers of organic substances), we would like to compare other lasers cleaning outcomes (e.g., Bertasa & Korenberg 2022). Thus, we further tested other lasers devices: the SmartClean 2 (used for the above-mentioned chlorides treatment) and an erbium laser (LIGHT BRUSH 2; https://www.lightforart.com/light-brush-2/). The former usually well suits the restoration of leaf-gilded bronze sculptures and bronze pieces with no gilding, especially using underwater ablation (see Rotondi et al. in this volume), while the latter is not usually applied on metal artifacts, where the most common lasers are the neodymium ones with an infrared wavelength of 1064 nm. The wavelength of the erbium laser (2940 nm), however, makes it suitable for being absorbed by oxygen-hydrogen bonds related to organic coatings and for this reason, it was included in the tests. Previous studies on erbium lasers report good results for removing shellac, linseed oil, and resins from overpainted artifacts, while its application to metal is not widespread (e.g., Boonrat et al. 2020; Teppo 2020).

On the three square areas of the relief *Arrest of John the Baptist*, different settings of the three laser devices were tested and compared. The cleaning was monitored in-situ by means of FTIR reflectance spectroscopy using a portable AGILENT handheld instrument, equipped with the external reflectance module. Owing to the highly reflective gold support and the limited thickness of the compounds on top of gold, the bands in the spectra show limited to no distortions, so their intensity can be consistently used to monitor the decrease of the amount of the compound in each cleaning trial (Unemura 2002).

Comparing the spectra of three areas of the frame of the Ghiberti panel before and after three different laser processes, the difference of the performance is evident. In the spectra, all the peaks related to concretions on the surface, basically gypsum, organic matter, and its alteration such as carboxylates and oxalates, are strongly reduced after the application of the laser. In this case, however, the EOS laser has proven the most effective as the after-treatment spectrum shows no other signals than those of the water vapour of the air (Figure 7).

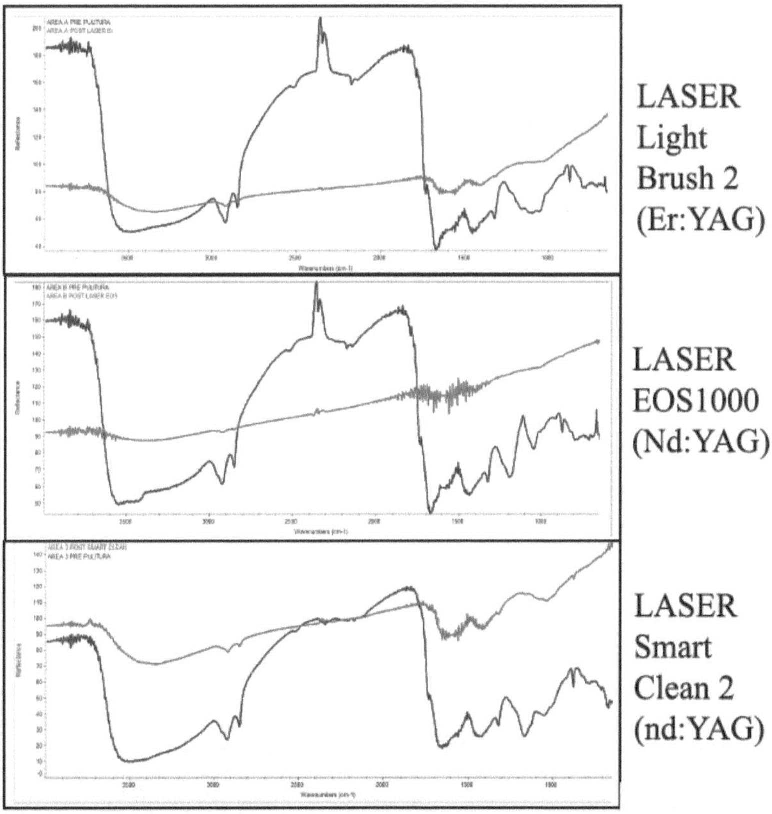

Figure 7. Reflectance FTIR for monitoring the impact of different cleaning processes on Ghiberti's panel. The blue lines are the spectra before cleaning, the red ones after cleaning. According to the spectra comparison, EOS1000 shows the greatest efficacy.

As for the capability of the isoctane-acetone to produce a surface more suitable for laser cleaning action, we tested the application of the oil-in-water emulsion (O/W 2) and the same solvent mixture with a swab before the laser treatment on three areas on the Giovanni da Turino *Birth of the Baptist* panel. The FTIR reflectance spectra show that the application of the solvents blends alone has nearly no effect: spectra still show the presence of characteristic peaks of calcium sulphate, C-H stretching of oils and waxes, metal carboxylates, oxalates and silicates with nearly the same intensity as in the uncleaned areas. Only a weak peak at about 1280 cm^{-1} attributed to the $Si-CH_3$ stretching, possibly due to a siloxane (Ding-Shan et al. 2010), is no more present after the treatment (Figure 8).

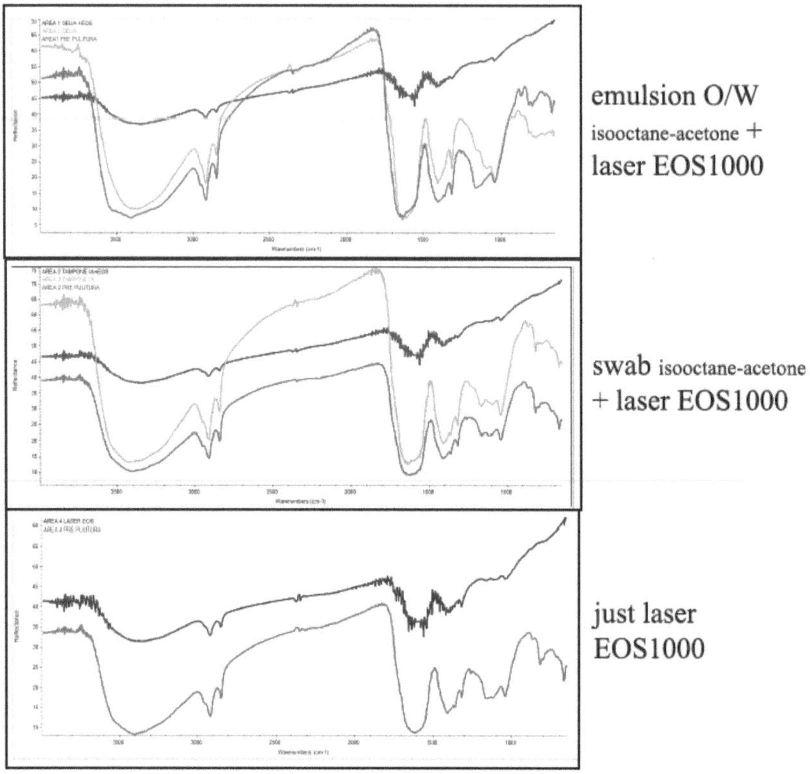

Figure 8. Reflectance FTIR for monitoring the impact of different cleaning processes on Birth's panel. The red spectra are before cleaning, the light blue ones are after the application of the isooctane-acetone blend either in the form of emulsion or with a swab, the dark blue spectra are after the application of laser.

However, if we compare the spectra of the three areas after the full processes (laser alone, solvents as gel+laser, solvents with a swab +laser) we can see that the laser seems to be slightly less effective where no solvents blend was applied before it (Figure 8). In particular, we can observe that the impact of the cleaning on the intensity of the C-H peaks of organic matter in the 2840-3000 cm^{-1} range, the oxalate peak at about 1320 cm^{-1} and the Si-O broad band (attributed to silicates) around 1000 cm^{-1}, is maximum where the combination of solvents blend in emulsion and laser was used and slightly poorer when the solvents were applied with the swab. In the case of no pre-treatment before the laser, we still observe significant peaks of oxalate and silicates in the final spectrum. This is evidence that the solvents, particularly the emulsion, helps the action of the laser.

The oil-in-water isooctane-acetone emulsion was always applied on the surface using soft brushes. After a few minutes it was stirred on the surface to enhance its action. Then, after 40 minutes, it was removed only by water with brushes and soft sponges. In this way the laser cleaning could take place on a less hard encrustation because of the partial softening of the organic coatings.

The outcomes derived from both FTIR instrument and visual inspection underline the effectiveness of laser cleaning based on the LQS system and the advantage of using the laser after a treatment with a mixture of solvents. The monitoring of laser cleaning on enamelled stripes is still in progress. Preliminary visual inspection of cleaned areas points out that the fluence rates set up for the cleaning of the gilded bronze elements are safe also for the perimetral stripes of the marble basin. Further analysis of the surface with SEM-EDS are in progress.

4 CONCLUSIONS

In conclusion, the restoration process of the fire-gilded bronzes from the Baptismal Font of Siena Baptistery represents a new challenge with respect to the gates of the Baptistery of Florence, for which the first laser system was developed many years ago. The massive presence of organic substances applied in past interventions requires new adapted solutions, possibly combining more cleaning options.

EOS1000 LQS has been confirmed as the best laser system, except when thick encrustations lie on uneven gilding layers (such as scraped areas and gaps). In this case restorers need to avoid or protect non-gilded portions, so as to prevent accidental laser ablation.

Attempts with the erbium laser resulted in the partial removal of the organic coatings, but a long time is required (about triple than with the other lasers), while organic solvents (especially with oil-in-water emulsion) allow a faster preliminary treatment to reduce organic coatings. Combining laser with oil-in-water emulsions, but also with ice blasting and mechanical tools, can certainly improve the cleaning efficacy and speed up the process, although some side effects, such as the turning of the colour of nude bronze into mat and purple hues, cannot be controlled on abraded areas where the gilded layer is thinner.

Smart Clean 2 outcomes on this amalgam gilding, instead, are strongly case-dependent but some side-effects suggested to prefer this device only for chloride treatment.

Selecting the type of laser and adjusting the operating parameters, along with scientific investigations aimed at the characterization of surface materials and the monitoring of ablation, have led to a better awareness on the effectiveness of some operating solutions. However, due to the issues discussed above about the use of laser on the most persistent and stubborn encrustations, the cleaning treatment applied up to now on such areas is based on organic solvents and mechanical tools, such as needles, porcupine spines, dental ultrasonic cleaner, and vibrating cutters.

ACKNOWLEDGEMENTS

We would like to acknowledge R. Gennaioli (Art historian and Head of both Stone and Jewellery Departments), C. Mancini (Head Conservator restorer of the Stone Department); L. Speranza (Art historian and Head of the Bronze Department), the restoration team of the Masterpiece owner (Opera Metropolitana di Siena) and the University of Siena; O. Caruso and M. Mercante for HMI, multispectral imaging and 3D scanning of the whole artwork; T. Radelet for the XR analysis of some panels and small figures; A. Giamello, A. Scala, S. Mugnaini (University of Siena) for the XRD on the casting core traces; Alessia Andreotti, Francesca Modugno (University of Pisa) for the GC/MS, that helped us in identifying residues of organic altered coatings; A. Pacini for technological and technical observations and studies.

REFERENCES

Bacci, P. 1929. Jacopo della Quercia: nuovi documenti e commenti. Stamperia di San Bernardino, Siena, Italy.

Bertasa, M. & Korenberg, C., 2022. Successes and challenges in laser cleaning metal artefacts: A review, in *Journal of Cultural Heritage* 53: 100–117.
Boonrat, P., Dickinson, M. & Cooper, M., 2020. Initial investigation into the effect of varying parameters in using an Er:YAG laser for the removal of brass-based overpaint from an oil-gilded frame, in *Journal of the Institute of Conservation* 43 (2020): 94–106.
Cremonesi, P., 2011. L'ambiente acquoso per la pulitura di opere policrome, I talenti. Padova, Italy: Il Prato.
Cremonesi, P., 2019. L'ambiente acquoso per il trattamento di manufatti artistici. Padova. Italy: Il Prato.
Di Virgilio, S., Agnoletti, S., Baruffetti, M., Ciseri, I., Galeotti, M. & Speranza, L. 2023, Due bronzetti raffiguranti Sansone uccide i filistei dal Museo Nazionale del Bargello. Proposte per la rimozione graduale di protettivi cerosi su vernici oleo-resinose. In *OPD Restauro* 34 (2023): 92–99.
Fattorini, G. 2005. Dalla "historia d'attone pel Battesimo" a "le porti di bronzo del Duomo": Donatello e gli inizi della scultura senese del Rinascimento". In *Pio II e le arti. La scoperta dell'antico da Federighi a Michelangelo*. Siena.
Mather, R. G., 1937. Donatello debitore oltre la tomba, in *Rivista d'Arte* 19 (2): 186.
Pucci, F. & Brini, A., 2021. L'utilizzo del laser nel restauro della Vittoria Alata. In A. Patera & F. Morandini (eds), *Necessitano alla Vittoria Alata le cure del restauratore. Studi, indagini e restauro del grande bronzo di Brescia*, Firenze, Italy: Edifir.
Ding-Shan, R., Li, Ya-Li, Wang, Lei, Su, Dong & Hou, Feng., 2010. Fabrication of Silicon Oxycarbide Fibers From Alkoxide Solutions Along the Sol–Gel Process. *Journal of Sol-Gel Science and Technology* 56: 184–190.
Siano, S. & Salimbeni, R. 2001. The Gate of Paradise: physical optimization of the laser cleaning approach. *Studies in Conservation* 46: 269–281.
Siano, S., Grazzi, F. & Parfenov, V. A. 2008. Laser cleaning of gilded bronze surfaces, in *Journal of Optical Technology* 75: 419–427.
Siano, S. & Salimbeni, R., 2010. Advances in laser cleaning of artwork and objects of historical interest: the optimized pulse duration approach, in *Accounts of chemical research* 43 (2010), 739–750.
Siano, S., Agresti, J., Cacciari, I., Ciofini, D., Mascalchi, M., Osticioli, I. & Mencaglia, A.A. 2012. Laser cleaning in conservation of stone, metal, and painted artifacts: state of the art and new insight on the use of the Nd:YAG laers, in *Applied Physics A* 106 (2012): 419–446.
Teppo, E., 2020. Introduction: Er:YAG lasers in the conservation of artworks, in *Journal of the Institute of Conservati*on 43: 2–11.
Unemura, J., 2002, Reflection-Absorption Spectroscopy of Thin Films on Metallic Substrates, in J. M. Chalmers, P. R. Griffiths (eds), *Handbook of Vibrational Spectroscopy*, vol. 2, John Wiley & Sons Ltd: 982–998.
Wolbers, R., 2004. Un approccio acquoso alla pulitura dei dipinti, Quaderni Cesmar7. Padova, Italy: Il Prato.

Lasers in the Conservation of Artworks XIII – Siano & Ciofini (eds)

Assessment of material-atmosphere interactions during scanning laser cleaning of archaeological bronze alloys: A Roman coin case study

E. Di Francia & S. Grassini
Dipartimento di Scienza Applicata e Tecnologia (DISAT), Politecnico di Torino, Torino, Italy

D. Neff
NIMBE/LAPA-IRAMAT, CEA/CNRS/Université Paris-Saclay, UMR3685, CEA Saclay, Gif/Yvette, France

R. Lahoz
Instituto de Nanociencia y Materiales de Aragón, (INMA), CSIC - Universidad de Zaragoza, Zaragoza, Spain

ABSTRACT: The main goal of this research was to assess the presence of possible re-oxidation phenomena during laser cleaning procedures carried out on Cu-based archaeological artefacts. Previous studies conducted on archaeological and artificial corrosion layers, highlighted that the laser ablation procedures (removal of unwanted materials) do not change the composition of the corrosion layers and that detectable re-oxidation phenomena occur on laser-treated surfaces for intense laser conditions, not applicable on Cultural Heritage artefacts. A novel approach, recently developed on Material Science field by these Authors, is here applied in the Conservation Science field: the use of a traceable isotope combined with Time-of-Flight Secondary Ion Mass Spectrometry (ToF-SIMS) on an archaeological bronze coin. This approach, applied for the first time on an archaeological artefact, allows assessing the presence of possible re-oxidation phenomena that might occur on object surfaces during optimised cleaning procedures.

1 INTRODUCTION

In recent years, many researches have been focused on laser applications on a wide range of artefacts (Landucci et al. 2000, Siano et al. 2012). Despite the very good performance of laser cleaning, the complex phenomena that take place during laser-material interaction are usually very difficult to describe. Nevertheless, very few systematic studies have been reported on the effects of the laser cleaning on highly heterogeneous materials such as Cu-based metal alloys. Generally they have been focused on discussions of laser-interaction mechanisms, chemical characterisation of laser effects on surfaces (Bertasa & Korenberg 2022; Di Francia et al. 2021), or systematic studies of the laser parameters to be applied on metal artefacts (Burmester et al. 2005, Di Francia et al. 2018, 2022). Indeed, knowing the laser-material interactions in depth will help to optimise the laser cleaning parameters and make it a more efficient procedure.

In addition to this, the complex stratified structure of the corrosion products grown on the Cu-based archaeological artefacts buried in soil should be considered, as model by Robbiola in a two-layer or a three-layer structure. In the two-layer model, a protective copper oxide layer (*noble* patina) grows in contact with the metal and then several types of unwanted, corrosion products might grow on it. These unwanted corrosion products are inhomogeneous

DOI: 10.1201/9781003386872-27

and polycrystalline structures which might differ in composition (due to oxygen, humidity and soil elements, e.g. Si, Fe, Ca), and so in their chemical and physical properties. Furthermore, these products can also degrade, transforming the cuprous products into cupric products and incorporating soil elements, forming new compounds as e.g. brochantite ($Cu_4(OH)_6(SO)_4$) and malachite ($Cu_2(OH)_2CO_3$). In the three-layer model, dangerous and reactive corrosion products form due to the presence of Cl^- ions in the environment: the Cl^- ions penetrate trough both the external porous layer of Cu(II) compounds (e.g. hydroxysilicates, hydroxyphosphates, hydroxychlorides) and the middle layer of cuprite (Cu_2O) up to the internal copper oxide layer, in contact with the metal. This concentration of Cl^- ions, at the metal-corrosion interface, can activate the formation of reactive cuprous chlorides (nantokite, CuCl), starting a cyclical and dangerous copper corrosion process, commonly noun as *bronze disease* (Robbiola et al. 1998, Soffritti et al. 2014).

The nature of these inhomogeneous and polycrystalline structures should be taken in consideration during the ablation process: indeed, the laser energy interacts with the material decaying into heat and generating a stress wave, which can cause disaggregation of the constituents and creation of reactive radicals. As a consequence, it is possible to assist at the formation of a plasma plume composed by free species (multiphoton/thermionic ionizations) of the ablated materials and the surrounding environment, forming a cloud (plume) on the treated surface. Eventually, if the substrate melts, it is possible to observe recombination reactions that latter might form by-products (Fotakis et al. 2007). A final parameter of paramount importance that poses an important role in a laser cleaning process is the surrounding atmosphere. Although most of the tests are usually performed in air, there are other works that have tested the cleaning conditions under controlled atmospheres (Gomes et al. 2018, Grigor'eva et al. 2017), to enhance the cleaning efficiency or prevent unexpected reactions on the surface, but never with the aim of studying the laser interaction mechanisms in the presence of a controlled atmosphere.

Due to the complexity of these interactions, several previous laser-cleaning tests had been conducted by our group on artificially-corroded layers simulating the protective patina to be preserved and the unwanted corrosion products to be removed. From these previous works, no changes had been detected in the chemical and microstructural composition of the laser-treated artificially-corroded layers (Di Francia et al. 2018, 2022). Our previous analyses also highlighted that high irradiance values and long pulse duration may cause a re-oxidation phenomenon on the surface during the ablation process. This demonstrated that, even if the composition found after the laser processes is the same, the corrosion compounds might be of new formation (Di Francia et al. 2021).

This work exposes the results obtained applying, for the first time on Cultural Heritage, the already explained procedure for detecting material-atmosphere recombination in controlled synthetic air (Di Francia et al. 2021). Details of the internal and complex stratified corrosion structures of the archaeological coin and of the optimal laser parameters assessment that allows the most effective ablation have been described in previous papers (Di Francia et al. 2018, 2022). The novelty here is that the present work assesses the laser-surface material interactions and, for the first time, the possible presence of re-oxidation phenomena during optimal laser-cleaning treatment conditions on archaeological bronze coin fragments. To achieve this goal, scanning laser experiments have been carried out in a controlled atmosphere enriched with ^{18}O isotope. Then, advanced Time-of-Flight Secondary Ion Mass Spectrometry (ToF-SIMS) analyses have provided the data necessary to detect the possible presence of re-oxidation phenomena.

2 MATERIALS AND METHODS

2.1 *Archaeological samples and their preparation*

The laser-cleaning procedure had been validated on real ancient samples with low archaeological value, a coin (*Follis Massenzio*, Roman Empire bronze coin) coming from a private collection. An extensive description of the coin and the sample preparation before and immediately after the cleaning treatments in synthetic air are reported in (Di Francia et al. 2021, 2022).

2.2 *Laser system and laser parameters*

The laser used for the treatments is a 1064 nm Nd:YAG fibre laser (model EasyMark-20 from Jeanologia). The system works in a Q-Switched (QS) regime, from 4 ns to 200 ns of pulse duration, which pulses are delivered by means of two-galvanic mirrors and are focused with a f-Theta lens with 160 mm of focal distance for a final beam diameter of ca. 30 μm. The laser system is coupled via computer with EzCAD 2.1 UNI, a vector graphic editor, with a CAD-*like* capability that enables users to perform, in a repeatable way, rapid, precise and complex surface scanning treatments.

Two values of energy were chosen for the laser treatments. Those values were optimised in previous works as the energy values suitable for the laser cleaning without damaging the substrate (Di Francia et al. 2018, 2022). Table 1 reports the optimised laser experimental parameters set as a function of the chemical composition of the corrosion products to be removed and applied on different coin fragments.

Table 1. Laser experimental parameters used on the archaeological samples. P (Power), F (Fluence), I (Irradiance), t_p (Pulse Duration), v_{scan} (Scanning Speed) and d_L (Interlining).

Laser parameters					Geometrical parameters	
Test	P (W)	F (J/cm^2)	I (GW/cm^2)	t_p (ns)	v_{scan} (mm/s)	d_L (mm)
1	0.23	1.63	0.41	4	300	0.015
2	0.73	5.16	1.29	4	300	0.015

2.3 *Experimental set-up and assessment of the ablation mechanism on archaeological samples*

The innovative procedure here reported has been already described in (Di Francia et al. 2021) for the assessment of the laser-surface material interaction mechanisms during ablation processes conducted with intense laser conditions on artificially corroded layers. Those conditions were selected to stress possible re-oxidation phenomena.

To assess if a re-oxidation phenomenon occurs during the ablation process, and so if the compounds analysed after the cleaning treatment are the same with the one previously present or they are compounds of new formation, the optimised laser cleaning treatments were performed in a controlled atmosphere chamber filled with synthetic air (20% ^{18}O + 80% N_2). This synthetic air is characterised by the absence of the ^{16}O, the most common oxygen isotope in the natural air: in this way, if re-oxidation phenomena occur, they will be traced by the mostly presence of the ^{18}O in the new compounds and detected by ToF-SIMS analyses (Figure 1).

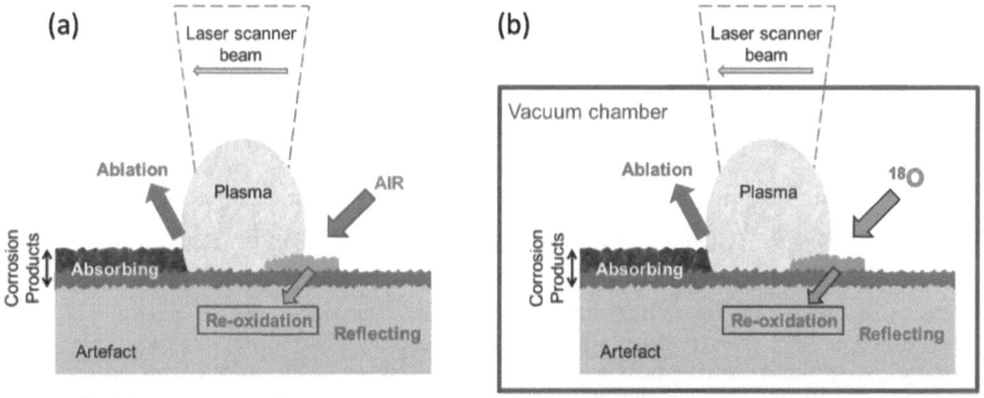

Figure 1. Scheme of laser-cleaning ablation model for Cu-based archaeological artefacts. *a)* Laser-cleaning in natural air *(left)*: possible re-oxidised products (*light green* in the figure) non-detectable; *b)* novel experimental set-up adopted using ^{18}O isotope *(right)*: possible re-oxidised products (*light blue*) detectable.

Laser cleaning tests were conducted positioning the coin fragments in a chamber presenting a window transparent to the infrared radiation. A depression of around 1 bar was done and then the chamber was filled back to recover 1 bar with synthetic air marked by the presence of ^{18}O isotope, an oxygen isotope presents in small amount in natural air, and so in every material (0.205(14) at.%) (Lide 2005).

Moreover, a coin fragment was laser-treated with Test 1 parameters in natural air and then analysed with ToF-SIMS in order to have a control sample.

2.4 *Sample characterisation*

The cross-sections of the non-treated and laser-treated coin samples were characterised by Optical Microscopy (Olympus BX51 microscope equipped with a Nikon EOS camera). It allows to acquire images in Bright and in Dark-Field modality (OM-BF and DF, respectively).

ToF-SIMS analysis (TOF-SIMS 5 instrument from IONTOF) was used to detect percentage amount of ^{18}O isotope on the ablated corrosion product layers of the coin samples; this analysis was performed on both the control sample, treated in natural air, and on the samples treated in synthetic air.

The high sensibility of the ToF-SIMS technique allows detecting very small fractions of the isotope that would be incorporated due to the laser interaction within the synthetic air atmosphere, enriched with the ^{18}O isotope. This will determine how the laser interaction occurs and how the recombination or re-oxidation processes take place.

The sample laser-treated in natural air (Test 1), considered as the control sample, was analysed on two randomly selected areas while on the samples irradiated in synthetic air, in the conditions of Test 1 and Test 2, were respectively acquired two and three random selected areas. Subsequently, for each selected area, the isotopic oxygen ratio percentage (O_r) was calculated as follows:

$$O_r = \frac{^{18}O}{^{18}O + ^{16}O} * 100 \tag{1}$$

where O_r is the isotopic oxygen ratio percentage; ^{18}O is the isotopic amount of ^{18}O; ^{16}O is the isotopic amount of ^{16}O.

Two sources of uncertainty, strictly connected with the applied measurement parameters, were also considered for each acquisition, as previously reported (Di Francia et al. 2021).

The arithmetic mean values obtained were used as the O_r estimation, while type-B uncertainties (u_B) were used to characterise the measurements. The type-B uncertainty was estimated according to the count number of the ^{18}O measurements.

The data and the results here presented are in accordance with the *Guide to the Expression of Uncertainty in Measurement* (GUM 2008).

3 RESULTS

3.1 *Cross-section analysis and composition of the laser-cleaned archaeological corrosion products*

Extensive discussion of the results are present in previous works (Di Francia et al. 2018, 2022), however here is extremely important to report that OM observations on the non-treated cross-section coin fragment had showed the presence of an outer corrosion products layer (*external* and *intermediate* layers) completely corroded, up to 200 µm-in thickness, and an internal and partially corroded layer, up to 800 µm-in thickness (Figure 2). From the picture, the heterogeneity in composition of the outer layer is shown by the presence of a *lighter grey* layer (intermediate layer) between the internal and the *darker grey* external corrosion products layers. The figure schematises also the structure.

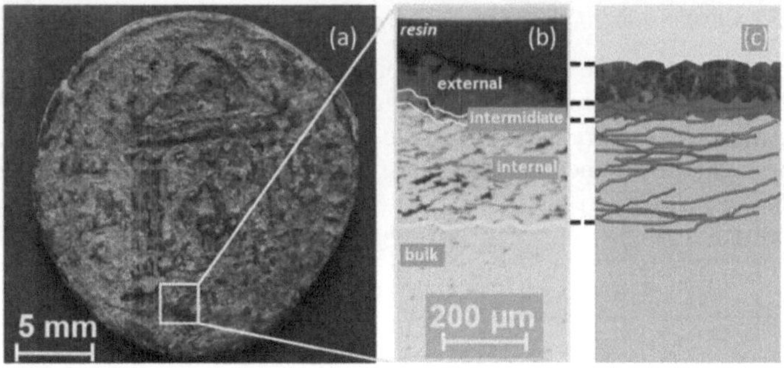

Figure 2. High-resolution coin photograph and OM cross-section image of the Roman Empire coin: verso side of the coin (left), cross-section acquired by optical microscope, bright field (centre) and scheme of the internal structure (right).

Deeper, Figure 3 compares the DF-OM observations of the cross-sections, for both laser-treated samples (Test 1 and Test 2), with the non-treated one. These analyses showed the complex nature of archaeological stratified corrosion products. Indeed the coin presented a turquoise green colouration (Cu^{++} compounds, e.g. hydroxychloride, hydroxycarbonates) overlapped to a thinner red-brown-in colour layer (Cu^{+} compounds, e.g. cuprous oxide) with spot-areas where only red-brown colouration was presented; yellowish-in colour (possible cuprous chloride or a mix of copper and nanometric tin oxides) areas or very thin layers were presented between the green and the red-brown layers (Soffritti et al. 2014, Robbiola et al. 1998). The Figure 3 shows that this structure was still present on both laser-treated samples: no qualitative differences in the composition of the outer corrosion products on the non-treated and the laser-treated samples were detectable.

Both laser treatments removed part of the corrosion products: Test 2 (I, 1.29 GW/cm^2) reduced the outer (external and intermediate) corrosion layer, in average, up to ≈41 μm while for Test 1 (I, 0.41 GW/cm^2) only a marginal average thickness reduction up to ≈15 μm was observed. Moreover, the laser probable interacted only with the surface of the outer corrosion products layers without reaching the internal layer and the artefact metal: at the highest value in irradiance (Test 2), the laser interacted with and partially ablated the external Cu^{++} and the intermediate Cu^{+} compounds layers. Due to a more compactness and thickness of the archaeological corrosion layers compared to the artificially-corroded ones, Test 2 resulted more efficient in terms of ablation of archaeological corrosion products respect to the optimal laser conditions discussed in (Di Francia et al., 2018).

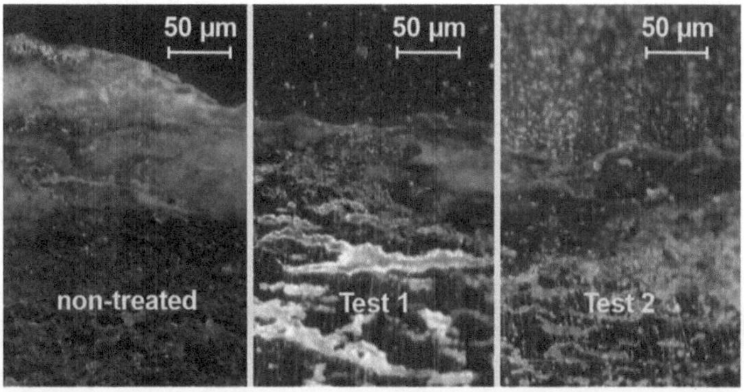

Figure 3. DF-OM observations of the cross-sections: comparison of the non-treated, Test 1 and Test 2 samples.

3.2 *ToF-SIMS identification of isotope incorporation*

ToF-SIMS maps supported the identification of an internal corrosion layer, in the shape as a net (characterised by the presence of chlorine isotopes), overlapped by a more corroded outer layer (identified by the presence of oxygen and copper isotopes). The structure was still present after the higher irradiance values of Test 2, as Figure 4 shows. The presence of chlorine in the internal layer, Cu^{+} in the intermediate layer and Cu^{++} in the external layer, might suggest the interaction of the coin with an aggressive burial environment (Robbiola et al. 1998).

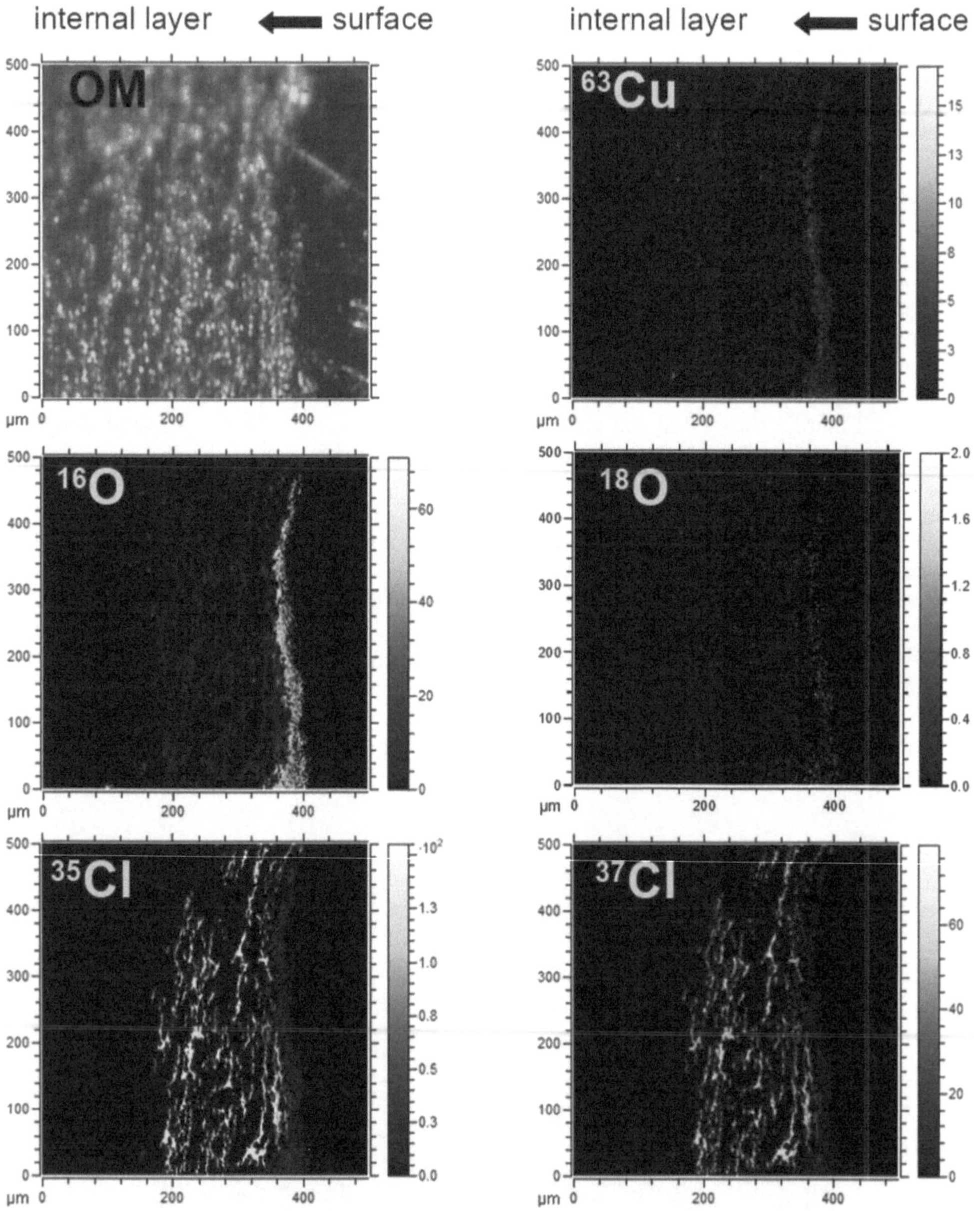

Figure 4. Stratifications of the corrosion layers: OM image (up left) and example of ToF-SIMS maps of different isotopes on Test 2-area cross-section.

4 DISCUSSION AND ASSESSMENT OF THE LASER-SURFACE MATERIAL INTERACTIONS DURING LASER ABLATION ON ARCHAEOLOGICAL SAMPLES

To assess if the optimised laser-cleaning parameters can induce surface re-oxidation phenomena on real corrosion product layers, several areas on the cross-section samples were considered: two areas of the cross-section controlling sample and, respectively, two and three areas of the cross-section samples treated in the chamber filled with synthetic air.

Table 2 shows the obtained figures of isotopic oxygen ratio percentage (O_r) and the corresponding type-B uncertainties (u_B) for each sample.

Table 2 O_r and u_B determined on the Follis Massenzio coin fragments laser-treated in natural and synthetic air with laser Test 1 parameters and Test 2 parameters.

Laser Test	Air	O_r *average*	u_B
1	natural	0.212	0.006
1	synthetic	0.218	0.007
2	synthetic	0.233	0.009

As said previously, Test 2 presents the best cleaning results on the coin and, as expected, this test has also the highest content of O_r, as shown in Table 2.

The data showed that the O_r of the controlling sample (and so of the coin) is close to the $^{18}O\%$ natural abundance of (0.205(14) at.%), as reported in (Lide, 2005): during the laser treatment, if re-oxidation occurred, the interaction would took place with the isotopes in the amounts of the natural air, so no differences would be observed with respect to the non-treated coin and to the natural isotope abundance.

Different the case of both laser-treated coin fragments in synthetic air: a possible increase of isotope ratios as a function of the irradiance values can be hypothesised. However, for Test 1 (I, 0.41 W/cm^2), laser-treated in synthetic air, knowing that the $^{18}O\%$ natural abundance value percentage is (0.205(14) at.%), it was not possible to assess the presence of re-oxidation phenomena since the O_r average values (0.218%) is in the range of the $^{18}O\%$ natural abundance uncertainty. Whereas for Test 2 (I, 1.29 GW/cm^2), laser-treated in synthetic air, the O_r average value (0.233%) is slightly above this range, at the limit of reliability to detect difference in abnormal amount of $^{18}O\%$.

Even though that, this kind of interaction was not excluded: it is possible that with these irradiance values, a re-oxidation occurred but in a very little amount.

This can be explained by the too low irradiance values of Test 1 and Test 2 used in this study, that lead to an insufficient interaction with the synthetic air, to be clearly detected by ToF-SIMS analyses. For both Tests, it can also be explained by a dilution effect: if laser induces re-oxidation phenomena, they occur at the surface layer (with a different depth penetration in dependence of the corrosion products present). Analysing the cross-sections, the *region of interest* selected probably had interacted with the synthetic air but, however, it is possible that part of the information come from an area not touched by the re-oxidation phenomenon.

In fact, at this pulse duration and irradiance values, the laser energy interacted with the heterogeneous corrosion layers, partially ablating them. The reactive radicals and the free electrons of the ablated material present either in the plasma or on the laser-treated surface might have interacted with the oxygen present in the atmosphere.

Even if, in synthetic air, slight differences of O_r averages of Test 1 and Test 2 were detected respect to $^{18}O\%$ natural abundance (0.205(14) at.%), it is highly probable that they are not due to normal fluctuation of the ratio: a re-oxidation trend is detectable at the increase of the irradiance values (Di Francia et al. 2021).

These optimised parameters probably generate re-oxidation phenomena but, in any case, in a little amount, barely outside the uncertainty value of the ^{18}O natural abundance (0.205(14) at.%).

5 CONCLUSIONS

The results here presented confirm the need of validating the optimised condition acquired on artificially-corroded samples on real archaeological corrosion products, due to their complexity in inhomogeneous and polycrystalline structure. Moreover, the use of a scanning laser system allows performing, in a repeatable way, surface homogeneous treatments.

In this work, an innovative approach was introduced, for the first time in the Conservation Field, in order to assess if a re-oxidation on the ablated surfaces occur during the laser cleaning of an archaeological Cu-based artefact by a NIR Q-switched Nd:YAG fibre laser.

ToF-SIMS technique applied on the archaeological artefact samples, treated in ^{18}O traced atmosphere, revealed an isotopic oxygen ratio percentage at the limit of reliability to detect difference in abnormal amount of ^{18}O%. Nevertheless, even if the detected amounts are slight above the range of the ^{18}O % natural abundance uncertainty, an interaction of the corrosion product layers with the air during the laser treatments was determined.

This validation experiment demonstrates that it is highly probable that re-oxidation phenomena occur even with the optimised laser-cleaning parameters used in conservation, as it may find reported with the applied laser-cleaning parameters used in other cleaning processes (Di Francia et al. 2021).

To conclude, at the optimised laser-cleaning conditions for archaeological corrosion products, re-oxidation phenomena would occur in a very slight amount. Therefore, at those laser conditions, laser-cleaning procedure can be considered as a low-invasive cleaning procedure which does not significantly modify the treated layers.

Laser-cleaning operators in the conservation field may find useful to know about this possible re-oxidation so they can keep attention especially on heterogeneous materials, where each point can behave in a different way towards the same combination of laser parameters.

ACKNOWLEDGEMENTS

The Authors would like to thanks the European Federation of Corrosion (EFC) for supporting the measurement campaign in the frame of the EUROCORR Young Scientist Grant 2016 and Dr Jacopo Corsi for donating the bronze coin. The author R. Lahoz wishes to acknowledge professional support of the CSIC Interdisciplinary Thematic Platform "Open Heritage: Research and Society (PTI-PAIS)".

FUNDING

This work was partially supported by the European Federation of Corrosion (EFC) [EUROCORR Young Scientist Grant 2016].

REFERENCES

Bertasa, M. & Korenberg, C. 2022. Successes and challenges in laser cleaning metal artefacts: A review. *J Cult Herit* 53:100–17. https://doi.org/10.1016/j.culher.2021.10.010.

Burmester, T., Meier, M., Haferkamp, H., Barcikowski, S., Bunte, J. & Ostendorf, A. 2005. Femtosecond Laser Cleaning of Metallic Cultural Heritage and Antique Artworks. In Dickmann K, Fotakis C, Asmus JF (eds) *Lasers Conserv. Artworks. Springer Proc. Physics, vol 100*. Berlin, Heidelberg: Springer. https://doi.org/10.1007/3-540-27176-7_8.

Fotakis, C., Anglos, D., Zafiropulos, V., Georgiou, S. & Tornari, V. 2007. *Lasers in the Preservation of Cultural Heritage: Principles and Applications* (Series in Optics and Optoelectronics). CRC Press.

Di Francia, E., Lahoz, R., Neff, D., Angelini, E. & Grassini, S. 2018. Laser cleaning of Cu-based artefacts: Laser/corrosion products interaction. *Acta IMEKO* 7. https://doi.org/10.21014/acta_imeko.v7i3.610.

Di Francia, E., Lahoz, R., Neff, D., de Caro, T., Angelini, E. & Grassini, S. 2022. Laser-cleaning effects induced on different types of bronze archaeological corrosion products: chemical-physical surface characterisation. *Appl Surf Sci* 573:150884. https://doi.org/10.1016/j.apsusc.2021.150884.

Di Francia E., Lahoz R., Neff D., Rico V., Nuns N., Angelini E. 2021. Novel procedure for studying laser-surface material interactions during scanning laser ablation cleaning processes on Cu-based alloys. *Appl Surf Sci* 544:148820. https://doi.org/10.1016/j.apsusc.

Gomes, V., Dionísio, A., Pozo-Antonio, J.S., Rivas, T. & Ramil, A. 2018. Mechanical and laser cleaning of spray graffiti paints on a granite subjected to a SO_2-rich atmosphere. *Construction and Building Materials* 188(10): 621–632.

Grigor'eva, I.A., Parfenov, V.A., Prokuratov, D.S. & Shakhmin, A.L. 2017. Laser cleaning of copper in air and nitrogen atmospheres. *Journal of Optical Technology* 84(1): 1–4.

Guide to the Expression of Uncertainty in Measurement (GUM). Int Bur Weight Meas n.d. https://www.bipm.org/en/ publications/guides/gum.html.

Landucci, F., Pini, R., Siano, S., Salimbeni. R. & Pecchioni. E. 2000. Laser cleaning of fossil vertebrates: A preliminary report. *Journal of Cultural Heritage* 1(Suppl. 1): S263–S267. https://www.sciencedirect.com/science/article/pii/S1296207400001412

Lide, D.R. 2005. CRC *Handbook of Chemistry and Physics*, Internet Version https://doi.org/10.1016/0165-9936(91)85111-4.

Robbiola, L., Blengino, J.M., Fiaud, C. 1998. Morphology and mechanisms of formation of natural patinas on archaeological Cu-Sn alloys. *Corrosion Science* 40: 2083–2111.

Siano, S., Agresti, J., Cacciari, I., Ciofini, D., Mascalchi, M., Osticioli, I. & Mencaglia, A.A. 2012. Laser cleaning in conservation of stone, metal, and painted artifacts: State of the art and new insights on the use of the Nd:YAG lasers. *Applied Physics A Mater Sci Process* 106: 419–46. https://link.springer.com/article/10.1007/s00339-011-6690-8.

Soffritti, C., Fabbri, E., Merlin, M., Garagnani, G.L. & Monticelli, C. 2014. On the degradation factors of an archaeological bronze bowl belonging to a private collection. *Applied Surface Science* 313:762–70. https://www.sciencedirect.com/science/article/pii/S016943321401352X.

Lasers in the Conservation of Artworks XIII – Siano & Ciofini (eds)
 ISBN 978-1-032-47995-8

Author Index